Paid Work Beyond Pension Age

Paid Work Beyond Pension Age
Comparative Perspectives

Edited by

Simone Scherger
University of Bremen, Germany

Selection and editorial matter © Simone Scherger 2015
Individual chapters © Respective authors 2015

All rights reserved. No reproduction, copy or transmission of this publication may be made without written permission.

No portion of this publication may be reproduced, copied or transmitted save with written permission or in accordance with the provisions of the Copyright, Designs and Patents Act 1988, or under the terms of any licence permitting limited copying issued by the Copyright Licensing Agency, Saffron House, 6–10 Kirby Street, London EC1N 8TS.

Any person who does any unauthorized act in relation to this publication may be liable to criminal prosecution and civil claims for damages.

The authors have asserted their rights to be identified as the authors of this work in accordance with the Copyright, Designs and Patents Act 1988.

First published 2015 by
PALGRAVE MACMILLAN

Palgrave Macmillan in the UK is an imprint of Macmillan Publishers Limited, registered in England, company number 785998, of Houndmills, Basingstoke, Hampshire RG21 6XS.

Palgrave Macmillan in the US is a division of St Martin's Press LLC, 175 Fifth Avenue, New York, NY 10010.

Palgrave Macmillan is the global academic imprint of the above companies and has companies and representatives throughout the world.

Palgrave® and Macmillan® are registered trademarks in the United States, the United Kingdom, Europe and other countries.

ISBN 978–1–137–43513–2

This book is printed on paper suitable for recycling and made from fully managed and sustained forest sources. Logging, pulping and manufacturing processes are expected to conform to the environmental regulations of the country of origin.

A catalogue record for this book is available from the British Library.

Library of Congress Cataloging-in-Publication Data
Paid work beyond pension age : comparative perspectives / [edited by]
  Simone Scherger, University of Bremen, Germany.
     pages   cm
  Includes bibliographical references.
  ISBN 978–1–137–43513–2
  1. Older people—Employment.   I. Scherger, Simone.
  HD6279.P35 2015
  331.3'98—dc23   2015014925

Contents

List of Figures and Tables vii

Acknowledgements x

Notes on Contributors xii

1 Introduction: Paid Work Beyond Pension Age – Causes, Contexts, Consequences 1
   Simone Scherger

Part I Country Cases

2 Work Beyond Age 65 in England and the USA 31
   David Lain

3 The Social Stratification of Work Beyond Pension Age in Germany and the UK: Quantitative and Qualitative Evidence 57
   Anna Hokema and Thomas Lux

4 Characteristics of Working Pensioners in Italy: Between Early Retirement Tradition and Reforms to Extend Working Life 81
   Andrea Principi, Pietro Checcucci, Mirko Di Rosa and Giovanni Lamura

5 Work Beyond Pension Age in Sweden: Does a Prolonged Work Life Lead to Increasing Class Inequalities Among Older People? 107
   Björn Halleröd

6 Work Beyond Pension Age in Russia: Labour Market Dynamics and Job Stability in a Turbulent Economy 129
   Jonas Radl and Theodore P. Gerber

7 Working Pensioners in China: Financial Necessity or Luxury of Choice? 151
   Ge Yu and Klaus Schömann

Part II Contexts

8 Pension Reform in Europe: Context, Drivers, Impact 177
   Karen M. Anderson

9 The Transition to Retirement: The Influence of Globalization, Public Policy and Company Policies 198
   Victor W. Marshall

10 Companies and Older Workers: Obstacles and Drivers of Labour Market Participation in Recruitment and at the Workplace 217
    Jutta Schmitz

11 Concepts of Retirement: Comparing Unions, Employers and Age-Related Non-Profit Organizations in Germany and the UK 237
    Steffen Hagemann and Simone Scherger

Part III Consequences

12 Later-Life Work, Health and Well-Being: Enduring Inequalities 259
    Katey Matthews and James Nazroo

13 The Decline of 'Late Freedom'? Work, Retirement and Activation – Comparative Insights from Germany and the USA 278
    Silke van Dyk

14 Open Questions and Future Prospects: Towards New Balances Between Work and Retirement? 298
    Harald Künemund and Simone Scherger

Index 316

Figures and Tables

Figures

1.1	Employment-population ratio for men aged 65+ in selected European and other countries	2
1.2	Influences on post-retirement work	9
2.1	Educational breakdown by employment status for those aged 65 to 74 in 2008	41
3.1	Employment ratio of people aged 65 to 69 in Germany and the UK (2001–13)	58
3.2	EGP class (reduced version) of current job of workers in pension age in 2008–09, workers aged 45 to 55 in 1991 and workers aged 45 to 55 in 2008–09 (Germany and the UK)	65
5.1	Employment rate of men and women aged 55 to 64 (1976–2013) and 65 to 74 (2005–2013)	109
5.2	Per cent of PSAE sample in paid work by age in 2002–03 and 2010–11	118
6.1	Percentage of pension-age RLMS respondents employed and unemployed, 2000–12, by gender	137
7.1	Age of first pension receipt by gender	160
7.2	Labour force participation of pensioners and non-pensioners aged 45 and over in 2011, by age group and gender	161
8.1	Public and private expenditure on pensions in selected European countries, 2012, as per cent of GDP	192
8.2	Public spending on pensions (as per cent of GDP) in 2011	193
9.1	US labour force participation by age and sex (1950–2013)	201
10.1	Employment rates of people aged 55 to 64 in selected EU countries, 2003 and 2013 (by sex)	219
11.1	Concepts of retirement and associated ideas	252
14.1	Life-course models	311

Tables

1.1	Combinations of age, work and pension receipt	5
2.1	'Capacity' and 'need' factors associated with working at age 65 to 74 in 2008	37
2.2	Household and other factors associated with working at age 65 to 74 in 2008	43
2.3	Logistic regression results for employment at age 65 to 74 in 2008 – average adjusted predictions (AAPs) and average marginal effects (AMEs)	45
2.4	Influence of job characteristics in 2002 on employment in 2008 at age 62 to 70 (excludes non-employed in 2002)	50
3.1	Characteristics of current job for those working beyond pension age and workers aged 45 to 55 in the UK and in Germany (2008–09)	63
3.2	Determinants of working beyond pension age in Germany and the UK in 2008–09 (binary logistic regression – average marginal effects)	66
3.3	Case overview	69
4.1	Working pensioners among pensioners by age group and gender, 2007–11, per cent (by column)	83
4.2	Main requirements for retirement before and after the 'Fornero reform'	88
4.3	Prevalence of combinations of working and pension receipt among the population aged 55+, different years, per cent (by row)	91
4.4	Combinations of working and receiving a pension among the population aged 55+, by individual characteristics (2011, row per cent)	94
4.5	Work characteristics of working pensioners and non-retired workers in 2011, per cent (by column)	97
4.6	Logistic regression analysis on working (1 = yes, 0 = no) among people aged 55 to 64 and 65+ receiving a pension (2011, average marginal effects)	99
5.1	Descriptive statistics for wave 1 and 2 – not working and working in the age group 65 to 74 (column per cent and averages)	117
5.2	The probability of working after pension age among 65 to 74-year-old Swedes (estimates of OLS regressions and confidence intervals)	120

5.3	Estimated incomes and probability of self-reported health problems among 65 to 74-year-old Swedes (estimates of OLS regressions and confidence intervals)	122
6.1	Employment status of older Russians, before and after reaching state pension age, by gender	138
6.2	Occupational class pre- and post-retirement (per cent, rounded), RLMS respondents who reached retirement age 2000–12	140
6.3	Employed RLMS respondents' job stability and changes within one year of reaching pension age (row per cent, rounded; only 'stayers')	142
6.4	Percentage of retired older persons who would have preferred to continue in paid work at the time they retired	144
7.1	Average working hours per week of working pensioners by gender	162
7.2	Distribution of working pensioners and workers across economic sectors	163
7.3	Characteristics of working pensioners and nonworking pensioners (aged 45 and older)	164
7.4	Binary logistic regression model for determinants of paid work among pensioners	166
12.1	Results of propensity score matching: The effect of later-life working on health outcomes (means differences and their standard errors)	264
12.2	Effects of route into retirement and retirement wealth on post-retirement depression (linear regression coefficients)	266
12.3	Change in well-being over two years: Volunteers compared with non-volunteers (linear regression coefficients)	270

Acknowledgements

The publication of this book would not have been possible without the support of many people and organizations to whom I am deeply indebted.

The idea for the book and its realization were part of the work of the Independent Junior Research Group 'Paid work beyond pension age in Germany and the UK', which has been generously funded by the German Research Foundation (DFG) under its Emmy Noether Programme. We are very grateful to Karin Gottschall, who has continuously supported the group in numerous ways. Chapters 1, 3, 11 and 14 of this book have been (co-)written by members of the Research Group.

The Hanse Wissenschaftskolleg (HWK), Institute for Advanced Studies in Delmenhorst, not only hosted but also provided financial and organizational support for the conference in April 2013 at which drafts of most contributions to this book were presented for the first time. I am particularly grateful to Sabine Friedrichs, Susanne Fuchs and Wolfgang Stenzel, who had a substantial part in the success of the event, and to all participants and commentators for the stimulating discussion.

Steffen Hagemann, Karl Hinrichs, Anna Hokema, Martin Kohli, David Lain, Thomas Lux and Jonas Radl served as additional commentators or reviewers for many of the contributions to the book in different stages of their preparation. I would like to thank them for their very useful and constructive remarks that helped to further improve the quality of the texts.

I would also like to thank Sandra Reinecke, who has done an incredibly precise and reliable job in editing the text formally, as well as the tables and the figures. I am also obliged to Gabriele Lumpp for her further support in the preparation of the final manuscript.

Katherine Bird deserves sincere thanks for her outstanding language editing, her patience and her thoughtful suggestions for language-related improvements to the texts. At Palgrave Macmillan, Amelia Derkatsch and Harriet Barker were always ready to answer questions regarding the manuscript quickly and reliably, for which I would like to thank them.

Of course I also owe thanks to the contributors to the book whose articles not only make the book what it is but who were also tremendously responsive and patient in the course of the completion of the texts. Finally, I would like to warmly thank Anna Hokema, Steffen Hagemann and Thomas Lux, who not only gave valuable advice during the preparation of the book. In the course of this project I have learnt a great deal from them, and without them, the project and the last four years would not have been half as much fun.

Simone Scherger

Contributors

Karen M. Anderson is Associate Professor of Social Policy at the University of Southampton. She holds a PhD in Political Science from the University of Washington (Seattle) and has held positions at Radboud University Nijmegen (NL), Leiden University (NL) and the University of Twente (NL). Her research focuses on comparative social policy development, the interaction of labour market policy and social policy and the impact of Europeanization on national welfare states. She is the author of *Social policy in the European Union* (Palgrave Macmillan, 2015) and the editor of the *Handbook of West European Pension Politics* (with Ellen M. Immergut and Isabelle Schulze, 2007). Her work has also appeared in *Comparative Political Studies, Comparative Politics, Zeitschrift für Sozialreform, West European Politics, Canadian Journal of Sociology* and the *Journal of Public Policy*.

Pietro Checcucci, sociologist, has been a researcher at the Institute for the Development of the Vocational Training of Workers (ISFOL), funded by the Italian Ministry of Labour, since 1995. He is currently head of the ISFOL research group on the ageing of the workforce and changes in the working life cycle. His research interests include age management policies and practices; the attitudes of employers towards an ageing workforce; the pension system; social policies for the ageing population; measures for facilitating the social and labour market integration of people with disabilities.

Mirko Di Rosa, economist, holds a PhD in Economics from Ancona University, Italy. His doctoral thesis specifically focused on the quality of public services and citizens' satisfaction. Since 2009, he has been collaborating at the Italian National Institute of Health and Science on Ageing (INRCA), where he has gained experience in international research projects in the fields of family care of older people; reconciling professional and caring responsibilities; migrant care workers; prevention of elder abuse and neglect; long-term care; older workers and the role of technology for improving the quality of life of older people.

Theodore P. Gerber is Director of the Center for Russia, East Europe and Central Asia, and Professor of Sociology at the University of

Wisconsin-Madison, USA. His research examines social stratification, demographic processes, public opinion and social change in contemporary Russia and other former Soviet republics. He is currently working on a comparative study of the relationship between housing and societal stability in four Eurasian countries.

Steffen Hagemann, political scientist, is a research assistant at the Research Center on Inequality and Social Policy (SOCIUM), and PhD candidate at the Bremen International Graduate School of Social Sciences (BIGSSS), both University of Bremen, Germany. He is currently working in the Emmy Noether Research Group 'Paid work beyond retirement age in Germany and the UK'. His research interests are pension reforms, ageing and the labour market, and the role of knowledge in welfare reform debates. In his PhD, he investigates the debate between sociopolitical actors on old age, work and pensions in Germany and the UK.

Björn Halleröd is Professor of Sociology at the Department of Sociology and Work Science and at the Centre for Ageing and Health (AGECAP), both at the University of Gothenburg, Sweden. During the past 25 years, he has worked on issues related to poverty, well-being, general living conditions and old age. As part of this work, he has been responsible for building up the Swedish Panel Survey of Ageing and the Elderly. Since the beginning of 2012, he has been leading an international project on child poverty in developing countries, focusing on the link between political institutions and children's living conditions; and since 2014, he has been involved in the AGECAP research centre. In tandem with his research, he has, as a representative for the Swedish research community, been heavily involved during the past decade in work at the Swedish Research Council, and he is currently Vice Chairman of the Swedish Council for Research Infrastructures.

Anna Hokema is a research assistant at the Research Center on Inequality and Social Policy (SOCIUM), and a PhD candidate in Sociology at the Bremen International Graduate School of Social Sciences (BIGSSS), both University of Bremen, Germany. She currently works in the Emmy Noether Research Group 'Paid work beyond retirement age in Germany and the UK'. In her PhD, she investigates the subjective experience of working beyond pension age from a comparative perspective. Her research interests are sociology of ageing, sociology of work and qualitative research methods.

Harald Künemund is Professor of Research Methods at the Institute for Gerontology, University of Vechta, Germany. He studied sociology at the Freie Universität Berlin and holds doctoral and post-doctoral degrees in sociology. He is currently Chair of the Section 'Ageing and Society' of the German Sociological Association (DGS) and member of several other associations, scientific and advisory boards. He was a co-ordinator of the Research Network 'Ageing in Europe' of the European Sociological Association (ESA) from 2007 to 2013. His main research interests are research design and methods, social participation and intergenerational relations; in particular, his recent research has dealt with productive ageing, social networks and support, political participation, intergenerational transfers, technology and ageing, the life course and social security.

David Lain is a senior research fellow at the University of Brighton Business School, UK. His research focuses primarily on older workers and policies influencing the finances and employment of older people. He has published on these topics in journals such as *Work, Employment and Society* and *Journal of Social Policy*. He has also co-edited journal special issues on pension reform in Anglo-Saxon countries (in *Social Policy and Society*) and changes to older peoples' employment (in *Employee Relations*). He held a Leverhulme Early Career Fellowship from 2011 to 2014. Research conducted under this fellowship is brought together in a forthcoming book, *Reconstructing Retirement? Work and Welfare Past Age 65 in the UK and USA*.

Giovanni Lamura is a social gerontologist with an international and interdisciplinary background, working at the Italian National Institute of Health and Science on Ageing (INRCA) since 1992. He graduated in economics in Italy in 1990, obtained a PhD from Bremen University in 1995, was a visiting fellow in 2006–07 at the University of Hamburg-Eppendorf and research director of the pillar 'Health and Care' of the European Centre for Social Welfare Policy and Research in Vienna in 2010–11. He has gained experience in international research projects mainly focused on family and long-term care of dependent older people, work-life balance, migrant care work, prevention of elder abuse and neglect, and intergenerational solidarity.

Thomas Lux is a research assistant at the Research Center on Inequality and Social Policy (SOCIUM), and a PhD candidate in Sociology at the Bremen International Graduate School of Social Sciences (BIGSSS), both

University of Bremen, Germany. He is working in the Emmy Noether Research Group, 'Paid work beyond retirement age in Germany and the UK'. His research interests are social stratification, life course sociology, action theory, sociology of work and quantitative methods. In his PhD, he investigates the social stratification of work after pension age in Germany and the UK.

Victor W. Marshall retired in 2013 from his position as Professor of Sociology at the University of North Carolina at Chapel Hill (USA), where he had also been a senior scientist at the UNC Institute on Aging, which he directed from 1999 to 2009. His research has focused on work and retirement and was, amongst others, supported by Employment and Immigration Canada and the Social Sciences and Humanities Research Council of Canada. His other research areas include health care, ageing and the life course; public policy in relation to ageing and health; veterans' health and well-being; and social theory of ageing and the life course. His publications include 13 books and 160 refereed journal articles and book chapters. He is a founding member of the Canadian Association on Gerontology, where he served as vice-president and in several further functions. In the USA, he has been a member of the Governor's Advisory Council on Aging (N.C.) and has served on various executive committees in the Gerontological Society of America, the American Sociological Association and the Southern Gerontological Society. He continues professional and scholarly involvements while in his retirement.

Katey Matthews is a research associate in Social Statistics at the University of Manchester, UK. Her key research interests focus on employment, retirement and social engagement in later life, particularly on issues concerning the current raising of the UK State Pension Age and the effects this may have on the health of older working populations. She recently completed her studies on the health and well-being of older workers in relation to work quality in later life, the potential impact of working for longer on the well-being of older volunteers and caregivers, and the effects of changes in life circumstances, such as entering retirement and the onset of frailty, on social behaviour among older people. She has also researched the consequences of changes in vision in later life on older people's mental and social well-being.

James Nazroo is Professor of Sociology at the University of Manchester, UK, Director of the ESRC Research Centre on Dynamics of Ethnicity

(CoDE) and Co-Director of the Manchester Institute for Collaborative Research on Ageing (MICRA). Issues of inequality, social justice and underlying processes of stratification have been the primary focus of his research activities, which have centred on gender, ethnicity, ageing and the interrelationships between these. His research on ageing has been concerned to understand the patterns and determinants of social and health inequalities within ageing populations, with a particular interest on the 'transmission' of inequalities across the life course, patterns of 'retirement', formal and informal social and civic participation, and how class operates post-retirement. He is a principal investigator of the fRaill Programme, an interdisciplinary study of inequalities in later life, and co-PI of the English Longitudinal Study of Aging (ELSA), which is a multidisciplinary panel study of those aged 50 and older.

Andrea Principi, sociologist, has been a researcher at the Italian National Institute of Health and Science on Ageing (INRCA), Ancona, Italy, since 2000. His main research interests and scientific publications relate to active ageing, that is, work, volunteering and education in older age and working carers' reconciliation of work for the labour market with informal care to older family members. He is currently involved in the following European projects: 'Mobilising the potential of active ageing in Europe' (MOPACT, 2013–17); 'Extending working lives – health and wellbeing implications and facilitators' (EWL, 2014–17).

Jonas Radl is Assistant Professor of Sociology at Carlos III University of Madrid. He holds a PhD from the European University Institute in Florence and has published extensively on the transition from work to retirement. His research interests also include social stratification, family, education and the life course. His work has appeared in journals such as *Social Forces, European Sociological Review, Social Science Research* and *Journal of Gerontology: Social Sciences*.

Simone Scherger is Leader of the Emmy Noether Research Group 'Paid work beyond pension age in Germany and the UK', based at the Research Center on Inequality and Social Policy (SOCIUM), University of Bremen, Germany, and funded by the German Research Foundation (DFG). She is a sociologist and holds a PhD from Freie Universität Berlin. Her research focuses on the life course, old age and ageing, social policy, social inequality and generations.

Jutta Schmitz is a research assistant at the Institute for Work and Qualification (IAQ) at the University Duisburg-Essen. Her focal points of research include employment and labour market policies, poverty and social exclusion, pension systems (in European comparison) and life courses within the welfare state. Since 2012, she has been directing the project 'Gainful employment and retirement in Germany – employees, companies and pension system' (funded by the Hans Böckler Foundation).

Klaus Schömann is Head of Programme at the German Institute for Adult Education, Leibniz Centre for Lifelong Learning in Bonn (DIE), and Professor of Sociology at Jacobs University Bremen, Germany. He worked at UNESCO in Paris from 1989 to 1991 and the Social Science Research Center in Berlin (WZB) from 1992 to 2004. His research focuses on employment and labour market structures as well as policy evaluations. He has published in, among others, the *American Sociological Review*, the *European Sociological Review* and recently in the *Journal of Health and Social Behavior*.

Silke van Dyk is Professor of Sociology at the University of Kassel, Germany. Her research focuses on comparative welfare state research, political sociology, discourse analysis and ageing studies. Her recent publications include 'The appraisal of difference: Critical gerontology and the active-ageing-paradigm' (*Journal of Aging Studies* 31, 2014) and the book *Leben im Ruhestand. Zur Neuverhandlung des Alters in der Aktivgesellschaft* (*Retirement Life. Renegotiating Old Age in the Active Society*, 2014). Together with Thomas Küpper she is guest editor of the special issue 'Theorizing Age. Postcolonial Perspectives in Aging Studies' of the *Journal of Aging Studies*, which will be published in winter 2015–16.

Ge Yu is University Lecturer at the Faculty of Philosophy and Public Administration at Liaoning University, China, and PhD fellow at the Bremen International Graduate School of Social Sciences, University of Bremen and Jacobs University Bremen, Germany. Her research focuses on social policy, age and the labour market, with special attention to China. Interests include the impact of institutional changes on the job mobility of older workers, individual decisions of work and retirement, and social inequality in China. She is the main editor of the book *Comparative Social Security – A Global Perspective* (in Chinese, co-edited with Shan Li, 2010). She is also the main translator of the book *Ageing Labour*

Forces – Promises and Prospects (edited by Philip Taylor) into simplified Chinese (co-translator Long Qin 2011). Her papers have been published in Chinese journals such as the *Journal of Gansu Social Sciences* and *Journal of Liaoning University (Philosophy and Social Sciences)*.

1
Introduction: Paid Work Beyond Pension Age – Causes, Contexts, Consequences

Simone Scherger

1.1 Paid work and retirement: A shifting relationship

The institution of retirement is a defining characteristic of modern and contemporary welfare states. After a long period of decreasing effective and, in part, statutory pension ages in many Western countries (see, for example, Blossfeld et al. 2006), this trend has started to reverse in (Western) Europe since around 2000. Connected to this and against the background of demographic ageing, retirement and its relationship to work have become contested issues (again).

Over and above these shifts in the timing of the transition into retirement, the boundary itself between working life and retirement has become more blurred. Partial retirement or partial pensions before normal pension age, flexible transitions into retirement, volunteering and other activities during retirement and paid work[1] beyond pension age, often whilst receiving an (old-age) pension, are the most important examples of these blurring boundaries. The interpretation of increasing post-retirement work varies widely: it can be seen as a deplorable exception from (the social right to) retirement, as a welcome flexibilization of the life course, as a 'solution' to problems connected with demographic ageing or as the result of a successful fight against age discrimination. One aim of this book is to achieve a more precise description and analysis of post-retirement work, in order to allow informed and well-based conclusions on the potential consequences of its growth.

To illustrate the recent trends, Figure 1.1 shows the employment-to-population ratio among men aged 65 and older in the 15 most populous countries of the EU, the average of the EU-28 countries and the three

2 Introduction

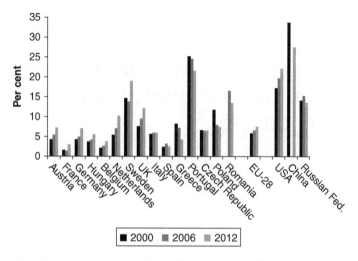

Figure 1.1 Employment-population ratio for men aged 65+ in selected European and other countries
Notes: Romania: data for 2000 unavailable; China: data for 2000 and 2010.
Source: OECD (2014) data.

other countries discussed in this book: the USA, China and the Russian Federation.

In Europe, the starting levels of employment of men aged 65 and older are low in 2000 (around five per cent or less), except for Sweden, Portugal and Poland and, to some extent, the UK and Greece. Until 2012, the ratio increases in most (North-)Western European countries, as well as across the EU-28 countries, whereas it tends to decrease or stay the same in (South-)Eastern and Southern Europe, where some countries have high starting levels of employment. While in the USA more and more older people are in employment, starting from an already high level of around 17 per cent in 2000, this does not apply to China and the Russian Federation which also have high employment ratios from the beginning. Trends for women (not shown) are, in many cases, roughly similar, but on a lower level. Although these shares still seem to be low in Northern and Western Europe, it has to be kept in mind that they cover all those aged 65 and over, with the rates for men aged 65 to 69 being much higher, for example around 25 per cent in the UK and in Sweden in 2012.

Regarding the employment wishes that people of main working age have for their retirement, according to recent Eurobarometer evidence a third of EU citizens (EU-27; European Commission/TNS Opinion & Social 2012: 74–6) say that they would like to continue working after they reach the age when they are entitled to a pension. The related percentages vary widely between EU countries, from 16 per cent (Slovenia) to 57 per cent (Denmark). Differentiating by age, the average share of those wishing to work beyond pension age is highest among those aged 55 and older (41 per cent).

Paid work after pension age, whether or not an old-age pension is received, is not a new phenomenon. Although public old-age pensions for state employees such as civil servants and soldiers were available from the 18th century onwards and early forms of occupational pensions were developed soon after in some industrial sectors or banking, the first more general public pension scheme was only introduced towards the end of the 19th century in Germany (Kohli 1987; Thane 2006). Such early public pension schemes only covered a small fraction of the population, and they were not aimed at completely substituting working incomes. They were only supposed to complement incomes from paid labour, which decline in old age and due to ill health. Pension recipients usually continued to work, be it in formal paid labour, often experiencing downward occupational mobility (Ransom and Sutch 1995), or in subsistence economy – or they were supported by their children, other relatives or poor relief. Their risk of being poor was high. In most Western welfare states, it was only sometime after the Second World War that retirement as a work-free and, consequently, distinct phase of life had become part of most people's lives (Thane 2006), as payments from public and/or occupational pensions were high enough and covered a majority of the population. Retirement had become part of the 'institutionalised life course' (Kohli 1986), the normative 'programme' (Kohli 1986: 291) consisting of education, employment and retirement which people, in particular men, expected to pass through during their lives. This normative programme also served and still serves – to differing degrees in different countries – as point of reference for regulations in social policy.

The increase in work beyond pension age and/or despite receiving an old-age pension thus not only raises questions associated with this late employment itself, its patterns, drivers and consequences on different levels. It also challenges the fundamental meaning of old age, retirement and old-age-related policies and, in a wider sense, also the institutionalized life course. In comparison to related topics, such as employment just before pension age, the transition into retirement and volunteering in old age, work past pension age has been much less in the focus of research. While work after pension age has been studied for at least two decades in the USA, notably because rates of post-retirement work have been higher there for a longer time, and for at least a decade now in the UK for similar reasons, the subject has not or only very recently been investigated more broadly in other European countries (with some early exceptions, for example in Germany, see Kohli and Künemund 1996; Wachtler and Wagner 1997). Therefore, much of the research summarized in this introduction focuses on the USA and the UK. The studies presented in Chapters 4, 5, 6 and 7 of this book can be counted among the first to investigate working pensioners in these countries. The relative lack of research applies even more to a comparative perspective (with the exception of Eurofound 2012 and Alcover et al. 2014). This edited book aims at examining post-retirement work in a systematic and comparative way and at discussing some of the broader issues raised by old-age work. The following sections of this introduction give an overview of the negotiated themes and issues, existing research and the contributions to this book. Although this synopsis cannot be complete, in particular with regard to existing research, it sets the scene for what follows.

1.2 Post-retirement work: Definition, types and relationship to earlier career

Retirement is the *transition into* as well as *the life phase itself* that marks the end of the working career. This life phase is characterized by the absence of paid work and usually implies receiving an old-age pension (or, in some cases of early retirement, a disability pension). 'Work in retirement' or 'work beyond pension age' thus mean combinations of working, receiving an old-age pension and having reached pension age which do not correspond to the institutionalized setup of this transition or life phase, that is, the expected and 'normal' combination of being of pension age, receiving a pension and not working. Consequently and in contrast to 'normal', nonworking pensioners, different subgroups of people working beyond pension age can be differentiated (see Table 1.1; also Scherger et al. 2012: 16–18).

Working pensioners in a strict sense combine pension receipt with paid employment, whereas those of pension age who are working and not receiving a pension either have deferred receiving their pension(s) or do not have any pension claims from their own employment record. Age further complicates the picture, as in many pension regimes certain old-age pensions can be drawn before statutory pension age, which is often combined with working, creating a category of pensioners

Table 1.1 Combinations of age, work and pension receipt

			Paid work	
			Yes	No
Younger than pension age	Pension receipt	No	Workers (of main working age)	Nonworking population of main working age
		Yes	Working early pensioners	Early pensioners
Of pension age	Pension receipt	No	Workers of pension age without pension claims Workers of pension age with pension claims (pension deferral)	Nonworking people of pension age without pension claims
		Yes	Working pensioners	Nonworking pensioners ('normal' retirement)

Source: Own table (extended version of Scherger et al. 2012: 16).

working before (regular) pension age. The definition of 'retirement', and thus the differentiation of post-retirement work, is further complicated by the fact that pensioners can receive payments from several pension schemes, possibly starting at different ages – which is particularly relevant in multi-pillar pension systems.

Distinguishing different forms of post-retirement work might seem academic. However, with these different forms, the experience and the consequences of post-retirement work will vary, and their incidence will differ between countries depending on the institutional context such as pension systems and labour markets. Although in most countries, working pensioners (pre- or post-pension age) dominate the picture of those combining work and pension receipt in unusual ways, in some pension systems (such as in the UK), pension deferral is not uncommon, amongst others because it is rewarded by higher pension payments later (Crawford and Tetlow 2010).

Further possible differentiation of post-retirement work concerns the relationship of the post-retirement job to the one before reaching pension age or starting to receive a pension. Most importantly, this relates to the questions as to whether the employer, the employment status (dependent employment or self-employment, for example) and the occupation are the same as before retirement and whether people restart working after some time of economic inactivity or simply continue working, in an unchanged or changed working arrangement, for example with regard to hours worked. Depending on the dimensions studied, 'stayers' or 'continuers' can thus be distinguished from 'movers' and 'recruits' (Smeaton and McKay 2003; Lain 2012), with the latter sometimes experiencing downward occupational mobility (Lain 2011: 90). Some of the more complex terms used to denominate post-retirement work, such as 'bridge employment' (for example Alcover et al. 2014), also imply ideas on what role this work plays in relation to the main career or to old-age provision (see the title of Parry and Wilson 2014).

1.3 Characteristics of paid work after pension age

The huge variety of pathways into and forms of post-retirement work defy simplifying descriptions or conclusions and also entail difficulties in empirically investigating and comparing post-retirement work. Nonetheless, regarding its basic features and structural characteristics, there are many similarities across European and other Western countries. In most countries, the majority of those working past pension age do so part-time (Banks and Tetlow 2008; Eurofound 2012: 37; Scherger et al. 2012). Although the occupations that are pursued cover a large spectrum, some are over- or under-represented in comparison to the employment structure of workers of main working age. The most clearly over-represented group in many countries are the self-employed or freelancers, often without or with only few employees (Hayward et al. 1994; Eurofound 2012: 39; Brenke 2013). While these might also include people who start self-employment late and in the prospect of retirement, or who give up dependent employment and continue their old occupation as a freelancer, there are indications that most of them were already self-employed before pension age. Furthermore, qualitative anecdotal as well as (potentially unreliable) quantitative evidence suggests that the share of post-retirement work that is done off the books is considerable (Eurofound 2012: 42).

Regarding sectors, jobs in manufacturing seem to be clearly under-represented in many countries, whereas professional jobs and sometimes those in retail and other services are more frequent among those of pension age (Smeaton and McKay 2003: 33; Eurofound 2012: 34). Information on the class profile of post-retirement jobs not only allows inferences about the pathways into and causes of late employment but is also associated with well-documented patterns of social stratification in retirement behaviour (Radl 2013) and a country's employment structures and job opportunities for older people. Hokema and Lux (Chapter 3 in this book) find that post-retirement jobs are somewhat shifted towards the classes of unskilled manual and low-routine jobs in the UK in comparison to younger workers (see also Lain 2012), whereas this is not the case in Germany where self-employment gains more relative importance. These under-researched characteristics of work post pension age also help to characterize the structural role that working pensioners play on the labour market (see below).

1.4 Theoretical approaches to post-retirement work

Whether and how people work beyond pension age depends on a multiplicity of influences which can be structured in different ways. Hayward, Hardy and Liu (1994: 84; see also Hardy 1991) describe work after retirement as the result of two selection processes: first, the retirees' self-selection into employment, and second, the selection of retirees wanting to work (or wanting to continue working) by the labour market. This distinction designates the two main areas of broader theoretical approaches to work beyond retirement: first, approaches focusing on the individual and his or her desire to work, and second, demand- or supply-based theories explaining what happens in the labour market. In the former area, often spelled out by scholars with a background in psychology, theories discussed include Atchley's continuity theory of ageing (Atchley 1989) or role theory in general (see Kim and Feldman 2000; von Bonsdorff et al. 2009). Here, post-retirement work can be seen as an attempt to maintain a certain daily routine and work-related contacts, or to preserve the occupational role in order to avoid the disruption and the role loss connected to (full) retirement. Some of these non-material incentives of post-retirement work also resemble the drivers of voluntary activities, which also help people maintain social contacts and are a source of social appreciation (see, for example, Griffin and Hesketh 2008).

In the theoretical approaches connected to the labour market, general models of supply and demand are usefully supplemented by models of occupational stratification or segmentation of the labour market and dual queuing in labour queues and job queues (Lain 2012: 80 – in referring to Reskin and Roos 1990). Furthermore, it can be assumed that the factors which influence labour market participation in old age are the opposite of what facilitates or delays retirement transitions – a very well-studied field (for example Blossfeld et al. 2006; Radl 2013). The same factors that increase the probability of (early) retirement, such as bad health or unemployment, clearly pose barriers to post-retirement work, and comparisons can also be drawn with explanations of labour market participation in general and earlier in the life course. However, structural and institutional specificities of employment in old age shift and partly change the interplay of labour demand and supply with regard to people of pension age. The age boundary institutionalized in the pension system and in related legislation implies that people beyond pension age are not expected to work and 'normally' do not need to work because their default status is being a pensioner and their main sources of income are their pensions. This shifts the weight of many other factors in such a way that much stronger incentives together with other favourable conditions must be at work for pensioners to actually pursue paid employment. Thus, the absence of the above-mentioned barriers to work (for example bad health) alone is not sufficient to explain post-retirement work: additional drivers, such as financial or social motivations for working, are required to effectively lead to engagement in paid work.

Both theoretical foci, the individual and the labour market, can be integrated into a broader life-course frame (see Wang et al. 2008: 820; von Bonsdorff et al. 2009: 82). This frame permits the analysis of post-retirement work as the result of a long-term interaction between individual lives and institutions that finds its expression (amongst others) in the socio-temporal schedules for typical life-course transitions. At the same time, the interconnectedness of individual lives is taken into account as well, as is the level of subjective experience (see, for example, Kohli 1986). The individual decision to restart or to continue working after having started to receive a pension is embedded in an individual's life course and accumulated experience in employment and family careers (see also Blekesaune et al. 2008), as well as in the institutional setting, in particular the pension system and the labour market.

1.5 Influences on post-retirement work

The single factors underlying this comprehensive theoretical approach can be summarized in a heuristic model of individual, meso- and macro-level influences on post-retirement work (see Figure 1.2).[2] The different levels of influence are interrelated, with, for example, meso- (especially company-level) influences being partly dependent on broader policy regulations.

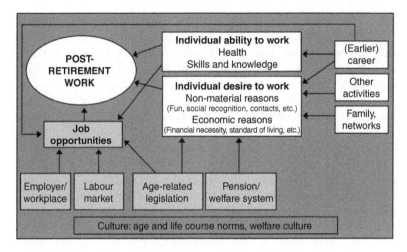

Figure 1.2 Influences on post-retirement work
Source: Based on Scherger et al. (2012).

1.5.1 Micro-level: Individual patterns of and reasons for post-retirement work

On the individual level, the ability to work, job opportunities and the desire to work affect whether someone participates in the labour market beyond pension age. The *ability to work* in particular comprises mental and physical health, and skills and knowledge (see, for example, Gould et al. 2008), whereas *job opportunities* derive from the interplay of the individual ability to work and the conditions on the labour market. Work ability and sufficient job opportunities are necessary, but in themselves not sufficient preconditions for working – their absence (poor health, low education and no opportunities for work) constitutes the most important barrier to working. The *individual desire* to work, by contrast, comprises the 'drivers' of working in a strict sense: financial and all kinds of other reasons for wanting post-retirement work. All three areas of influences – the individual desire to work, the ability to do so and job opportunities – are shaped by the earlier life course of older people, in particular by their employment career.

In relation to these individual influences on post-retirement work, there are a number of similarities among those working past pension age in different countries. Men, younger pensioners, those who have retired more recently and those of better health have been found to be more likely to continue or restart work after pension age in most studies that investigate work after pension age (Lain 2011; Eurofound

2012: 30–3; Pleau and Shauman 2013; Hofäcker and Naumann 2015). The gender difference indicates that, by and large, the career and employment patterns before pension age continue to be effective in old age. Furthermore, in most countries, higher educational qualifications go together with a higher probability of working after pension age (see also Lain in Chapter 2 of this book), just as better-educated people usually enter retirement later and have more continuous working careers.

In comparison to age, health and education, the influence of class and financial resources on post-retirement work varies more widely across countries. Related to the question as to which classes and sectors are under- or over-represented among jobs that are pursued in retirement is the propensity of those working in different classes (or sectors) *before* retirement to go on working or restart work after pension age. Whereas the small self-employed (without or with few employees) tend to continue working much more often than other classes in most countries under study (Scherger et al. 2012), the pattern for the other classes is more country-specific. While higher professionals often have a higher propensity to work beyond pension age, for example in Germany (Scherger et al. 2012; Brenke 2013), this association is not visible in others.

The influence of financial resources on taking up post-retirement work is more complex to study. Those who defer their pension (usually a minority) might have low household incomes if the working income is not included and might nonetheless have high pensions once they start drawing them, while for working pension recipients, low pensions and other nonwork incomes might indicate financial motives for working. There are indications that those with lower financial resources are more likely to work after retirement, for example those with high household debt and outstanding mortgages (Scherger 2013), low (nonwork) household or retirement incomes (Pleau and Shauman 2013) or low wealth. At the same time, receiving means-tested benefits and the resulting earnings threshold seem to constitute a disincentive for working in the very lowest income groups. This might be one reason for poor people seeming to be less likely to work after pension age, at least in some countries (Crawford and Tetlow 2010; Lain 2011), and when their health- and qualification-related barriers are not taken into account, for example through multivariate statistical approaches. These cursory remarks illustrate that the relationship between inequalities and late employment is not unidirectional: on the one hand, poor financial resources seem to be a driver of post-retirement work; on the other hand, higher education and class appear to be associated with better abilities, opportunities and probably also a stronger desire to work after pension age.

The probability of working after pension age is also affected by marital status, the family and private living arrangements and the related obligations of people of pension age. This can often be traced back to the underlying income dynamics but can by no means be reduced to these. For example, in many countries, divorced women show clearly increased rates of post-retirement work (for example Pleau 2010), which is often due to interrupted careers, low retirement incomes and the inability to pool resources. Furthermore, people with a working partner are more likely to work (Banks and Tetlow 2008), as are people with a partner who has a long-term illness (Crawford and Tetlow 2010). The complexity of domestic arrangements and negotiations related to working (or not working) in old age is also revealed by qualitative research (for example, Loretto and Vickerstaff 2013), which shows that these arrangements can be challenged and changed abruptly by unexpected events like health shocks.

Many of these patterns can directly or indirectly be explained by the individual desire to work. Quantitative and qualitative research on the reasons for working post-retirement reveal a broad range of individual motives for working. In quantitative evidence, both financial reasons and non-financial reasons are mentioned by considerable shares of respondents (see, for example, Deller and Maxin 2009; Eurofound 2012: 26; Scherger et al. 2012: 58). Among non-financial reasons, enjoying the job and similar reasons seem to be paramount, followed by reasons such as contact to other people, continuing to do something useful or wanting to stay active. If possible, people often choose several reasons, and the combination of financial and non-financial reasons is very common. This variety of individual reasons for working is also corroborated by qualitative research which finds even more motives for working (apart from the already mentioned ones), such as wanting to pass on skills and knowledge, having been asked by one's employer to continue, or family-related reasons such as the work or economic inactivity of a partner (Barnes et al. 2004; see also Hokema and Lux in Chapter 3 of this book). According to this strand of research, financial reasons themselves imply a whole range of motives which have to be differentiated: the wish to earn extra money in order to be able to afford special activities or goods, like holidays, the wish to support others (such as grandchildren) or the financial need to earn money, because without that money the working person and/or the household would be in serious financial need or poverty.

1.5.2 The meso-level: Employers and the workplace

Although they are not often studied systematically, meso-level influences on post-retirement work should not be overlooked – particularly, the role of employers and the workplace cannot be underestimated in their importance. It is in companies and other employing organizations where labour market policies and age-related regulations are translated into practice, where older workers are recruited, retained and dismissed (or not), and where the work is organized in temporal and spatial form, as well as in contents (see Schmitz in Chapter 10 of this book).

While older workers do not seem to be at higher risk of becoming unemployed than younger ones, finding a job again is more difficult for them than for middle-aged and young workers (see, for example, Lain 2012; Brussig 2011), but not necessarily labour market entrants. Correspondingly, research on age discrimination has demonstrated that age stereotypes and age norms in particular affect processes of recruitment (see, for example, Perry and Parlamis 2006), and this also applies to work after pension age (Karpinska et al. 2011a, 2011b). However, although they probably do not determine recruitment practices, there are also positive stereotypes of older workers, and young people can also be affected by ageist recruitment practices. Within the workplace, older workers are also less likely to be promoted and take part in training measures (Canduela et al. 2012).

On a more general level, the organization of the work affects the ability of the employed to continue working: health management, ageing-friendly workplaces, stress management and supportive human resource management (Naegele and Walker 2006; Eurofound 2012: 57–68; see also Schmitz in Chapter 10 of this book) can all facilitate employment until or beyond pension age, as can measures for flexible or phased retirement (Reday-Mulvey 2005; Eurofound 2007). The fact that in some countries and studies post-retirement workers are more likely to work in small companies (Smeaton and McKay 2003: 35; Eurofound 2012: 35) might indicate that the latter are more flexible or willing to retain or employ workers beyond pension age; it might, however, also be an effect of more generous (early) retirement or dismissal arrangements and occupational pension provision for older workers in larger companies.

1.5.3 Macro-level influences: Structures and institutions

How exactly the individual and workplace characteristics just summarized affect work after pension age is at least in part the result of macro-level institutions and their interplay. There are three important macro-level areas of influence which impact post-retirement work: pension systems and related welfare regulations, labour markets and legal provisions dealing with old age and (old) age discrimination. They are all predominantly shaped by national legislation and structures, which in turn 'filter' supra-national influences such as globalization (see Buchholz et al. 2006; also Blossfeld et al. 2011).

The regulations contained in *pension systems* (for more details see Anderson in Chapter 8 of this book), on the one hand, shape the timing of retirement. On the other hand, they determine the degree of decommodification of old age and thus affect the financial incentives for working. Higher inequalities and/or higher old-age poverty possibly imply more people wanting to work post-pension age. Furthermore, the exact setup of the pension system – importance of the three pillars, degree of privatization, financing, coverage, outcomes – has a direct effect on the income distribution in old age and on who is at risk of old-age poverty. In multi-pillar systems such as the American and the British, for example, income risks in old age are shifted towards the individual, and working can be one possibility to achieve a decent income in old age despite low pension income.

Regulations for flexible or stepwise retirement, pension deferral and rules for combining pension receipt and working also have a direct impact on the incentives for working (see, for example, Kantarci and van Soest 2008): In many countries, if pensions are drawn before statutory pension age (for example in the case of incapacity) or if means-tested old-age benefits are claimed, public pension or benefits payments are reduced if the recipient earns above a certain threshold.[3] Only in a minority of (European) countries does this also apply to pension payments in regular pension age (for an overview, see Eurofound 2012: 51–2). This can be a disincentive to work for the poorest people. However, non-take-up rates of old-age benefits can be high, as a good share of older people will rather work and voluntarily abstain from claiming benefits (Radford et al. 2012). Furthermore, working longer and deferring (public) pension payments is rewarded in many pension systems (Eurofound 2012: 51–2), and the possibility to claim partial pensions or to retire gradually can facilitate extended careers and 'bridge' employment (Reday-Mulvey 2005).

Many Western countries have seen pension reforms in the last one or two decades; in the member countries of the EU, these reforms were amongst others a consequence of the EU recommendations for increasing the share of older workers. Most of these policy changes involved increasing statutory pension ages, closing early retirement routes and often also changing the balance between public and private pillars in the pension system (for more details, see Anderson in Chapter 8 of this book). Often, this has resulted in a weakened public pension provision and an increasing need for private provision, as well as a tightened link between lifetime earnings and level of later pensions (see, for example, Kohli and Arza 2011, and Anderson in this book). The stepwise substitution of defined benefit by defined contribution schemes in occupational pensions, such as in the UK and the USA, has similar consequences. Most of these reforms are likely to have contributed to an increased number of people working after pension age or despite receiving a pension. Over and above pensions and means-tested benefits in old age, other social policy regulations may influence work after pension age as well, such as the availability of health insurance.

The general situation on the *labour market* also bears on the probability of older people working, so that, for example, employment rates of people of pension age are higher in regions and countries with lower unemployment (Crawford and Tetlow 2010) or with skills shortages, or, by contrast, lower in times of economic crisis. Over and above this general influence, the structure and flexibility of labour markets also have effects on late employment. Regarding older workers *before* pension age, less flexible, more regulated and protected labour markets with higher wages and possibly seniority wages for older workers tend to follow a 'logic of employment exit' (Buchholz et al. 2006), often framed by possibilities of early retirement. Correspondingly, the barriers for working beyond pension age are also higher. By contrast, the generally higher job mobility and higher occupational mobility in more deregulated regimes also bear on older workers before and after pension age. Following a logic of 'employment maintenance' (Buchholz et al. 2006), at least for those with sufficient qualifications, there are lower barriers to staying employed and to changing job in old age, and this also applies to the time after pension age. An extended flexible service sector (with potentially low wages) also offers jobs for less well-qualified people of pension age. This also implies that higher employment rates in old age might come with the cost of downward occupational mobility, lower wages and less advantageous working conditions.

Age-related legislation but also regulations at the level of collective agreements, for example, provide the legal framework for working after pension age. If 'default' or mandatory retirement ages are still legal, this will also hinder post-retirement work. Where, by contrast, strong legislation against age-based recruitment and retention practices exists and is implemented, it is easier for older workers to stay employed beyond pension age or to find another job. Whereas default retirement ages have been illegal since 1986 in the USA (Lain 2011; see also Lain in Chapter 2 of this book), the EU Employment Equality Framework Directive 2000/78/EC, which aims at combating (amongst others) age-based employment practices, has only been implemented hesitantly by some EU countries. This especially applies to countries where regulation is relatively de-centralized, for example on the basis of collective agreements (such as in Germany).

Connected to age-related legislation, *cultural* factors related to age (for example, de Vroom 2004; Jansen 2013), the life course and the general welfare culture (van Pfau-Effinger 2005; Oorschot 2007) also impact the individual wish to work, as well as employers' recruitment practices. These ideational influences probably change more slowly and vary in loose accordance with national welfare regimes. Although they are difficult to measure independently from the 'hard' institutional factors, they are important in shaping people's expectations and images regarding old age.

When comparing prevalence, forms and determinants of work past pension age between countries, typologies can be helpful tools for summarizing and systematizing the described institutional and structural influences, as they reduce and order the multiplicity of these macro-characteristics and describe their underlying logic and genesis. The most important examples of this would be Esping-Andersen's typology of welfare states (Esping-Andersen 1990) and Hall and Soskice's (2001) 'Varieties of Capitalism'.

Although comparative evidence on work after pension age has not yet been systematically analysed in terms of welfare typologies, the following cursory assumptions are in line with the evidence in this field. Lower degrees of old-age decommodification, higher degrees of privatization of welfare provision and less protected and more flexible labour markets characterize liberal welfare regimes. These countries follow the logic of 'market-induced' 'employment maintenance' (Buchholz et al. 2006: 17); thus, the shares of older workers before pension age are higher, and of those after pension age as well. This is in part the consequence of higher inequalities in old age and also earlier in the life course. The USA and the UK can serve as examples here, in which retirement and other life-course norms have always been less clearly institutionalized and more flexible (see Chapters 2, 9 and 11, respectively, for the contributions of Lain, Marshall and Hagemann and Scherger). Most social democratic welfare regimes also have higher proportions of older workers before and after retirement. This is due to relatively flexible labour markets, in their case combined with high degrees of labour market activation measures and universal public old-age benefits. Late careers in these countries also follow a 'logic of maintenance'; however, it is 'public-induced' rather than being governed by markets (Buchholz et al. 2006: 17; see the example of Halleröd in Chapter 5 of this book). By contrast, most conservative welfare regimes have for a long time promoted early retirement as a way of managing labour market problems, following the rationale of 'employment exit'; they usually have or had more generous, but also more stratified (public) pension provision, and (much) more protected labour markets with insider-outsider tendencies. Thus, at least in their traditional setup, employment shortly before and also after regular pension age was much less probable than in liberal and social democratic regimes. Southern European countries, not covered by Esping-Anderson's typology in the first place, tend to be similar to conservative regimes in many aspects discussed here and have also favoured the logic of early exit for a long time (see Principi et al. in Chapter 4 of this book). The European post-socialist countries are heterogeneous and hard to classify (see Buchholz et al. 2006); they have also been neglected by the related research on pensions, pension reforms and their outcomes (for an exception see Guardiancich 2013).

As is the case with all typologies, not all specifics of the single countries in question and their institutions are captured; in particular, typologies provide a rather static picture of what is going on. This is especially true for old-age employment and pensions. For example, many countries traditionally classified as conservative (or Southern European) have undergone reforms of their pension systems and labour markets that moved them closer to liberal regimes in some aspects and to social democratic ones in others (see Anderson in Chapter 8 of this book). One example is Germany whose old-age employment rates (before and after pension age) have increased considerably in the course of such reforms, although these reforms are only one of several drivers of this development.

Beside the post-socialist countries facing their very own constellations of challenges (see, for example, Radl and Gerber in Chapter 6 of this book, and Gerber and Radl 2014), countries outside Western Europe and the Western world also escape easy classification. Depending on the development of the welfare state, economic structures and policies (including the role of subsistence economy and informal work) and demographic ageing, these countries in most cases are in a very different situation with regard to old age and employment. Most developing countries do not yet face the problems of rapid ageing, and where there is no developed pension system, retirement as a work-free phase of life will not be institutionalized. Older people will work as long as they can, reduce their work hours according to their health and often be poor and/or dependent on the support of their family.

Emerging economies with high growth rates like China and Russia, which are covered by the country studies in this book, again encounter very specific challenges. In China (Yu and Schömann in Chapter 7 of this book), the current transformation of the traditionally very fragmentary pension system to one which also covers the large rural population coincides with rapid population ageing. Actual pension ages are very low, and the share of (relatively young) working pensioners is high. In Russia (Radl and Gerber in Chapter 6 of this book), by contrast, population ageing is currently not a huge problem, while the low pension ages, many possibilities for early retirement and low pension levels go together with high shares of working pensioners before and after statutory pension ages.

1.6 Pulling the threads together: Why is the number of people working after pension age increasing in many Western countries?

The discussion so far implies a cross-sectional perspective – it answers the question of which individual, workplace level, institutional and structural factors increase the probability of older people working beyond pension age at one moment in time. Although not the focus of the book, these factors also give explanations for the increasing shares of post-retirement workers in many Western countries over time. While population ageing leads or will lead to a growth in absolute numbers of people working after pension age even if everything else remains equal, the increase of the respective relative shares can be attributed to at least five interrelated changes. Although not all apply in every country, they provide reasons to expect further increases in the proportions of working pensioners in the future.

First, the birth cohorts who have entered retirement recently and those who will retire in the future are, on average, of better health and better educated than every cohort before them – which increases their ability to work, their employment opportunities and perhaps also the desire to work. Second, and relatedly, due to recent pension reforms, statutory and effective retirement ages have been increasing in many countries, and the difference between both has been partly reduced. Thus, increasing proportions of people working after pension age are to a certain degree a kind of spillover effect; if the end of the working career was not a long time ago, taking up work again becomes more probable. Third, institutional changes leading to lower average pension payments, more unequally distributed pension incomes and higher old-age poverty in many countries go together with a more widespread desire to continue working (or take up work again) for financial reasons. Together with these institutional changes in the pension system, structural influences have similar effects, in particular more discontinuous (earlier) employment careers and – depending on the exact institutional setting – deregulated labour markets. Fourth, a more widespread desire to work past pension age might also be due to changed attitudes towards old age. For example, some scholars argue that the culture of ageing has changed considerably and that the current generation of pensioners differs from earlier ones in that they want more from their retirement (Gilleard and Higgs 2000). Together with emerging ideologies of 'active ageing' (Moulaert and Biggs 2013) which ascribe the responsibility for a 'good' old age to older people themselves, this might also be an explanation of working pensioners' continued identification with their paid work or their desire for the social appreciation and the greater financial resources that come with it. Put differently, ideational change and changed attitudes have led to different expectations for one's own old age, for example with regard to individual fulfilment, and result in changed behaviour. Fifth and finally, changing labour market structures, the demographic transformation, the resulting decrease in labour supply and/or skills shortages (will) entail a higher demand for older workers. This potentially includes those of pension age and, amongst other factors, legitimizes some of the described policy changes. Additionally, higher flexibilization, anti-ageist legislation and changed attitudes of employers also (will) contribute to better labour market opportunities for older workers. This brief overview illustrates that behavioural changes, the changed composition of the population (especially with regard to health and education), macro-economic and institutional factors (Pleau and Shauman 2013), as well as ideological antecedents to the relationship of paid work, retirement and pensions are closely interrelated and hard to disentangle in their effects.

1.7 Consequences of post-retirement work for well-being and inequalities

In order to come to a balanced and realistic evaluation of post-retirement work, its ramifications must be known. On an individual level and beyond its income-increasing outcomes, the consequences of work after pension age are not well known, amongst others because they are hard to disentangle from their causes: it is a very selective, at least in some countries privileged group of people who work after pension age, and this selection effect needs to be taken into account methodologically. It is plausible, however, that the effects of working after pension age on well-being are probably as heterogeneous as the reasons and conditions of working are. Corresponding to the broad spectrum of reasons for post-retirement work, at least some people, particularly those for whom the earned money does not play a central role, (might) experience work beyond pension age as more pleasant in comparison to their pre-retirement employment. In these cases, paid work can become more similar to voluntary activities (Hagemann et al. forthcoming).

The few studies that examine this question in a methodologically adequate way speak in favour of none to moderately positive effects, for example with regard to psychological health and well-being (Matthews and Nazroo in Chapter 12 of this book; see also Lux and Scherger, forthcoming). Matthews and Nazroo (in this book) also show that there are no positive effects if the subjective rewards for working do not match the subjective effort of work. These moderately optimistic results, however, point to the importance of considering social inequalities in post-retirement work. They are also only a snapshot of the current situation of working retirees in the studied countries and might be different in other countries. The dominating effect on well-being is bound to change in the future if the composition of people working beyond pension age and their motivations change.

Patterns and explanations of post-retirement work can furthermore be condensed to make assumptions about the relation of this work to social inequalities and the structural role of these workers on labour markets. On the individual level, there are two opposing relationships at work with regard to inequalities: economic motives for paid work, which are often negatively linked to pension income, qualifications, health, occupational class and the continuity of earlier careers, on the one hand; and better job opportunities (including more attractive jobs) for those with higher qualifications, better health, better occupations and more continuous careers, on the other hand, which are in turn positively associated with pension income. Whether this results in a u-shaped distribution of paid work beyond pension age (in terms of individual material resources and qualifications) or whether one relationship outweighs the other depends on the precise and nation-specific interplay of the individual, workplace-related and institutional factors described above. Paid work in retirement can thus be a privilege for some and a burden for others (Scherger et al. 2012).

With regard to the level of inequalities in general, the extension of paid work into pension age has not often been assessed so far (but see Halleröd in Chapter 5 of this book for Sweden), and different scenarios are conceivable. If post-retirement work is mostly done by those who are well-off anyway, the inequality-related consequences of work after pension age bring about a Matthew effect: those in advantageous positions will be able to accumulate more resources which will in turn increase overall (income) inequality in old age. If, by contrast, many of those who are poor and disadvantaged (for example due to interrupted working careers) manage to work beyond pension age, this might offset some of their disadvantages (and, to a degree, their discontinuous careers) in old age, resulting in less inequality in old age. As income-related inequalities are closely connected to lower qualifications and ill health, this is so far hardly realistic except for a very specific population of those on low pension incomes, while for the others in this group employment barriers remain substantial. On the whole, this makes the optimistic scenario of decreasing old-age inequalities due to post-retirement work less plausible, and post-retirement work cannot be seen as an easy way to diminish inequalities in old age. This is even truer if health and unequal life expectations are included into the speculations systematically. The institutional framing of work after pension age, in particular pension and labour market regulations, including access to other (means-tested) benefits, will play a key role in shaping the exact outcomes of working lives extended into pension age.

Corresponding to these speculations, one can also reflect on the structural role of post-retirement workers for the labour market: at least some of them seem to serve as an industrial 'reserve army' (see also Kohli and Künemund 1996: 28) who have sometimes undergone downward mobility (Lain 2012) and can be fobbed off with low wages and at best mediocre working conditions, at times exploiting the fact that for some retirees, working resembles a paid hobby or volunteering. Working pensioners can also be a much-needed and highly qualified resource in ageing societies or even the vanguard of a future flexible life-course regime. Especially with regard to those deferring their pensions, they can also be 'normal' workers who are just a bit older than others but have similar wages and working conditions. Again, the exact structural role of post-retirement work depends on the point in time observed, the institutional framing and how the group of working pensioners is composed.

1.8 The evaluation of work past pension age and the reinvention of retirement

In brief, the individual and structural significance of post-retirement work can vary significantly between countries, but also within countries and for different parts of the population of retirees. The political evaluation of post-retirement work correspondingly varies widely – it is, for example, seen as a deplorable exception from (the social right to) retirement which destroys its protective function, as a welcome flexibilization of the life course and 'activation' of old age, as a 'solution' to problems connected with demographic ageing and the sustainability of pensions or as the result of a successful fight against age discrimination. The question as to which of these evaluative interpretations prevail depends on the sociopolitical context, that is, the framing institutions and the welfare culture, and on the political position of the individual or collective actor evaluating them (see Hagemann and Scherger in Chapter 11 of this book; Scherger and Hagemann 2014).

The social right to retire and not to work or the right to work even after pension age are only two very clear positions on post-retirement work which are systematically related to positions on what should happen to the institution of retirement and pensions in the future. The outcomes of the political reform of retirement (Kohli and Arza 2011) are difficult to anticipate and will depend on many factors: on national policies and circumstances, but also on policies at the level of the EU and on global economic development. What is debated here is not only the institutional relationship between paid work, on the one hand, and old age and retirement, on the other, but also the question as to how people want to or should live in old age and the meaning that is attached to this. In this respect, how binding pension ages are and their coincidence with actually stopping work, and how (that is, with which activities) the phase of retirement is spent, are the most important dimensions of observation. A mere shift of pension ages in contrast to their flexibilization, individualization or even abolition constitutes the horizon of what might happen in the future. While work after pension age is clearly an indicator of change and challenges traditional notions of retirement, it remains open whether this change will be limited to some features of retirement or to specific parts of the population or lead to a more hybrid transition phase in between the end of the main career and full retirement ('bridge employment') – or whether the reinvention of retirement will be a more fundamental one, contesting the very core of retirement, and thus parts of the institutionalized life course

(Sargent et al. 2013; see also Künemund and Scherger in Chapter 14 of this book).

1.9 The structure of the book

In light of the questions just outlined, the contributions in this book serve a threefold purpose, corresponding to the three parts of the book. First, they contribute to a deeper understanding of work after pension age, its forms, conditions and explanations. The country studies in Part I of the book give an overview of the profiles of working pensioners after and, in part, also before regular pension age. With the UK (Lain in Chapter 2, Hokema and Lux in Chapter 3), the USA (Lain; see also Marshall in Chapter 9 in Part II of the book), Germany (Hokema and Lux), Italy (Principi et al. in Chapter 4) and Sweden (Halleröd in Chapter 5), they cover examples of welfare regimes which differ considerably with regard to the framing institutions. Consequently, the average length of working lives and the prevalence of working while receiving a pension varies, as well as the underlying inequality-related processes. The cases of Russia (Radl and Gerber in Chapter 6) and China (Yu and Schömann in Chapter 7) further broaden the horizon of observation to include institutional settings that are completely different from European ones and, in the case of China, go together with very many young pensioners working full-time. Despite these differences, at least some of the individual factors influencing post-retirement work are shown to be similar across most countries investigated.

Part II of the book aims at spelling out in more detail at least parts of the institutional and structural context of what is described in the country studies. The dynamics of pension systems (Anderson in Chapter 8, Marshall in Chapter 9) and how older workers are treated at the workplace level (Schmitz in Chapter 10, Marshall) both directly affect employment before and after pension age. Whereas Marshall also gives an overview of public and company policies and their impact on the transition into retirement for the USA, Hagemann and Scherger (Chapter 11) develop a deeper understanding of the conceptions of retirement that are negotiated when work after pension age is debated, thus directing attention to the 'soft' cultural and ideological influences on old-age policies.

Finally, Part III of the book focuses on possible consequences of late-life work on retirement in general and on individual well-being. Whereas Matthews and Nazroo (Chapter 12) compare the individual well-being of workers and nonworkers, and in part of those volunteering, van Dyk (Chapter 13) focuses on the more general trend towards the 'activation' of old age, not only on the level of employment but also with regard to activities in general and normative discourses on old age. Both contributions critically reflect the consequences of a changed balance between work and retirement in more detail. The concluding Chapter 14 (Künemund and Scherger) sketches the gaps in the book that future research should fill, pursues some of the critical perspectives discussed earlier and challenges dominating views on the life course and retirement by proposing a potential alternative to the known tripartite life course.

Notes

1. Whenever 'work' is mentioned in the following, this refers to (all forms of) paid employment. Voluntary and unpaid activities are not included (unless explicitly mentioned).
2. The model does not claim to be complete. For reasons of simplification and readability, only the most important relationships between different influences are marked by an arrow.
3. Similar limitations for earning extra money apply to derived pension benefits which are not based on the recipient's own employment record (such as dependent's or survivor's pensions). Occupational pensions are also subject to rules with regard to continuing work.

References

Alcover, C.-M., Topa, G., Parry, E., Fraccaroli, F. and Depolo, M. (eds.) (2014), *Bridge employment. A research handbook*, New York: Routledge.
Atchley, R. C. (1989), 'A continuity theory of normal aging', *The Gerontologist*, 29 (2), 183–90.
Banks, J. and Tetlow, G. (2008), 'Extending working lives', in: J. Banks, E. Breeze, C. Lessof and J. Nazroo (eds.), *Living in the 21st century: Older people in England. The 2006 English Longitudinal Study of Ageing (Wave 3)*, London: The Institute for Fiscal Studies, 19–56.
Barnes, H., Parry, J. and Taylor, R. (2004), *Working after state pension age: Qualitative research, Department for Work and Pensions Research Report No 208*, Leeds: Department for Work and Pensions.
Blekesaune, M., Bryan, M. and Taylor, M. (2008), *Life-course events and later-life employment, Department for Work and Pensions Research Report No 502*, London: Department for Work and Pensions.
Blossfeld, H.-P., Buchholz, S. and Hofäcker, D. (eds.) (2006), *Globalization, uncertainty and late careers in society*, London/New York: Routledge.
Blossfeld, H.-P., Buchholz, S. and Kurz, K. (eds.) (2011), *Aging populations, globalization and the labor market. Comparing late working life and retirement in modern societies*, Cheltenham: Edward Elgar.
Brenke, K. (2013), 'Immer mehr Menschen im Rentenalter sind berufstätig', *DIW Wochenbericht*, 2013 (6), 3–12.

Brussig, M. (2011), *Neueinstellungen im Alter: Tragen sie zu verlängerten Erwerbsbiografien bei?*, Altersübergangs-Report 2011–03, Düsseldorf/Berlin/ Duisburg: Hans-Böckler-Stiftung/Forschungsnetzwerk Alterssicherung/Institut Arbeit und Qualifikation.

Buchholz, S., Hofäcker, D. and Blossfeld, H.-P. (2006), 'Globalization, accelerating economic change and late careers. A theoretical framework', in: H.-P. Blossfeld, S. Buchholz and D. Hofäcker (eds.), *Globalization, uncertainty and late careers in society*, London/New York: Routledge, 1–23.

Canduela, J., Dutton, M., Johnson, S., Lindsay, C., McQuaid, R. W. and Raeside, R. (2012), 'Ageing, skills and participation in work-related training in Britain: Assessing the position of older workers', *Work, Employment and Society*, 26 (1), 42–60.

Crawford, R. and Tetlow, G. (2010), 'Employment, retirement and pensions', in: J. Banks, C. Lessof, J. Nazroo, N. Rogers, M. Stafford and A. Steptoe (eds.), *Financial circumstances, health and well-being of the older population in England. The 2008 English Longitudinal Study of Ageing (Wave 4)*, London: Institute for Fiscal Studies, 11–75.

Deller, J. and Maxin, L. (2009), 'Berufliche Aktivität von Ruheständlern', *Zeitschrift für Gerontologie und Geriatrie*, 42 (4), 305–10.

de Vroom, B. (2004), 'Age-arrangements, age-culture and social citizenship: A conceptual framework for an institutional and social analysis', in: T. Maltby, M. L. Mirabile, E. Øverbye and B. de Vroom (eds.), *Ageing and the transition to retirement. A comparative analysis of European welfare states*, Burlington: Ashgate, 6–17.

Esping-Andersen, G. (1990), *The three worlds of welfare capitalism*, Cambridge: Polity Press.

Eurofound (European Foundation for the Improvement of Living and Working Conditions) (2007), *Early and phased retirement in European companies. Establishment survey on working time 2004–2005*, Luxembourg: Publications Office of the European Union.

Eurofound (2012), *Income from work after retirement in the EU*, Luxembourg: Publications Office of the European Union.

European Commission/TNS Opinion & Social (2012), *Active ageing. Special eurobarometer 378/Wave EB76.2*, Brussels: European Commission (Directorate General Communication).

Gerber, T. P. and Radl, J. (2014), 'Pushed, pulled, or blocked? The elderly and the labor market in Post-Soviet Russia', *Social Science Research*, 45, 152–69.

Gilleard, C. and Higgs, P. (2000), *Cultures of ageing. Self, citizen and the body*, Harlow: Prentice Hall/Pearson Education.

Griffin, B. and Hesketh, B. (2008), 'Post-retirement work: The individual determinants of paid and volunteer work', *Journal of Occupational and Organizational Psychology*, 81 (1), 101–21.

Gould, R., Ilmarinen, J., Järvisalo, J. and Koskinen, S. (eds.) (2008), *Dimensions of work ability. Results of the Health 2000 Survey*, Helsinki: Finnish Centre for Pensions/The Social Insurance Institution/National Public Health Institute/Finnish Institute of Occupational Health.

Guardiancich, I. (2013), *Pension reforms in Central, Eastern and Southeastern Europe. From postsocialist transition to the global financial crisis*, London: Routledge.

Hagemann, S., Hokema, A. and Scherger, S. (forthcoming), 'Gründe für Erwerbstätigkeit jenseits der Rentengrenze aus Sicht von arbeitenden Älteren und Sozialpolitik-Akteuren' (journal article).
Hall, P. A. and Soskice, D. (eds.) (2001), *Varieties of capitalism*, Oxford: Oxford University Press.
Hardy, M. A. (1991), 'Employment after retirement: Who gets back in?', *Research on Aging*, 13 (3), 267–88.
Hayward, M. D., Hardy, M. A. and Liu, M.-C. (1994), 'Work after retirement: The experiences of older men in the United States', *Social Science Research*, 23, 82–107.
Hofäcker, D. and Naumann, E. (2015/online first), 'The emerging trend of work beyond retirement age in Germany. Increasing social inequality?', *Zeitschrift für Gerontologie und Geriatrie*.
Jansen, A. (2013), 'Kulturelle Muster des Altersübergangs: Der Einfluss kultureller Normen und Werte auf die Erwerbsbeteiligung älterer Menschen in Europa', *Kölner Zeitschrift für Soziologie und Sozialpsychologie*, 65 (2), 223–51.
Kantarci, T. and van Soest, A. (2008), 'Gradual retirement: Preferences and limitations', *De Economist*, 156 (2), 113–44.
Karpinska, K., Henkens, K. and Schippers, J. (2011a), 'Hiring retirees: Impact of age norms and stereotypes', *Journal of Managerial Psychology*, 28 (7/8), 886–906.
Karpinska, K., Henkens, K. and Schippers, J. (2011b), 'The recruitment of early retirees: A vignette study of the factors that affect managers' decisions', *Ageing and Society*, 31 (4), 570–89.
Kim, S. and Feldman, D. (2000), 'Working in retirement: The antecedents of bridge employment and its consequences for quality of life in retirement', *Academy of Management Journal*, 43 (6), 1195–210.
Kohli, M. (1986), 'The world we forgot: A historical review of the life course', in: V. W. Marshall (ed.), *Later life. The social psychology of aging*, Beverly Hills/London/New Delhi: Sage, 271–303.
Kohli, M. (1987), 'Retirement and the moral economy: An historical interpretation of the German case', *Journal of Aging Studies*, 1 (2), 125–44.
Kohli, M. and Künemund, H. (1996), *Nachberufliche Tätigkeitsfelder – Konzepte, Forschungslage, Empirie*, Stuttgart: Kohlhammer.
Kohli, M. and Arza, C. (2011), 'The political economy of pension reform in Europe', in: R. H. Binstock and L. K. George (eds.), *Handbook of aging and the social sciences*, New York: Elsevier, 251–64.
Lain, D. (2011), 'Helping the poorest help themselves? Encouraging employment past 65 in England and the USA', *Journal of Social Policy*, 40 (3), 493–512.
Lain, D. (2012), 'Working past 65 in the UK and USA: Occupational segregation or integration', *Work, Employment and Society*, 26 (1), 78–94.
Lux, T. and Scherger, S. (forthcoming). 'In the sweat of their brow? The effects of starting work again after pension age on life satisfaction in Germany and the UK' (journal article).
Loretto, W. and Vickerstaff, S. (2013), 'The domestic and gendered context for retirement', *Human Relations*, 66 (1), 65–86.
Naegele, G. and Walker, A. (2006), *A guide to good practice in age management*, Dublin: Eurofound.
Moulaert, T. and Biggs, S. (2013), 'International and European policy on work and retirement: Reinventing critical perspectives on active ageing and mature subjectivity', *Human Relations*, 66 (1), 23–43.

OECD (Organisation of Economic Co-operation and Development) (2014), *Labour force statistics by sex and age – Indicators: Employment-population ratios (DOI:10.1787/data-00310-en)*, Paris: OECD, http://stats.oecd.org/, date accessed 6 January 2015.

Parry, E. and Wilson, D. B. (2014), 'Career transitions at retirement age in the United Kingdom: Bridge employment or continued career progression?', in: C.-M. Alcover, G. Topa, E. Parry, F. Fraccaroli and M. Depolo (eds.), *Bridge employment. A research handbook*, New York: Routledge, 138–53.

Perry, E. L. and Parlamis, J. D. (2006), 'Age and ageism in organizations: A review and consideration of national culture', in: A. M. Konrad, P. Prasad and J. K. Pringle (eds.), *Handbook of workplace diversity*, London: Sage, 345–70.

Pleau, R. (2010), 'Gender differences in postretirement employment', *Research on Aging*, 32 (3), 267–303.

Pleau, R. and Shauman, K. (2013), 'Trends and correlates of post-retirement employment, 1977–2009', *Human Relations*, 66 (1), 113–41.

Pfau-Effinger, B. (2005), 'Culture and welfare state policies: Reflections on a complex interrelation', *Journal of Social Policy*, 34 (1), 3–20.

Radford, L., Taylor, L. and Wilkie, C. (2012), *Pension credit eligible non-recipients: Barriers to claiming, Department for Work and Pensions Research Report No 819*, London: Department for Work and Pensions.

Radl, J. (2013), 'Labour market exit and social stratification in Europe. The effects of social class and gender on the timing of retirement', *European Sociological Review*, 29 (3), 654–68.

Ransom, R. L. and Sutch, R. (1995), 'The impact of ageing on the employment of men in American working-class communities at the end of the nineteenth century', in: D. Kertzer and P. Laslett (eds.), *Aging in the past: Demography, society, and old age*, Berkeley: University of California Press, 303–27.

Reday-Mulvey, G. (2005), *Working beyond 60. Key policies and practices in Europe*, Basingstoke: Palgrave Macmillan.

Reskin, B. F. and Roos, P. A. (1990), *Job queues, gender queues: Explaining women's inroads into male occupations*, Philadelphia: Temple University Press.

Sargent, L. D., Lee, M. D., Martin, B. and Zikic, J. (2013), 'Reinventing retirement: New pathways, new arrangements, new meanings', *Human Relations*, 66 (1), 3–21.

Scherger, S. (2013), 'Zwischen Privileg und Bürde. Erwerbstätigkeit jenseits der Rentengrenze in Deutschland und Großbritannien', *Zeitschrift für Sozialreform*, 59 (2), 137–66.

Scherger, S. and Hagemann, S. (2014), *Concepts of retirement and the evaluation of post-retirement work. Positions of political actors in Germany and the UK, ZeS-Working paper no. 04/2014*, Bremen: Centre for Social Policy Research.

Scherger, S., Hagemann, S., Hokema, A. and Lux, T. (2012), *Between privilege and burden. Work past retirement age in Germany and the UK, ZeS-Working Paper Nno. 04/2012*, Bremen: Centre for Social Policy Research.

Smeaton, D. and McKay, S. (2003), *Working after state pension age: Quantitative analysis, Department for Work and Pensions Research Report No 182*, Leeds: Department for Work and Pensions/Policy Studies Institute/Personal Finance Research Centre.

Thane, P. (2006), 'The history of retirement', in: G. L. Clark and A. H. Munnell (eds.), *Oxford Handbook of pensions and retirement income*, Oxford: Oxford University Press, 33–51.

van Oorschot, W. (2007), 'Culture and social policy: A developing field of study', *International Journal of Social Welfare*, 16 (2), 129–39.
von Bonsdorff, M. E., Shultz, K. S., Leskinen, E. and Tansky, J. (2009), 'The choice between retirement and bridge employment: A continuity theory and life course perspective', *International Journal of Aging and Human Development*, 69 (2), 79–100.
Wachtler, G. and Wagner, P. S. (1997), *Arbeit im Ruhestand. Betriebliche Strategien und persönliche Motive zur Erwerbsarbeit im Alter*, Opladen: Leske + Budrich.
Wang, M., Zhan, Y., Liu, S. and Shultz, K. S. (2008), 'Antecedents of bridge employment: A longitudinal investigation', *Journal of Applied Psychology*, 93 (4), 818–30.

Part I
Country Cases

2
Work Beyond Age 65 in England and the USA

David Lain

2.1 Introduction

During the 20th century, the age of 65 became institutionalized as the most common male state pension age across OECD countries (Ebbinghaus 2006). Women in countries such as the UK received earlier state pensions, and men on average also retired below age 65 for much of the post-war period (Ebbinghaus 2006). However, pressures to work past 65 are now growing considerably. In the UK and USA, state pension ages are rising to 67 for both men and women by 2027, reaching age 69/70 over time if they are linked to life expectancy projections as currently proposed (Béland and Waddan 2012; *The Guardian* 2013b). Cash benefits for retirees below age 65 will decline, and opportunities for early exit via salary-related occupational pensions have diminished considerably (Friedberg and Webb 2005; Clark 2006).

Given the rising need to work, this chapter considers prospects for employment beyond age 65 in the USA and England, the largest country of the UK. Both the UK and the USA are typically labelled liberal residual welfare states (following Esping-Andersen 1990) with relatively unregulated competitive liberal market economies (after Hall and Soskice 2001). However, as we discuss in the first section, US policy has done more historically to promote employment beyond age 65; this includes disallowing most employers from setting mandatory retirement ages in 1986. UK policy is, however, catching up with that of the USA. In 2011, the 'default retirement age' of 65 was abolished, greatly reducing the ability of organizations to automatically retire off people at this age. Following a policy discussion, the main part of this chapter assesses the factors influencing employment at age 65 to 74 using the English Longitudinal Study of Ageing and the US Health and Retirement Study. The final section discusses the findings of the chapter, namely that socio-economic characteristics built up over the life course will constrain employment past age 65 for significant numbers. We therefore need positive policies that promote equitable and realistic employment/retirement outcomes for older people.

2.2 The institutional context and increasing pressures to work past 65

Until the early 2000s, the principles underpinning US retirement policy arguably differed in emphasis from those of the UK (Lain 2011). The USA has promoted 'self-reliance' amongst older people to a greater degree: access to benefits for those with low pensions was limited, but protection from forced retirement via age discrimination legislation was strong. In the UK, a more 'paternalistic' policy-logic was in operation in the past: older people were viewed as a more vulnerable group, with a more extensive safety net for poorer pensioners but few employment rights (Lain 2011). To summarize these differences, we might draw on Leisering's (2003: 222) assessment that 'Americans view retirement as a matter of civil rights while Europeans view it as a matter of social and political rights'. Americans have been more opposed to legal mandatory retirement ages than their British counterparts in the past (Hayes and Vandenheuvel 1994), while Britons have been more likely than Americans to say that it is the government's responsibility to ensure 'a decent standard of living for the old' (Hicks 2001: 7). This greater UK concern for poorer older people has been reflected in the means-tested safety net made available, rather than through generous social insurance pensions.

Before we examine retirement policy, it is important to acknowledge that a US policy focus on self-reliance actually affected older cohorts throughout their life courses. Leisering (2003: 206) argues that in the USA 'Government activities are more designed to secure equal opportunities [for example through education] [...] than to promote security across the lifecourse'. The USA was a laggard in developing social insurance but was a leader in providing education as a means of promoting self-reliance (Lindert 2004). Older Americans consequently continue to have higher levels of education than their UK counterparts (OECD 2004: 58), placing them in a potentially stronger position with regard to obtaining work (Kanabar 2012).

2.2.1 Policies affecting older people up to the early 2000s

UK state pension provision developed out of concern for poverty in older age identified in surveys (Rimlinger 1971: 229; Williamson and Pampel 1993: 43); it was initially flat-rate and intended to provide a subsistence income, rather than replicating market positions. The original state pension was set too low to do this, however, and a supplementary second pension was introduced in the 1970s; this typically provided very modest pensions to those without occupational pensions. While a significant proportion had occupational pensions, in 2005 around a third of pensioner households received income-tested benefits in the form of social assistance-based Pension Credit, Housing Benefit or benefits to cover local taxes (DWP 2007). These benefits would be largely lost as a result of working, but employment of this group was not expected (Lain 2011).

In line with the general thrust of promoting self-reliance, the US Social Security pension was designed to be 'consistent with the dominant values of self-help and rewards from individual effort' (Rimlinger 1971: 229). Like occupational pensions of the time, Social Security was earnings related to reward previous work (Williamson and Pampel 1993). Over time the formula became redistributive, replacing lower levels of earnings at higher replacement rates, and providing supplements for partners (usually wives) with low pensions in their own right. However, low earnings resulted in a low pension, even when a person had a full contribution record. Someone on half average earnings throughout their working life would receive a pension worth 25 per cent of average earnings in the mid-2000s (Lain et al. 2013). In the UK, the pension level for those on half average earnings would be comparable to that in the USA (Lain et al. 2013), but there would also be a greater safety net of means-tested benefits available. For those with low pension income in the USA, means-tested benefits were even less generous than in the UK and harder to get (Lain 2011). US housing benefits covered only part of the rent and were only received by a minority of eligible individuals, given budget constraints (Priemus et al. 2005; Lain 2011). US Supplemental Security Income (SSI) was more meagre than its UK equivalent, Pension Credit, and strict capital limits on eligibility restricted access (Lain 2011: 496). SSI became increasingly difficult to get over time (Elder and Powers 2006), given a lack of political will and public support to protect 'unearned' benefits and perceived abuse by non-contributing immigrants (Berkowitz and DeWitt 2013). This is in contrast to 'earned' Social Security, which enjoys much higher levels of public support (Harrington Meyer and Herd 2007).

In the context of a declining safety net, US employment rights for those over 65 were increased to encourage people to take financial responsibility for their retirement and its timing (Macnicol 2006). Mandatory retirement ages were outlawed in 1986,[1] following federal age discrimination legislation in 1967 and 1978 and state-level initiatives dating back as early as 1960 (Neumark and Stock 1999). The evidence suggests that 'age discrimination legislation has succeeded at boosting the employment of older individuals through allowing them to remain in the workforce longer' (Adams 2004: 240). In the UK, on the other hand, people have historically had few employment rights beyond age 65. In the recent past, around half of individuals worked for organizations with mandatory retirement ages (Metcalf and Meadows 2006: 65). However, irrespective of mandatory retirement ages it appears that line managers were important in deciding who was allowed to work past age 65 (Metcalf and Meadows 2006: 75–6; Vickerstaff 2006).

2.2.2 Reforms in the 1990s and 2000s

In both countries, reforms will increase both the future need and potential opportunities to work past 65. State pension ages in both countries will rise to 67 by 2027 for men and women. In 2008 – the year this chapter examines – the state pension age in England was 65 for men and 60 for women. The increase to age 67 is therefore rapid compared with the USA, where 'normal' state pension age reached 66 by 2008. A US option of reduced pension at 62 will continue, but the reductions for early receipt will increase considerably as state pension ages rise (SSA 2014). The means-tested safety net for older people exiting employment before state pension age is also declining. The age at which those on low incomes can receive UK Pension Credit, 60 in 2008, is to rise to 67 in line with state pension increases. In the USA, access to means-tested Supplementary Security Income at 65 will decline further due to strict and unchanging asset requirements, leading Elder and Powers (2006) to dub it 'the incredible shrinking program'.

Other pension reforms encourage employment beyond 65. Limits on earnings that can be received whilst receiving a state pension have been lifted in the UK (in 1989) and the USA (in 2000). Most people working past 65 in both countries receive their pension at the same time (Lain and Vickerstaff 2014). UK reforms in 2008 also make it easier for employers to allow older employees to continue working while receiving their occupational pension (Thurley 2011). The USA has alternatively mandated employers to continue contributing to an employee's pension if they work beyond normal pension age (Quadagno and Hardy 1991).

Wider pension changes also reduce opportunities to exit work before 65. Historically, in both countries, defined benefit (DB) salary-related occupational pensions were a common route to early retirement. However, these pensions have declined sharply, first in the USA (Friedberg and Webb 2005) and then in the UK (Pensions Commission 2004). US employers replaced DB pensions with defined contribution (DC) pensions, which are effectively investment accounts that pay out a lump sum. In the UK, DC pensions did not increase sufficiently in number to replace closed DB pensions (Pensions Commission 2004). Starting in 2012, the UK government has therefore mandated employers without an approved occupational pension to automatically enrol employees into a DC scheme (unless the employee decides not to join). US and UK evidence suggests that people with DC pensions retire later on average than those with DB pensions (Friedberg and Webb 2005; Arkani and Gough 2007; Banks et al. 2007). This reflects greater financial incentives to continue working and contributing to a DC pension (see Lain and Vickerstaff 2014) and the fact that these pensions typically provide less generous and secure retirement incomes.[2]

A final area of convergence between the two countries over time is in relation to employment rights. In 2006, UK age discrimination legislation was introduced for the first time, including the right to request continued employment beyond a 'default retirement age' of 65. In 2011, this 'default retirement age' was abolished, meaning that employers can no longer force retirement on the basis of age unless they can provide a legally defensible justification.

In summary, in both countries, the needs and potential opportunities to work past age 65 are increasing. However, it remains an open question as to whether employment will be realistic for many of those most in need of earnings; we turn to this issue now.

2.3 Data

We examine the factors influencing employment at ages 65 to 74 in 2008, in order to assess the groups likely to remain in work in future. This is a useful time to compare the countries. US mandatory retirement had been abolished for decades by this point, but UK employers had a largely free hand in deciding whether someone could continue working beyond 65. In the UK, women could access a state pension at 60, but employers could not force them to retire until they reached 65. In the USA, 'normal' state pension age reached 66 in 2008. However, most of those in the 65 to 74 age band in 2008 had been eligible for an unreduced pension when they were 65.

Surveys with a high degree of comparability were analysed: the English Longitudinal Study of Ageing (ELSA) and the US Health and Retirement Survey (HRS). We used comparable data files created by the RAND organization (see Phillips et al. 2012) alongside the original data. England has 84 per cent of the UK population (ONS 2012), so the results should be broadly reflective of the UK as a whole. The analysis was conducted using STATA 13, and weights provided were used as advised by both surveys to increase the representativeness of the samples; it was particularly important to weight the HRS analysis because of its complex sample structure (Aneshensel 2013: 167–96). The unweighted sample of people aged 65 to 74 was 2,941 for England (including 405 workers) and 5,891 for the USA (1,668 workers). The non-regression results include 95 per cent confidence intervals. If the confidence intervals do not overlap, we can say that the employment rate for one group (for example men) was different from that of another group (women) at the statistically significant level.

2.4 Results

We identify four important sets of factors influencing employment at age 65:

- the *need* to work (including pensions, home ownership status and wealth),
- the *capacity* to work (including education and health),
- *household factors* (related to partnership status) and
- *workplace* factors.

We present analyses on each of these factors in turn, discussing the results in the context of the broader literature.

2.4.1 Need factors

As noted above, the lack of employment rights past age 65 in the UK has constrained the ability of many individuals to respond to financial needs or incentives by working. As Table 2.1 shows, in 2008, prior to the UK abolition of mandatory retirement ages, less than one-sixth of the English worked at age 65 to 74 compared with just under a third of Americans. Women were less likely to work past age 65 than men in both countries; apart from the generally lower labour market participation

Table 2.1 'Capacity' and 'need' factors associated with working at age 65 to 74 in 2008

	USA			England		
	% working[1]	Confidence interval (95%)		% working[1]	Confidence interval (95%)	
All people	30.9	29.4	32.6	13.6	12.3	14.9
Male	38.0	38.0	35.4	17.3	15.3	19.5
Female	24.8	24.8	23.2	10.2	8.7	11.8
Need factors						
Highest wealth quartile	37.8	34.7	41.0	18.4	15.7	21.4
2nd wealth quartile	29.9	27.4	32.6	14.4	12.0	17.2
3rd wealth quartile	32.8	30.4	35.2	12.2	10.0	14.9
Lowest quartile	23.3	20.8	26.0	8.8	6.8	11.3
Has private pension	34.7	32.8	36.7	14.4	12.9	16.0
No private pension	25.1	22.9	27.5	11.7	9.7	14.1
Own house outright	27.5	25.4	29.7	13.2	11.8	14.7
Buying house	41.6	38.5	44.9	25.3	19.9	31.6
Renting	24.9	21.2	28.9	9.5	6.9	12.8
Other	19.0	13.2	26.5	(16.4)	(9.5)	(27.1)
Capacity factors						
Good health	36.4	34.6	38.3	16.7	15.1	18.4
Fair/poor health	15.6	13.4	18.0	5.6	4.1	7.5
College and above	43.2	39.9	46.6	22.0	18.3	26.3
Some college	33.0	29.8	36.3	16.3	13.5	19.5
High school	27.4	25.7	29.2	16.4	13.3	20.0
Below high school	21.2	19.2	23.5	9.6	8.1	11.3
Mean ADL[2] limits –						
all	2.4	2.3	2.5	2.1	2.0	2.2
Workers	1.7	1.6	1.8	1.1	0.9	1.3
Nonworkers	2.8	2.6	2.9	2.2	2.1	2.3
Base N	5,891			2,941		

Notes: Results are weighted. Results in brackets denote an estimate based on less than 30 cases.
[1]Percentage of those working within each category; [2]Mean limits in Activities of Daily Living.
Source: Own analysis of the ELSA Wave 4 and the HRS Wave 9.

of women, this in part is likely to reflect the fact that some couples coincide their retirement and the female partner is likely to be younger (Pienta 2003; Loretto and Vickerstaff 2013). However, US *women* were more likely to work than English *men*, indicating the greater extent to which Americans worked.

Given the financial rationale for abolishing US mandatory retirement, this raises the question as to whether the high US employment rate relates to Americans with financial needs or desires continuing in employment. People work for a range of financial and non-financial reasons, including an interest in the work itself or social reasons (Parry and Taylor 2007; Scherger et al. 2012). Nevertheless, financial considerations appear to be an important reason for working among many Americans. Analysis of a one-off self-completion questionnaire given to half the HRS sample reveals that 64.1 per cent of workers aged 65 to 74 said that they would ideally like to leave work now but they needed the money (supplementary analysis, not shown).[3] In England, workers over pension age were asked why they worked; among 65- to 74-year-olds, 32.0 per cent stated they worked either to improve their finances or because they could not afford to retire. Although the questions were not asked identically, this does suggest that Americans were more likely than their English counterparts to be in employment for financial reasons. This may reflect a greater financial need to work in the USA compared with England to a degree, but it should be noted that average UK retirement incomes are relatively modest compared with working-age incomes (Sefton et al. 2007). It is therefore likely that Britons with modest retirement incomes were retiring with little sense that there was an alternative (see Vickerstaff 2006). With the UK abolishing mandatory retirement, we are likely to see a growth in financially motivated employment past age 65. UK survey evidence suggests that around three-quarters (72 per cent) of older individuals *expecting* to work past state pension age give financial reasons (Smeaton et al. 2009: 80).

While Americans are much more likely to say they were working for financial reasons, the relationship between financial resources and employment is complex. The descriptive analysis in Table 2.1 suggests that in both countries those with the lowest resources appear to be least likely to work, which is consistent with previous research (Haider and Loughran 2001; Crawford and Tetlow 2010: 21; Lain 2011). If we look at household wealth (excluding housing and pensions) equivalized to the individual level, in both countries the poorest quartile was less likely to work than the quartiles above it. These differences were statistically significant with one exception (the higher rate of employment for the second lowest English quartile was not significant at the five per cent level). Americans in the lowest quartile were nevertheless more likely to be in employment than their counterparts in England (23.3 vs. 8.8 per cent). The wider availability of UK means-tested benefits, such as Pension Credit, relative to the USA may discourage employment *to a degree*

(Lain 2011). It is likely, however, that in both countries lower levels of health and education limited employment prospects for poorer people (see regression analysis below). At the other end of the wealth spectrum, in both countries the largest estimated employment rate was for the highest quartile; this was significantly higher than the quartiles below it, with the exception of the third US wealth quartile and second English quartile.

These results only present a partial picture, however, as they exclude pensions and housing. Table 2.1 gives the proportions working among those with a private pension. This includes pensions the individual has received or is entitled to receive in the future. Pensions may be from a current or former employer or held by the individual. For the USA, we include Individual Retirement Accounts, because these are significant pension vehicles for the self-employed. Following the previous results, we might expect the absence of a pension to reduce the likelihood of working, because it signifies wider disadvantage. In England, however, there was no significant difference in employment between those with and without pensions. Private pension coverage in England is, however, a weaker indicator of advantage and disadvantage than in the USA. Most people have at least a marginal pension from a previous employer, because of the ability to opt out of (some) state pension provision into private schemes. In the USA, on the other hand, having a pension significantly increased the likelihood of working: a third of those with a pension worked compared with a quarter of those without a pension.

This evidence suggesting employment is *most* concentrated amongst the advantaged should, however, be put in a broader context to help understand why so many Americans say they are working for financial reasons. First, it should be noted that employment was relatively high at all wealth quintiles compared with England, including the poorest. Second, as we will see below, US employment is associated with having a defined contribution pension; these typically provide less secure financial pathways out of work than defined benefit schemes (Lain and Vickerstaff 2014). Third, wealth and pensions ignore the importance of debts, in particular outstanding mortgages. Table 2.1 shows that in both countries having an outstanding mortgage significantly increased the likelihood of working, which is consistent with previous research (Smeaton and McKay 2003; Mann 2011; Butrica and Karamcheva 2013). In total, 41.6 per cent of those with an outstanding mortgage at age 65 to 74 were still working in the USA, compared with 27.5 per cent of outright homeowners. Likewise, the employment rates for England were 25.3 per cent for homebuyers compared with 13.2 per cent for outright homeowners. This is a particularly important issue in the USA because of a large increase in the proportion of people aged 62 upwards with mortgage debt (Butrica and Karamcheva 2013). As a result, 29 per cent of those aged 65 to 74 in the USA had an outstanding mortgage, rising to 38.9 per cent among workers (supplementary analysis not shown). In England, the proportions with an outstanding mortgage were much lower (6.7 per cent overall vs. 12.7 per cent among workers).

To summarize, there is convincing evidence that the poorest groups were least likely to work. However, it is also evident that financial considerations were more generally an important influence on employment, particularly in the USA. In a context of greater financial uncertainty, people in both countries are increasingly required to take a decision as to whether to continue working for financial reasons. The question remains, however, whether they face barriers to employment.

2.4.2 Capacity factors

One set of explanations for lower employment among the poorest segment relates to what we call here 'capacity factors'; the evidence clearly demonstrates that better health and education increase the likelihood of employment. Health conditions are fairly common amongst older people even before they reach age 65. In the UK, around a quarter of people aged 50 to 69 in 2008–09 had some kind of work disability limiting the kind or amount of work they could do (Crawford and Tetlow 2010: 32). The poorest are most likely to have a health condition (Banks et al. 2007) and least likely to work if they have one (Crawford and Tetlow 2010), presumably because of wider disadvantage and the lack of less physically strenuous jobs.

Table 2.1 presents two measures of health, both of which demonstrate its importance on employment. First, in both countries those rating their health as 'good' were significantly more likely to be employed, although Americans with fair or poor health were more likely to work than their English counterparts. Second, Table 2.1 shows the mean number of difficulties individuals report with Activities of Daily Living (ADLs). ADLs include up to eight activities involving lifting, climbing, stooping, walking, raising limbs, pushing and rising from a chair (see Lain 2011). Table 2.1 shows that in both countries workers had significantly fewer ADL limitations than nonworkers.

In addition to health, education is a capacity factor known to exert a strong influence on employment (Haider and Loughran 2001; Smeaton and McKay 2003; Lain 2011). Qualifications may facilitate more stable careers with fewer involuntary exits (Blekesaune et al. 2008). Qualifications may also make it easier for older people to move into new work if required or desired. US research suggests the likelihood of entering new 'bridge' employment, rather than retiring fully, increases with the number of years of education a person has received (Wang et al. 2008). Likewise, Kanabar (2012) found that in England the likelihood of returning to work after retirement was higher for the higher educated.

Education levels for England have been harmonized with the measure used in the HRS in Table 2.1 (Phillips et al. 2012). The bottom two categories relate to whether or not the individual has attained 'high school' secondary-level qualifications. 'Some college' refers to individuals with qualifications between high school and higher education; this includes, for example, English A levels. Finally, 'college+' includes undergraduate higher education degrees and above. In both countries, having a higher education degree roughly doubles the likelihood of working relative to someone with below secondary qualifications. Americans with degrees were nevertheless around twice as likely to work as their English counterparts, with 43.2 per cent in work compared with 22 per cent in England.

It is clear that highly educated Americans were able to take advantage of opportunities to work past age 65 in large numbers. Furthermore, relative to England, these highly educated Americans were a comparatively large group. Figure 2.1 presents educational profiles for each country. Just over a fifth of Americans of this cohort had a college degree, around

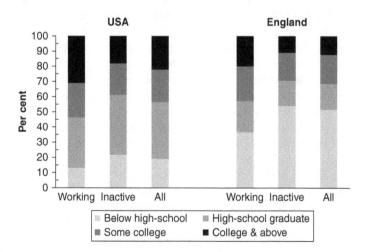

Figure 2.1 Educational breakdown by employment status for those aged 65 to 74 in 2008
Note: Results are weighted.
Source: Own analysis of the ELSA Wave 4 and the HRS Wave 9.

twice that of England. At the other end of the qualification spectrum, less than one-fifth of Americans had qualifications below secondary level compared with more than half in England (see OECD 2004). Purely because of their size, the modal group of 'workers' in England – around a third in total – were those with below secondary qualifications. Low qualifications place particular constraints on older people working up to and beyond age 65 in the UK, given the need to compete for jobs with more highly qualified younger people (Kanabar 2012). Qualification levels have risen markedly in the UK in more recent cohorts, but have not caught up with the USA, so the problem will remain (OECD 2004).

2.4.3 Household factors

While individual factors are important, people often make decisions about work and retirement from within a household and family context (Loretto and Vickerstaff 2013). To explore this, Table 2.2 breaks down employment rates by partnership status for men and women. 'Partnered' individuals are married or are part of a cohabiting couple; employment rates are given for those with and without a working partner. For a small third category, we do not know the employment status of their partner; in both survey samples, around a hundred individuals aged 65 to 74 were in this category (118 in England and 117 in the USA). We also include 'single' people in the table; these are individuals without a partner, often due to divorce, separation, widowhood or never having been married.

Table 2.2 shows that in both countries people in couples were much more likely to be employed if they had a working partner. For those with a working partner, employment rates were similarly high in the USA (44.1 per cent) and England (38.7 per cent). The proportions employed with a nonworking partner were much lower: 26.1 per cent in the USA and 9.2 per cent in England. This big disparity in employment between those with and without working partners in England suggests that household influences are particularly strong in the UK. In England, 46.7 per cent of men with a working partner were employed, compared with only 12.5 per cent of those with a partner not in employment. Likewise, 28.5 per cent of women with a working partner were employed, compared with only 5.7 per cent of those with an inactive partner. The corresponding differences for US men and women were important, but less pronounced (see Table 2.2). It should also be noted that in the USA employment was high among those with non-responding partners, although it is unclear why from this analysis.

Table 2.2 Household and other factors associated with working at age 65 to 74 in 2008

	USA			England		
	% working[1]	Confidence interval (95%)		% working[1]	Confidence interval (95%)	
Partnership status (all)						
Partnered, p. not working	26.1	23.9	28.3	9.2	7.8	10.8
Partnered, p. working	44.1	40.8	47.5	38.7	33.5	44.2
Partnered, partner's work status unknown	40.3	31.2	50.0	(20.0)	(13.6)	(28.4)
Single	27.9	25.5	30.3	10.9	9.0	13.2
All	30.9	29.4	32.6	13.6	12.3	14.9
Male partnership status						
Partnered, p. not working	33.2	29.9	36.6	12.5	10.3	15.1
Partnered, p. working	51.1	46.2	56.1	46.7	39.4	54.1
Partnered, partner's work status unknown	48.9	35.2	62.7	(20.3)	(12.2)	(31.7)
Single	31.2	25.5	37.4	11.4	8.2	15.6
All men	38.0	38.0	35.4	17.3	15.3	19.5
Female partnership status						
Partnered, p. not working	17.6	15.2	20.3	5.7	4.3	7.6
Partnered, p. working	35.1	31.3	39.1	28.5	21.7	36.4
Partnered, partner's work status unknown	(29.3)	(16.1)	(47.1)	(19.7)	(10.7)	(33.5)
Single	26.4	23.7	29.2	10.6	8.3	13.4
All women	24.8	24.8	23.2	10.2	8.7	11.8
Race/ethnicity						
White	30.9	29.2	32.7	13.6	12.4	14.9
Black (USA)	30.6	26.4	35.2	–	–	–
Other (USA)	33.2	25.3	42.1	–	–	–
Black and other (England)	–	–	–	(12.6)	(6.2)	(23.8)
All	30.9	29.4	32.6	13.6	12.3	14.9
Mean age of workers	68.3	68.1	68.4	67.9	67.7	68.2
Mean age of nonworkers	69.4	69.3	69.5	69.4	69.3	69.6
Base N	5,891			2,941		

Notes: Results are weighted. Results in brackets denote an estimate based on less than 30 cases.
[1] Percentage of those working within each category.
Source: Own analysis of the ELSA Wave 4 and the HRS Wave 9.

Moving on to single people, it is evident that employment rates did not vary dramatically on the basis of gender in either country. In the USA, 31.2 per cent of single men worked compared with 26.4 per cent of single women. Likewise, in England 11.4 per cent of single men worked compared with 10.6 per cent of single women. In both countries, single women were more likely to work than women with a nonworking partner, but less likely to work than partnered women with an employed partner. Unfortunately, due to sample sizes it is not possible to disaggregate employment levels of single women by whether they were divorced, separated, widowed or never married. However, we would expect different employment rates across these groups. For example, previous research suggests that in both countries employment is relatively high for older female divorcees, presumably for financial and social reasons (Smeaton and McKay 2003; Pleau 2010). The results suggest that partnership status is an important employment influence, although more research is needed on single people (Lain 2011).

The final section of Table 2.2 provides data on ethnicity and age. In neither country were there statistically significant employment differences between ethnic groups, although this may be due to the broad categories used. Table 2.2 also shows that workers were on average aged 68, compared with 69 for nonworkers.

2.4.4 Logistic regression of capacity and need factors

The above discussion suggests that capacity factors such as education and health are important influences on employment past age 65. However, it is important to disentangle these relationships; highly educated individuals may, for example, only be more likely to work because of better health. We therefore present a logistic regression analysis on whether an individual aged 65 to 74 is working (a score of one) or inactive (score of zero). This allows us to see the way in which one variable (education, for example) increases or decreases the likelihood of employment, controlling for the effects of other variables in the regression. The independent variables are those from the previous analysis.

To make the results more meaningful, we present average adjusted predictions (AAPs), as per the suggestion of Williams (2012). These show the likelihood of working of an 'average' person within a particular category (for example men). 'Average' equates to the average effect across all observed cases when only changing the variable in question. For example, the column titled AAPs in Table 2.3 shows that the 'average' man in the USA has a likelihood of working of 36.1 per cent (an AAP of 0.361), compared with 26.3 per cent for the 'average' women (an AAP of 0.263). This controls for the effects of other independent variables in the regression that might account for higher employment among men than women.

In addition to AAPs, we present average marginal effects (AMEs). These represent how much the likelihood of working differs between

Table 2.3 Logistic regression results for employment at age 65 to 74 in 2008 – average adjusted predictions (AAPs) and average marginal effects (AMEs)

	USA		England	
	AAPs	AMEs	AAPs	AMEs
Male	0.361*** (0.012)		0.157*** (0.010)	
Female	0.263*** (0.008)	−0.098*** (0.016)	0.109*** (0.008)	−0.048** (0.014)
Age	included		included	
Capacity factors				
College +	0.355*** (0.016)		0.163*** (0.017)	
Some college	0.314*** (0.014)	−0.041* (0.019)	0.141*** (0.014)	−0.021 (0.021)
High school graduate	0.294*** (0.009)	−0.061** (0.019)	0.151*** (0.015)	−0.012 (0.023)
Below high school	0.271*** (0.014)	−0.084** (0.025)	0.111*** (0.009)	−0.052* (0.020)
Fair/poor health	0.204*** (0.016)		0.078*** (0.012)	
Good health	0.341*** (0.009)	0.138*** (0.020)	0.149*** (0.008)	0.071*** (0.015)
ADL limits	included		included	
Need factors				
Highest wealth quartile	0.324*** (0.013)		0.136*** (0.012)	
2nd wealth quartile	0.282*** (0.011)	−0.042** (0.016)	0.140*** (0.012)	0.004 (0.016)
3rd wealth quartile	0.337*** (0.011)	0.012 (0.016)	0.136*** (0.013)	0.000 (0.018)
Lowest quartile	0.293*** (0.015)	−0.032 (0.020)	0.116*** (0.015)	−0.020 (0.020)
No private pension	0.291*** (0.012)		0.150*** (0.015)	
Has private pension	0.320*** (0.008)	0.029° (0.015)	0.128*** (0.007)	−0.022 (0.017)
Own house	0.278*** (0.010)		0.125*** (0.007)	
Buying house	0.366*** (0.013)	0.088*** (0.019)	0.192*** (0.025)	0.067* (0.026)
Renting/other	0.313*** (0.018)	0.035° (0.021)	0.141*** (0.019)	0.015 (0.021)

Table 2.3 (Continued)

	USA				England			
	AAPs		AMEs		AAPs		AMEs	
Household								
Partner not working	0.257***	(0.010)			0.090***	(0.007)		
Partner working	0.375***	(0.015)	0.118***	(0.019)	0.296***	(0.024)	0.207***	(0.025)
Partner's work status unknown	0.331***	(0.037)	0.074*	(0.037)	0.160***	(0.036)	0.071°	(0.037)
Single	0.330***	(0.014)	0.073***	(0.017)	0.133***	(0.013)	0.043*	(0.015)
Base N	*5,851*				2,825			

Notes: Regression analysis weighted; standard errors in brackets; 'included' denotes regression results from continuous variables that cannot be presented as AAPs or AMEs (see text); ***p < 0.001, **p < 0.01, *p < 0.05, °p < 0.10.
Source: Own analysis of the ELSA Wave 4 and the HRS Wave 9.

a particular group (for example women) and the reference category group (men). Table 2.3 shows that the AME for American women is −0.098, which indicates that their likelihood of working was 9.8 percentage points lower than that of American men. This AME is the difference between the AAP for women and the AAP for men (−0.098 = 0.263 − 0.361); whether or not the difference is statistically significant is shown. In this case, American women were significantly less likely to work than men, taking into account other variables in the regression. Similarly, the AME for English women shows that they had a reduced likelihood of working of around 5 percentage points (an AME of –0.048).

Starting with capacity factors, qualifications were associated with an increased employment likelihood in both countries, after taking into account the effects of other variables. The US AAPs show that over a third of those with college and above education were predicted to be employed compared with a quarter of those with below high school qualifications. The corresponding AMEs suggest that somebody in the USA with below secondary high school qualifications was 8.4 percentage points less likely to work than somebody with a college degree; this difference was significant. Indeed, the college-educated in the USA were significantly more likely to work than all of the qualification groups below them. In England, after having taken into account other factors, the results indicate that 16.3 per cent of college-educated people worked, compared with 11.1 per cent of those with below high school qualifications. The corresponding AME results show that this low educated group was 5.2 percentage points less likely to work than their college-educated counterparts, a result that was significant. For the middle three qualification groups, the AME signs were in the expected negative direction, which suggests a lower likelihood of working than for the highly educated. However, these results were not significant, perhaps partly due to smaller sample sizes. Nevertheless, in both countries having higher qualifications increased the likelihood of working significantly relative to the low educated, after taking into account other factors such as health.

As expected, poor health significantly reduced the likelihood of working in both countries. Somebody with good health was 13.8 percentage points more likely to be in employment in the USA than someone with fair or poor health; the corresponding figure for England was 7.1 percentage points. Limitations to activities of daily living were included in the regression and had the expected significant effects, but these results cannot be presented as AAPs because they are continuous variables.

Moving on to need factors, in both countries those in lower wealth quartiles were no longer less likely to work once we take into account capacity, household and other factors. This result is different from that in Lain (2011), who discovered that those with the lowest wealth in England were still significantly *less* likely to work after controlling for health, education and age in 2002. This may reflect the different time period examined. However, this chapter also uses a different measure of wealth (excluding housing) and is able to control for a broader range of factors that might explain lower rates of employment among those with the least wealth. In any case, the regression analysis suggests that lower levels of employment among the least wealthy can be attributed, in part, to lower work capacity and household factors.

The only aspect of relative advantage increasing the likelihood of working is having a private pension in the USA (albeit at the weak ten per cent level of significance). This may be due to the unmeasured relative advantage of those with pensions increasing their likelihood of working (for example related to job quality). Alternatively, people might be working as a response to financial incentives within DC pensions to increase their retirement incomes. We examine the influence of pension type on employment later.

The final measure related to need in Table 2.3 is housing status. In both countries, people buying their house were significantly more likely to still be working than those owning the house outright. 'Buyers' had an increased likelihood of working of 8.8 percentage points in the USA and 6.7 percentage points in England. In the USA, but not in England, being a renter also increased the likelihood of working (albeit at the ten per cent level of significance), taking into account other factors. Overall, then, need in the form of outstanding housing costs remained an important influence on employment in the regression.

The final segment of Table 2.3 confirms the importance of the household for employment. People in couples with a working partner had an increased likelihood of working of 11.8 percentage points in the USA and 20.7 percentage points in England; the employment status of the partner was therefore particularly important in England, as we saw earlier. In the USA, people with a non-responding partner were also significantly more likely to work than those reporting that their partner was not employed. Taking into account other variables in the regression, single people were more likely to work than married people with an inactive partner, presumably for a combination of reasons related to social contact and financial need.

In summary, the logistic regression analysis confirms the importance of capacity factors (health and education) and household factors on employment. Need factors in the form of outstanding mortgages also continued to be important. However, the lower likelihood of working for those in lowest wealth quartiles disappeared once we took into account other capacity and need factors. This supports previous research showing that low levels of health and education are impediments to working for the poorest (Lain 2011).

2.4.5 Workplace factors

A final set of factors examined relates to the workplace and the job held. This analysis focuses on individuals employed in 2002, drawing on work-related information from 2002 to examine how this influenced their employment in 2008. A slightly younger age cohort is examined, aged 62 to 70 in 2008, to ensure everyone was below 65 in 2002. This results in a sample size of 2,579 in the USA (with 1,474 workers in 2008) and 1,142 in England (516 workers).

Table 2.4 first presents employment outcomes in 2008 descriptively, broken down by the type of pension held with their employer, if any, by workers in 2002. This could be a salary-related DB pension, a DC pension providing a pension lump sum or a hybrid scheme. Hybrid schemes are similar to DC schemes in that they pay a lump sum upon retirement; however, like DB schemes they also retain some degree of predictability because the sponsor guarantees the interest rate (Johnson and Steuerle 2004). Only a minority of schemes are hybrids; these have been combined with DC schemes here because they provide similar incentives to continue working (Johnson and Steuerle 2004). Individuals could also be classed as being unsure of their pension type, or of having no pension at all with their current employer. The literature discussed above would suggest that people with DB pensions were less likely to work past age 65; Table 2.4 provides some evidence for this. In both countries, people with DB pensions in 2002 were around 10 percentage points less likely to be working than those with DC or hybrid pensions. However, this difference is only significant at the 95 per cent level in the USA. For England, the wide confidence intervals indicate that the lack of significance may be due to the small sample size for those with private pensions.

The second workplace factor, also related to need, is medical insurance in the USA. US employers have taken an important role in providing medical insurance due to the lack of a universal national health service of the kind found in England. Table 2.4, however, shows that being covered by employer-provided medical insurance in 2002 had no significant

Table 2.4 Influence of job characteristics in 2002 on employment in 2008 at age 62 to 70 (excludes non-employed in 2002)

	USA			England		
	% working in 2008[1]	Confidence interval (95%)		% working in 2008[1]	Confidence interval (95%)	
All workers from 2002	60.7	58.1	63.4	44.6	45.3	59.8
Job characteristics in 2002						
DB pension[2]	54.1	49.8	58.4	41.2	35.6	46.9
DC or hybrid pension	64.3	59.6	68.7	50.6	42.3	58.8
Unsure of type	(76.7)	(58.4)	(88.5)	(33.9)	(21.3)	(49.4)
No pension	61.0	57.1	64.7	45.4	41.5	49.5
Medical insurance	61.1	61.1	58.0			
No medical insurance	59.5	59.5	55.3			
Self-employed	75.3	68.2	81.3	52.6	45.3	59.8
Up to 19/24 employees[3]	59.2	51.4	66.6	40.7	33.4	48.4
20/25–99 employees	60.3	52.3	67.7	53.4	44.9	61.8
100–499 employees	58.5	52.4	64.4	42.6	32.8	53.0
500+ employees	57.5	47.4	67.0	41.7	37.4	46.2
Not answered	56.9	53.3	60.4	(31.9)	(15.6)	(54.3)
Strenuous job	58.3	54.5	62.0	45.0	39.7	50.4
Non-strenuous job	61.6	58.5	64.5	44.4	40.8	48.0
Mean working hours 2002 (only employed)	39.2	38.5	39.9	34.5	33.5	35.5
Mean working hours 2008 (only employed)	35.0	33.8	36.1	29.3	27.8	30.8
Base N	2,579			1,142		

Notes: Results are weighted. Results in brackets denote an estimate based on less than 30 cases.
[1] Percentage of those working within each category; [2] DB = defined benefit, DC = defined contribution; [3] ELSA: up to 19 employees, HRS: up to 24 employees.
Source: Own analysis of the ELSA Waves 1 and 4 and the HRS Waves 6 and 9.

bivariate effect on the likelihood of working in 2008. This is partly due to the fact that over 65s are covered by universal Medicare health insurance, which helps facilitate employment exit. However, Medicare insurance has had gaps in coverage (for example in relation to medication costs) and as a result it is likely to have complex influences on employment undetected here (see Green 2006).

Other workplace factors in Table 2.4 relate to the job itself. Individuals are categorized as being employees or self-employed in 2002; employees are disaggregated by workplace size. Previous research indicates that the majority of workers past age 65 are employees in the UK and the USA (Lain and Vickerstaff 2014). Nevertheless, in both countries workers over state pension age are more likely to be self-employed than younger workers (Smeaton and McKay 2003: 29; Karoly and Zissimopoulos 2004: 26). The self-employed may find it easier to continue working because they are less constrained by organizational policies and cannot be dismissed (Lain and Vickerstaff 2014); they are also less likely to have defined benefit pensions that encourage exit. Table 2.4 shows that in the USA self-employed individuals in 2002 were significantly more likely than employees to be still working in 2008.

In England, employment estimates were highest for the self-employed and those in smaller workplaces (with 25 to 99 staff). The results lack statistical significance but are consistent with previous research (Smeaton and McKay 2003; Cebulla et al. 2007). Small employers are less likely to provide defined benefit pensions that encourage earlier retirement. They may also have found it easier to accommodate requests for continued employment and for reduced or flexible working hours (Cebulla et al. 2007). This is because they typically have fewer bureaucratic rules, including fixed retirement ages (Metcalf and Meadows 2006: 74). With the abolition of mandatory retirement for employers of all sizes in 2011, we may see a reduced concentration in small organizations. In the USA, where large employers have been forbidden from setting compulsory retirement ages, there is no detectable difference in employment on the basis of workplace size in 2002. As Table 2.4 shows, mean working hours for those employed in 2002 and 2008 declined significantly, reflecting a preference among a significant number of older people for reduced hours (Lain and Vickerstaff 2014).

Finally, Table 2.4 shows employment rates in 2008 for people in strenuous and non-strenuous jobs in 2002. An individual had a strenuous job if they rated it as 'physical' or 'heavy labour' in England, or if it required physical effort 'all' or 'most' of the time in the USA. Despite the different underlying questions, the proportions defined as having strenuous jobs were similarly high in both countries at just under a third (30.3 per cent in England and 30.1 per cent in the USA, not shown in table). Perhaps surprisingly, having a strenuous job in 2002 did not increase the likelihood of leaving work by 2008 in either country. This is probably because people of relatively older age in strenuous jobs in 2002 were already a self-selecting group with a high capacity to work; this is something for future research to examine further. Nevertheless, the physical nature of many jobs past age 65 highlights the potential importance of capacity factors for realistically enabling employment in older age.

2.5 Conclusions

In future, the need to work past age 65 will increase considerably in the USA and the UK. In this context, the UK abolished mandatory retirement in 2011. In terms of considering the possible consequences, it may be insightful to look to the USA, where mandatory retirement ages were outlawed in 1986 to encourage those with modest pensions to continue working. American men and women in 2008 were more likely than those in England to work at all wealth, health and educational levels, and their employment was determined to a lesser degree by their partner's employment. While financial influences on employment were present in both countries, for example in relation to mortgage debt and pension type, these were stronger in the USA than in England, as would be expected given the US policy legacy.

Given the shifting policy context, we might expect the UK to move closer to the USA in terms of witnessing a continued increase in employment past 65 for financial and non-financial reasons. However, close examination of the US results indicates that disadvantages are likely to constrain employment past age 65 for many people. Capacity factors – low levels of health and education – significantly reduce the likelihood of working in the USA and England. In both countries, the descriptive analysis shows that somebody with a degree was twice as likely to work as somebody with below secondary qualifications. However, older people in England are less highly educated than their counterparts in the USA; low qualifications will make it harder for many to remain in employment, particularly as they will be competing with more highly educated younger cohorts. The descriptive results also show that those in the lowest wealth quartile were least likely to remain in work in both countries, a reduced likelihood that was no longer there once capacity and other factors were controlled for in the regression analysis. This suggests that lower employment among the poorest is, in part, explained by disadvantages, including low education and bad health built up earlier in the life course.

Further challenges to extending working life emerge when we recognize that employment behaviour is influenced by external factors. People in both countries make decisions about employment from within a particular domestic situation and are less likely to work if they have an inactive partner. This may be to coincide leisure activities or perhaps to care for a partner if they are ill (McGeary 2009). In addition, workplace factors are likely to influence whether people remain in employment, such as whether they are self-employed or an employee or perhaps have access to part-time work.

Capacity, household and workplace factors therefore influence and constrain prospects for work beyond age 65. Abolishing mandatory retirement, although broadly positive, does little to help those with poor employment prospects, some of whom have paid pension contributions from an early age and will have to rely on diminishing state financial support. We therefore need policies that recognize long contribution periods and the fact that not everyone will realistically work up to and beyond age 65.

Notes

1. The exceptions are where a fixed retirement age exists for safety reasons (Macnicol 2006: 237), or where the employer has fewer than 20 staff.
2. Employers contribute less on average to DC schemes (ONS 2013; Ghilarducci 2008). Furthermore, there has been a sharp decline in the regular income an annuity will buy using a DC pension pot (*The Guardian* 2013a). This has led the UK to join the USA in not requiring 'DC pensioners' to buy an annuity.
3. The weight provided for the self-completion survey was used.

References

Adams, S. J. (2004), 'Age discrimination legislation and the employment of older workers', *Labour Economics*, 11 (2), 219–41.

Aneshensel, C. (2013), *Theory-based data analysis for the social sciences*, Thousand Oaks, CA: Sage Publications.

Arkani, S. and Gough, O. (2007), 'The impact of occupational pensions on retirement age', *Journal of Social Policy*, 36 (2), 297–318.

Banks, J., Emmerson, C. and Tetlow, G. (2007), *Healthy retirement or unhealthy inactivity: How important are financial incentives in explaining retirement*, London: Institute for Fiscal Studies, http://www.ifs.org.uk/publications/3972, date accessed 27 November 2014.

Béland, D. and Waddan, A. (2012), *The politics of policy change: Welfare, medicare, and social security reform in the United States*, Washington, DC: Georgetown University Press.

Berkowitz, E. and DeWitt, L. (2013), *The other welfare: Supplemental security income and U.S. social policy*, New York: Cornell University Press.

Blekesaune, M., Bryan, M., Taylor, M. and Britain, G. (2008), *Life-course events and later-life employment*, DWP Research Report No 502, Norwich: Department for Work and Pensions.

Butrica, B. A. and Karamcheva, N. S. (2013), *Does household debt influence the labor supply and benefit claiming decisions of older Americans?*, CRR WP 2013–22, Chestnut Hill, MA: Center for Retirement Research at Boston College.

Cebulla, A., Butt, S. and Lyon, N. (2007), 'Working beyond the state pension age in the United Kingdom: The role of working time flexibility and the effects on the home', *Ageing and Society*, 27 (6), 849–68.

Clark, G. L. (2006), 'The UK occupational pension system in crisis', in: H. Pemberton, P. Thane and N. Whiteside (eds.), *Britain's pensions crisis: History and policy*, Oxford: Oxford University Press, 145–68.

Crawford, R. and Tetlow, G. (2010), 'Employment, retirement and pensions', in: J. Banks, C. Lessof, J. Nazroo, N. Rogers, M. Stafford and A. Steptoe (eds.), *Financial circumstances, health and well-being of the older population in England*, London: Institute for Fiscal Studies.

DWP (2007), *The pensioners' incomes series 2005/6 (revised)*, London: Department for Work and Pensions, http://web archive.nationalarchives.gov.uk/20100407010852/http://dwp.gov.uk /docs/pens024-220507.pdf, date accessed 12 March 2015.

Ebbinghaus, B. (2006), *Reforming early retirement in Europe, Japan and the USA*, Oxford: Oxford University Press.

Elder, T. E. and Powers, E. T. (2006), 'The incredible shrinking program', *Research on Aging*, 28 (3), 341–58.

Esping-Andersen, G. (1990), *The three worlds of welfare capitalism*, Princeton: Princeton University Press.

Friedberg, L. and Webb, A. (2005), 'Retirement and the evolution of pension structure', *Journal of Human Resources*, 40 (2), 281–308.

Ghilarducci, T. (2008), *When I'm sixty-four: The plot against pensions and the plan to save them*, Princeton: Princeton University Press.

Green, C. A. (2006), 'The unexpected impact of health on the labor supply of the oldest Americans', *Journal of Labor Research*, 27 (3), 361–79.

The Guardian (2013a), *Annuity rates rise but reprieve may be short-lived*, by J. Pappworth, 16 March 2013, http://www.theguardian.com/money/2013 /mar/16/annuity-rates-rise-reprieve-short-lived, date accessed 20 January 2015.

The Guardian (2013b), *A pension age of 70? That's what is in store for overburdened generation Y*, by T. Clark, 5 December 2013, http://www.theguardian .com/society/2013/dec/05/pension-age-70-in-store-generation-y, date accessed 27 November 2014.

Haider, S. and Loughran, D. (2001), *Elderly labor supply: Work or play?*, CRR WP 2001–04, Boston: Center for Retirement Research at Boston College.

Hall, P. A. and Soskice, D. W. (eds.) (2001), *Varieties of capitalism: The institutional foundations of comparative advantage*, Oxford: Oxford University Press.

Harrington Meyer, M. and Herd, P. (2007), *Market friendly or family friendly? The state and gender inequality in old age*, New York: Russell Sage Foundation.

Hayes, B. C. and Vandenheuvel, A. (1994), 'Attitudes toward mandatory retirement – An international comparison', *International Journal of Aging & Human Development*, 39 (3), 209–31.

Hicks, P. (2001), *Public support for retirement income reform*, OECD Labour Market and Social Policy Occasional Papers No. 55, Paris: OECD.

Johnson, R. W. and Steuerle, E. (2004), 'Promoting work at older ages: The role of hybrid pension plans in an aging population', *Journal of Pension Economics and Finance*, 3 (3), 315–37.

Kanabar, R. (2012), *Unretirement in England: An empirical perspective*, Discussion Paper Series No. 12/31, York: Department of Economics and Related Studies, University of York.

Karoly, L. A. and Zissimopoulos, J. (2004), 'Self-employment among older US workers', *Monthly Labor Review*, 127 (7), 24–47.

Lain, D. (2011), 'Helping the poorest help themselves? Encouraging employment past 65 in England and the USA', *Journal of Social Policy*, 40 (3), 493–512.

Lain, D., Vickerstaff, S. and Loretto, W. (2013), 'Reforming state pension provision in "liberal" anglo-saxon countries: Re-commodification, cost-containment or recalibration?', *Social Policy and Society*, 12 (1), 77–90.
Lain, D. and Vickerstaff, S. (2014), 'Working beyond retirement age: Lessons for policy', in: S. Harper and K. Hamblin (eds.), *International handbook on ageing and public policy*, Cheltenham: Edward Elgar, 242–55.
Leisering, L. (2003), 'Government and the life course', in: J. Mortimer and M. Shanahan (eds.), *Handbook of the life course*, New York: Kluwer Academic/Plenum, 205–25.
Lindert, P. H. (2004), *Growing public: Volume 1, the story: Social spending and economic growth since the eighteenth century*, Cambridge: Cambridge University Press.
Loretto, W. and Vickerstaff, S. (2013), 'The domestic and gendered context for retirement', *Human Relations*, 66 (1), 65–86.
Macnicol, J. (2006), *Age discrimination: An historical and contemporary analysis*, Cambridge: Cambridge University Press.
Mann, A. (2011), 'The effect of late-life debt use on retirement decisions', *Social Science Research*, 40 (6), 1623–37.
McGeary, K. A. (2009), 'How do health shocks influence retirement decisions?', *Review of Economics of the Household*, 7 (3), 307–21.
Metcalf, H. and Meadows, P. (2006), *Survey of employers' policies, practices and preferences relating to age, DWP Research Report No. 325*, London: Department for Work and Pensions.
Neumark, D. and Stock, W. A. (1999), 'Age discrimination laws and labor market efficiency', *Journal of Political Economy*, 107 (5), 1081–125.
OECD (Organisation of Economic Co-operation and Development) (2004), *Education at a glance 2004*, Paris: OECD.
ONS (2012), *2011 Census: Population estimates for the United Kingdom, 27 March 2011*, London: Office for National Statistics, http://www.ons.gov.uk/ons/dcp171778_292378.pdf, date accessed 28 January 2014.
ONS (2013), *Pension trends chapter 8: Pension contributions, 2013 edition*, Newport: Office for National Statistics, http://www.ons.gov.uk/ons/dcp171766_310458.pdf, date accessed 27 November 2014.
Parry, J. and Taylor, R. (2007), 'Orientation, opportunity and autonomy: Why people work after state pension age in three areas of England', *Ageing & Society*, 27 (4), 579–98.
Pensions Commission (2004), *Pensions: Challenges and choices. The first report of the pensions commission*, Norwich: T. S. Office.
Phillips, D., Chien, S., Moldoff, M., Lee, J. and Zamarro, G. (2012), *RAND ELSA data documentation, version B*, Bethesda, MD: Labor & Population Program, National Institute on Aging/National Institutes of Health, http://doc.ukdataservice.ac.uk/doc/5050/mrdoc/pdf/5050_RAND_ELSA_B.pdf, date accessed 4 November 2014.
Pienta, A. M. (2003), 'Partners in marriage: An analysis of husbands' and wives' retirement behavior', *Journal of Applied Gerontology*, 22 (3), 340–58.
Pleau, R. L. (2010), 'Gender differences in postretirement employment', *Research on Aging*, 32 (3), 267–303.

Priemus, H., Kemp, P. A. and Varady, D. P. (2005), 'Housing vouchers in the United States, Great Britain, and the Netherlands: Current issues and future perspectives', *Housing Policy Debate*, 16 (3-4), 575-609.
Quadagno, J. S. and Hardy, M. (1991), 'Regulating retirement through the age-discrimination in employment act', *Research on Aging*, 13 (4), 470-5.
Rimlinger, G. V. (1971), *Welfare policy and industrialization in Europe, America and Russia*, New York: Wiley.
Scherger, S., Hagemann, S., Hokema, A. and Lux, T. (2012), *Between privilege and burden: Work past retirement age in Germany and the UK, ZeS-Working Paper No. 4/2012*, Bremen: Centre for Social Policy Research.
Sefton, T., Evandrou, M. and Falkingham, J. (2007), *Mapping the incomes of older people in the UK, US and Germany, CRA Discussion Paper No. 0704*, Southampton: Centre for Research on Ageing.
Smeaton, D. and McKay, S. (2003), *Working after state pension age: Quantitative analysis, DWP Research Report No. 182*, Leeds: Department for Work and Pensions.
Smeaton, D., Vegeris, S., Sahin-Dikmen, M. and Britain, G. (2009), *Older workers: Employment preferences, barriers and solutions*, Manchester: Equality and Human Rights Commission.
SSA (2014), *Retirement planner: Benefits by year of birth*, Baltimore: Social Security Administration, http://www.ssa.gov/retire2/agereduction.htm, date accessed 27 November 2014.
Thurley, D. (2011), *Pension age: Occupational and personal pensions, Standard Note SN/05847*, London: House of Commons Library.
Vickerstaff, S. (2006), ' "I'd rather keep running to the end and then jump off the cliff." Retirement decisions: Who decides?', *Journal of Social Policy*, 35 (3), 455-72.
Wang, M., Zhan, Y. J., Liu, S. Q. and Shultz, K. S. (2008), 'Antecedents of bridge employment: A longitudinal investigation', *Journal of Applied Psychology*, 93 (4), 818-30.
Williams, R. (2012), 'Using the margins command to estimate and interpret adjusted predictions and marginal effects', *Stata Journal*, 12 (2), 308-31.
Williamson, J. B. and Pampel, F. C. (1993), *Old-age security in comparative perspective*, Oxford: Oxford University Press.

3
The Social Stratification of Work Beyond Pension Age in Germany and the UK: Quantitative and Qualitative Evidence

Anna Hokema and Thomas Lux

3.1 Introduction

In both Germany and the UK, labour force participation of people past state pension age is increasing. The employment ratio for those aged 65 to 69 has risen strongly in both countries, and it nearly doubled from 2001 to 2013 (Figure 3.1). This holds true for both men and women. Rising figures of working pensioners are also reported regularly by the media such as the BBC ('The pensioners who choose to work' – BBC News 2013) or the German Spiegel Online in a more provocative fashion ('Work despite retirement: "I'm condemned to clean"' – own translation, Spiegel Online 2012). The question often underlying these reports is whether these pensioners work past state pension age because of financial need or for other motives. This is also debated in academia (see, for example, Komp et al. 2010; Lain 2011; Brenke 2013; Hochfellner and Burkert 2013), a discussion that we join in. In order to investigate the influences of very divergent institutional and structural settings, we compare working beyond pension age in Germany and the UK. Furthermore, we use quantitative and qualitative data, allowing us to analyse objective circumstances in combination with subjective experiences of people working beyond pension age.[1]

The chapter is structured as follows: In Section 3.2, we describe the institutional and structural contexts that influence working beyond pension age. Then we derive, in Section 3.3, guiding questions for the subsequent analyses. In Section 3.4, we introduce our quantitative and qualitative data. After that, in Section 3.5, we quantitatively analyse the job characteristics of post-retirement work and differences

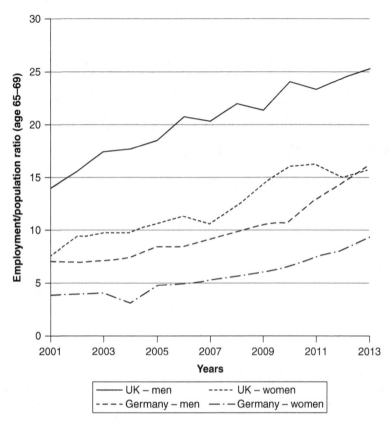

Figure 3.1 Employment ratio of people aged 65 to 69 in Germany and the UK (2001–13)
Source: OECD 2014.

between working and nonworking people beyond retirement age. In Section 3.6, we illustrate these findings with four qualitative case studies to exemplify the subjective perspective on work in retirement. Lastly, in Section 3.7, we summarize and discuss our findings and evaluate how fruitful the integration of qualitative and quantitative data is with regard to this topic.

3.2 The institutional and structural context of post-retirement work in Germany and the UK

Germany and the UK differ strongly with respect to their welfare state and their labour market. The German welfare state is usually characterized as conservative (Esping-Andersen 1990), with the principle of public insurance being very important. With respect to the pension system, this means that there are comprehensive and generous earnings-related social insurance pensions for the majority of employees, while occupational and private pensions are less widespread (Schulze and Jochem 2007).[2] Thus, the vast majority of current German pensioners receive the main share of their retirement income from public pensions. Important exceptions are the self-employed, whose main income sources are often private pensions, and civil servants for whom an extra system exists. It is important to note that in recent decades this system has undergone important structural reforms, amongst others the abolition of early retirement routes, which have created considerable pressures to prolong individual working lives (Buchholz et al. 2011). Furthermore, but more important for future retirees, retirement age has been increased to 67, private pensions have been strengthened through subsidies and a decrease of the level of (future) public pensions has been decided.

The UK is classified as a liberal welfare state, which relies more strongly than Germany on private welfare provision (Esping-Andersen 1990). With respect to old-age provision, this means that every pillar of the pension system is important. The comprehensive but modest flat-rate basic State Pension covers employees and self-employed and is based on prior National Insurance contributions. It is complemented by the earnings-related additional public pension (State Second Pension).[3] In addition, occupational and private pensions are relatively widespread but unequally distributed (Schulze and Moran 2007). This institutional setup results in higher income inequality in old age in the UK than in Germany as many employees do not have sufficient income from occupational and private pensions (Möhring 2013: 301).[4] Furthermore, it also results in higher financial uncertainty for old age because private pensions are generally more volatile in their outcome due to fluctuations in the financial markets such as the financial crisis 2008–09 (Möhring 2013: 297).

With regard to the regulation of labour, Germany is classified as a coordinated market economy (Hall and Soskice 2001) with strong employment protection. This results in a rigid labour market that distinguishes heavily between insiders – those in secure employment – and outsiders. Recent policy changes have led to a liberalization and flexibilization of the labour market which has increased the prevalence of insecure work. Furthermore, the German occupational structure is characterized by strong boundaries between qualifications and occupations because of highly standardized educational and vocational training (Buchholz et al. 2011). The UK, by contrast, is classified as a liberal market economy (Hall and Soskice 2001) with weak employment protection and high importance of on-the-job training which makes insider/outsider markets less important and occupational boundaries more permeable (Buchholz et al. 2011). As a result, job mobility and occupational mobility over the life course are much lower in Germany than in the UK (Allmendinger and Hinz 1998), which might also result in lower rates of job change in later life. Furthermore, the rigid labour market in Germany creates stronger barriers for reintegration of older workers who – voluntarily or involuntarily – left the labour market. This is reflected in higher unemployment rates for the German population aged 55 to 64 compared to the UK (OECD 2013a: 245).[5]

3.3 Guiding questions for the empirical analyses

Explanations for working beyond pension age can refer to the cultural level (such as age and life-course norms), the institutional level (such as pension legislation and labour market regulation), the structural level (such as the degree of inequality and unemployment), the organizational level (such as personnel policy of specific companies) and the individual level (such as individual income and education) (Scherger et al. 2012). Furthermore, explanations at the individual level can be differentiated into objective characteristics, on the one hand, and subjective experiences, on the other hand. In the following analyses, we will focus on influences on working beyond pension age at the institutional, structural and (objective and subjective) individual level.

The guiding questions for this are as follows:

(1a) *Structural and institutional level:* How does the higher income inequality in old age in the UK affect the job characteristics of those working beyond pension age, compared to middle-aged workers?

(1b) *Structural and institutional level:* How do the rigid labour market and the strong occupational boundaries in Germany affect the job characteristics of those working beyond pension age, compared to middle-aged workers?

(2) *Objective individual level:* Which socio-economic differences can be found between those working beyond pension age and those not working? Are there indications of different groups who have different reasons for such work?

(3) *Subjective individual level:* Do different socio-economic groups reflect differently on their post-retirement work? Do the subjective accounts contain explicit references to institutional regulations?

3.4 Data and methods

For the quantitative analyses, we use the German Socio-Economic Panel (GSOEP) and the British Household Panel Survey (BHPS). These data sets are representative yearly surveys of private households in Germany and the UK, collecting information on respondents aged 16 years and older who are re-interviewed every year (Haisken-DeNew and Frick 2005; Taylor et al. 2010). For our analyses, we use the data from 2009 (Germany) and 2008 (UK) and focus on respondents beyond state pension age, which at that time was 65 in Germany, 65 for British men and 60 for British women.[6] In the quantitative analyses, respondents are defined as working (1) if they state that their current working status is full-time employed, part-time employed, self-employed or marginally/irregularly employed, (2) if they state that they currently have some kind of second job or (3) if they have been engaged in paid work in the last seven days.

For the qualitative data analysis, we drew from a sample of 49 problem-centred interviews (Witzel and Reiter 2012) with persons past state pension age and in paid work. The interviews were conducted in Germany and the UK in 2011 and 2012. The sample was stratified according to gender, country and the qualification level of the job. Based on the transcribed interviews, both a theory-driven and a data-driven coding technique were applied. For the purpose of this chapter, we present four individual case studies to show how the subjective experiences of working and individual biographies are embedded in institutional regulations. The qualitative and quantitative approaches are integrated at several stages: the overall research objective applies to the analysis of both kinds of data, and the selection of qualitative cases is based on the quantitative results. In this way, quantitative and qualitative results are integrated to mutually validate or complement each other (Erzberger and Kelle 2003).

3.5 Quantitative findings

First, we examine the particular features of post-retirement work in both countries by comparing the hours worked, the work income (quartiles) and the class of jobs people beyond pension age work in with those of workers aged 45 to 55.[7] For this purpose, we use the class scheme of Erikson and Goldthorpe (EGP scheme). It distinguishes jobs based on the ownership of the means of production, the employment relations and human capital specificity. Accordingly, the scheme differentiates between the high service class and the low service class (who enjoy the best socio-economic conditions in terms of income, job security and work autonomy, for example managers or teachers), on the one hand, and, on the other, the semi- and unskilled manual workers, the low-routine non-manuals (who face difficult socio-economic conditions, for example cleaners or salespersons) and the skilled manual workers (who also face difficult socio-economic conditions, but to a lesser degree, for example electricians). The class of high-routine non-manuals (for example office clerks) falls in between the classes just mentioned, and two additional classes include those who are self-employed (Erikson and Goldthorpe 1993; Goldthorpe 2007).

In a second step, we examine the differences between those working and those not working beyond pension age in both countries. For these multivariate analyses, we primarily focus on the respondents' educational level and income and control for age, health, gender, marital status, region and pension receipt.[8] The educational level is recoded on the basis of the CASMIN classification (Brauns and Steinmann 1999). It is 'very low' for those with at most general elementary education (CASMIN 1a, 1b), 'low' for those with general elementary education and vocational qualification (CASMIN 1c) and 'medium' for those with general maturity certificate and vocational qualification at the most (CASMIN 2a, 2b, 2c_gen, 2c_voc). And lastly, it indicates 'high' education for those with lower or higher tertiary education (CASMIN 3a, 3b). For income, we use the total gross monthly household income (which includes income from work, pensions, social benefits, transfers and investments of all household members). This is reduced by the monthly gross labour earnings of the person working beyond pension age.[9] The reduced income is equalized on the basis of the new OECD scale to account for differences in size and composition of households. Finally, this reduced equivalent income is divided into five income quintiles. The first quintile includes those people with the lowest reduced gross equivalent income (not exceeding £730 in the UK and €1015 in Germany), while the fifth quintile contains those with the highest incomes (at least £1626 and €2110).[10] It is important to note that when constructed in this way the first quintile also includes individuals who have a high earned income but no other household income sources, as may be the case with single persons who defer pension receipt. The variable 'pension receipt' indicates that the respondent currently receives the basic State Pension in the UK, and any kind of old age pension in Germany (for example state pension, civil service pension, occupational pension, private pension). In both cases, this only refers to pensions which are based on the respondent's own employment record.

In Table 3.1, we compare people working beyond retirement age with workers aged 45 to 55 (referred to in the following as the middle-aged). With respect to the class position, those past pension age in the UK work less frequently in the high service class than the middle-aged,

Table 3.1 Characteristics of current job for those working beyond pension age and workers aged 45 to 55 in the UK and in Germany (2008–09)

	UK		Germany	
	45–55	60+/65+	45–55	65+
EGP class (column percentages)				
High service	21.5	8.4	12.9	12.7
Low service	25.0	24.2	21.8	22.0
High routine non-manual	12.4	11.7	9.9	2.3
Self-employed with employees	2.8	4.9	3.1	6.8
Self-employed w/o employees	6.1	8.5	4.7	18.1
Skilled manual	10.7	8.5	14.6	5.3
Low routine non-manual	7.5	14.7	11.5	7.5
Semi-, unskilled manual	14.0	19.1	21.5	25.3
n (unweighted)	1,839	395	3,737	399
Actual weekly hours worked (column percentages)				
1–10 hours	2.5	23.9	5.3	52.3
11–20 hours	8.3	28.3	8.4	22.1
21–40 hours	53.4	38.0	42.8	14.3
40+ hours	35.8	9.8	43.5	11.3
n (unweighted)	2,004	402	3,631	355
Monthly gross work income (in Euro)[1]				
P25	1,365	473	1,373	200
P50 (median)	2,364	921	2,558	390
P75	3,637	1,785	3,530	1,032
n (unweighted)	1,888	395	3,708	359

Notes: [1]Only incomes greater than zero; UK incomes transformed into € according to exchange rate at beginning of 2008.
Source: GSOEP 2009 and BHPS 2008, weighted.

but there is no such difference in Germany. Concerning the low service class, there is hardly any difference between the two age groups in both countries. At the lower end of the class structure, jobs in the semi- and unskilled manual class are more widespread among those beyond pension age than among the middle-aged in both countries, but this difference is more pronounced in the UK. Furthermore, the share of the low routine non-manual class is also larger for those beyond pension age than for the middle-aged in the UK, whereas the contrary is true in Germany. At the middle of the class structure, jobs in the skilled manual class and the high routine non-manual class are less common among those working beyond pension age compared to middle-aged workers in both countries, but to a greater extent in Germany. Finally, self-employment with and without employees is more frequent among those beyond pension age in both countries. However, for the self-employed without employees, this difference is much more pronounced in Germany than in the UK.

In order to show the country differences more clearly, we used a reduced 5-class EGP scheme in Figure 3.2.[11] Furthermore, to account for the cohort-related structural changes in the labour market, we also show the class positions of the middle-aged group in 1991, which is the year in which a considerable number of people beyond pension age in 2008–09 were aged 45 to 55 years.

The figure shows that, in comparison to the class profile of middle-aged workers (in 1991 and in 2008–09), the class profile of those working beyond pension age is shifted towards low-qualified and low-income jobs in the UK. Such a shift is not visible in Germany where, by contrast, self-employment (and especially self-employment without employees) shows the highest increase.[12] Furthermore, due to longer working hours, the earnings of those working beyond pension age are much higher in the UK than in Germany (see Table 3.1), which might speak for a higher need for earned income during retirement age in the UK.

Taken together, this pattern of job characteristics (class, income and hours worked, as shown in Table 3.1) is in line with the interpretation that the higher post-retirement employment rates in the UK result (at least partly) from higher inequalities in old age (which increase the economic need for post-retirement work with long working hours and high working incomes, especially for employees in low class occupations) and from a more flexible labour market in the UK (which increases the opportunities for paid employment in old age). In Germany, opportunities for dependent employment beyond pension age seem to

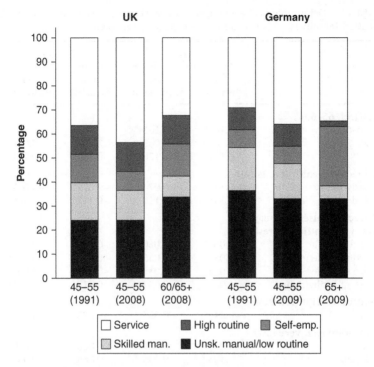

Figure 3.2 EGP class (reduced version) of current job of workers in pension age in 2008–09, workers aged 45 to 55 in 1991 and workers aged 45 to 55 in 2008–09 (Germany and the UK)
Note: UK 1991 is based only on the English sample.
Source: GSOEP 1991 and 2009; BHPS 1991 and 2008, weighted.

be limited, which is reflected by the high share of self-employment in that age group.[13]

In order to analyse the differences between people beyond pension age who work and those who do not work, we calculated a set of logistic regressions. Their results are shown in Table 3.2. Columns one and three show the average marginal effects resulting from several country-specific bivariate regressions, and columns two and four show the average marginal effects of the country-specific multivariate regressions.

In the bivariate regression for the UK with the reduced household equivalent income as the only explanatory variable (Table 3.2, column 1), we observe a significantly higher probability of working for those in the lowest (the poorest) and in the two highest (the richest) quintiles. In more detail, the probability of working beyond pension

Table 3.2 Determinants of working beyond pension age in Germany and the UK in 2008–09 (binary logistic regression – average marginal effects)

	UK		Germany	
	(1) Bivariate	(2) Multivariate	(3) Bivariate	(4) Multivariate
Age (ref. 65–69)				
60–64	0.165**	0.185**	–	–
70–74	−0.105**	−0.083**	−0.093**	−0.074**
75+	−0.151**	−0.120**	−0.132**	−0.111**
Subjective health (ref. good)				
Satisfactory	−0.111**	−0.074**	−0.077**	−0.037**
Bad	−0.145**	−0.117**	−0.112**	−0.060**
Reduced household income (ref. 1st quintile: poorest)				
2nd quintile	−0.061**	−0.050**	−0.021	−0.032+
3rd quintile	−0.056**	−0.055**	−0.036**	−0.061**
4th quintile	−0.012	−0.035+	−0.025+	−0.069**
5th quintile (richest)	0.032	−0.034+	0.018	−0.071**
Pension receipt (ref. none)				
Yes	−0.246**	−0.095**	−0.020	−0.025
Education (ref. very low)				
Low	0.056**	0.039*	0.050**	0.040**
Middle	0.105**	0.044**	0.064**	0.053**
High	0.110**	0.044**	0.127**	0.111**
Gender (ref. male)				
Female	0.024*	−0.056**	−0.062**	−0.038**
Family status (ref. married)				
Divorced/separated	0.051*	0.037+	0.050*	0.028+
Single	−0.051*	−0.029	−0.026	−0.013
Widowed	−0.103**	−0.025	−0.059**	−0.017
Region UK (ref. North England)				
South England	0.059**	0.041*	–	–
Wales	−0.029+	−0.025	–	–
Scotland	0.005	0.005	–	–
Northern Ireland	−0.016	−0.010	–	–
Region Germany (ref. West)				
East Germany			−0.055**	−0.061**
Nagelkerkes R^2		0.31		0.19
McKelvey & Zavoinas R^2		0.43		0.31
Observations		3,241		4,769

Notes: Men older than 64 (UK and Germany), women older than 64 (Germany) and than 59 (UK); dependent variable: working beyond retirement age (yes/no); $^+p < 0.10$, $^*p < 0.05$, $^{**}p < 0.01$.
Source: GSOEP 2009 and BHPS 2008, unweighted.

age in the lowest income quintile is 6.1 percentage points higher than in the second quintile and 5.6 percentage points higher than in the third quintile, while the difference between the first quintile, on the one hand, and the fourth and fifth quintile, on the other, is not statistically significant. However, if we control for age, health, pension receipt, education, marital status, gender and region (Table 3.2, column 2), the probability of working beyond pension age in the lowest income quintile is significantly higher than in all other quintiles. Additional stepwise regressions show that this reduction of the probability in the two highest quintiles (and the resulting significant difference to the lowest) is mainly the result of controlling for education (not shown).[14] This means that the high probability of the two quintiles with the highest income in the bivariate model is mostly due to the higher level of education in these groups, resulting in better labour market opportunities and probably also in a higher work motivation, for example because of larger non-financial gains from work such as having interesting work tasks or high work-related prestige.

For Germany, we find a similar pattern: a u-shaped influence of income in the bivariate model which becomes a higher probability only for the lowest income quintile after controlling for additional aspects (especially education). In both countries, this probably speaks for different reasons for working beyond pension age for different groups: financial needs on the one hand and non-financial gains on the other. However, one could argue that respondents in the lowest quintile are not really or not only those with limited finances because what is included here is the household equivalence income without the income from working. This means that the lowest quintile probably also includes people who earn a lot of money and have no other sources of income as a household because they defer pension receipt. As we control for pension receipt in the multivariate models (Table 3.2, models 2 and 4) and we can replicate the u-shaped pattern of income in bivariate regressions which are restricted to respondents with pension receipt (basic State Pension in the UK, any pension in Germany; table not shown), this objection does not invalidate the above interpretation that individuals in the lowest quintile mainly work for financial reasons.[15]

The multivariate results for the control variables are – in both countries – in line with what one might expect (Table 3.2, columns 2 and 4): men, the 'younger old', those with good health, those with high education, divorced people and those living in regions with low unemployment (South England and West Germany) are more likely to work beyond pension age. Pension receipt only has an effect in the UK: not receiving the basic State Pension is related to a higher probability of post-retirement work, which is not the case for the receipt of a pension in Germany. This might indicate that there is indeed a distinct group of pension deferrers in the UK, but not in Germany.

In a last step, we estimate the probability of working beyond pension age for specific groups of people (specific combinations of independent variables) on the basis of the multivariate regressions. We focus on married men receiving a pension as this group is often regarded as the most important group with continuous work biographies (male breadwinner model).[16] In more detail, we focus on married men receiving a pension, aged 65 to 69 (as this is the age at which working beyond pension age is most likely), and living in West Germany or North England (as these are the regions where the qualitative interviews were carried out). Within the group defined in this way, we differentiate between men in good socio-economic circumstances (high education, highest quintile of reduced household income, good health), men in medium circumstances (medium education, third income quintile, satisfactory health) and men in poor socio-economic circumstances (low education, first income quintile, satisfactory health). In both countries, men living in good circumstances are most likely to work beyond pension age (0.371 in Germany vs. 0.291 in the UK), while men in poor circumstances are somewhat less likely to do so (0.287 vs. 0.187). However, they are still more likely to work than men living in medium socio-economic circumstances (0.189 vs. 0.117).[17] As the analyses above suggest that the two extreme groups (good and poor socio-economic circumstances) in both countries might have different reasons for working, we will now use the qualitative interviews to reconstruct the work biographies of exemplary actors in these groups and examine their subjective motives with respect to post-retirement work.

3.6 Qualitative findings

In this section, we present one exemplary case from each of the two just mentioned groups of working pensioners in Germany and the UK: Henry Bell[18] and Gerd Schmitt are in poor socio-economic circumstances, and Michael Turner and Holger Weiss in good circumstances (see Table 3.3).

Finding comparable cases in the qualitative sample which correspond exactly to the combination of characteristics mentioned above proved

Table 3.3 Case overview

	Henry Bell (UK, 66)	Michael Turner (UK, 68)	Gerd Schmitt (DE, 69)	Holger Weiss (DE, 66)
Education	Low (in-house-training)	High (university)	Low-middle (vocational training)	High (university)
Health status	Satisfactory	Good	Bad[4]	Good
Monthly household income[1]	£1.800+	£1.800+	€1.500–1999[4]	€2.500–2.999
Income from work[2]	£25.000/year	£25.000/year	€313/month	€150–300/month
Pension	State p.	State p., occupational p.	State p., occupational p.	State p., occupational p.
Socio-economic circumstances	Poor	Good	Poor	Good
Marital status	Single[4]	Married	Separated[4]	Married
Probability[3]	0.138	0.291	0.278	0.380

Notes: [1]Household income including income from work; [2]Before tax; [3]Probability of working beyond pension age estimated according to the combination of individual characteristics on the basis of the multivariate logistic regressions in Table 3.2; [4]Indicates attributes that do not correspond to groups described quantitatively above.
Source: Own compilation based on cases presented in the text.

difficult. Nevertheless, despite some deviations, the chosen cases can be regarded as examples of the two socio-economic groups. The deviations are as follows: Henry Bell is single instead of married and is in the highest predefined (open-ended) household income bracket provided by our questionnaire. However, Henry currently only receives his working wage and his basic State Pension, and can thus be assigned with relative certainty to the lowest income quintile (not including income from work). Gerd Schmitt has poor health, is separated instead of married and belongs to the second income quintile. However, he pays alimony to his wife, which worsens his financial situation. Holger Weiss is in the fourth instead of the fifth income quintile.

3.6.1 Henry Bell

Henry Bell's work career has been discontinuous; he left school at 15 without any formal qualifications and worked in different engineering firms in his hometown. In his 40s, after a number of in-house training courses, he had worked his way up to the position of distribution manager. However, in his late 40s, he lost this job when his employer relocated to a different part of the country, and he could not move because he was looking after his elderly mother at the time. Through an acquaintance, he heard about a job offer as a porter at the university. He applied and got this job in the – for him unknown – sector of service work. In the following years, he moved positions and up the career ladder inside his workplace until he reached his current position as building superintendent in 2001, aged 56. In this full-time job, Henry supervises a group of cleaners and is the general caretaker of one of the university buildings. He has worked past state pension age, without any changes to his work arrangement; he started to receive the basic State Pension when he turned 65. He clearly states that his strongest motivation to keep working is his perceived difficult financial situation without an earned income:

> [...] well the question why I'm still working past 65 is just pure and simple, well not altogether but mostly financial. The state pension in England is appalling you have to work [...]. And the [other] side to it obviously is I have had a routine, [...] I worked for up to now 51 years and so it would be trying to find things [to do] in that work period during the week from Monday to Friday for eight hours each day [...].

The quote makes clear that Henry also works because it structures his day and gives him something to do. His leisure activities are organized in a way that is rather typical for a working person's life, where everything revolves around work. Living by himself and in rented accommodation also adds to the fact that he perceives his financial situation in old age as vulnerable. His income is made up of the basic State Pension (£115 a week) and his working wage (£25,000 a year before tax). Additionally, he only began paying into an occupational pension scheme when he started working for the university; before that, he says, he neither had the possibility nor the knowledge to do that. He will start receiving his occupational pension once he stops working. Henry also started paying regularly into a savings account in midlife. Despite these efforts to save, he reports that these three different sources of income will only allow him a modest lifestyle when he stops working. In conclusion, Henry still lives the life of an employee, even though he is older than the state pension age. For him the meaning of work is centred around the income he gains from it. His discontinuous work career, his late start of pension saving and his single marital status without pooling resources with anyone further increase the need to prolong his working life.

3.6.2 Michael Turner

Michael Turner's work career was much more continuous than Henry's. However, it was influenced by economic changes too. Michael completed his A levels and then continued his education at university to become a mechanical engineer. Right after university, he started working at the Central Electricity Generating Board, which was responsible for the national electricity supply. In 1993, his state-owned employer was privatized. His office was moved to a different part of the country, but Michael decided against moving with it, because he did not want to uproot. Both he and his wife played and still play instruments in amateur orchestras, and they did not want to give that up. The employer offered him a generous redundancy package including the occupational pension being topped up by the equivalent of eight years' worth of contributions and the possibility to work as a sole trader for the new private company. Aged 50, he started to receive his occupational pension and has been working in his current job for the past 15 years with no change. In his work, he carries out the maintenance and security checks of electricity plants. He drives up and down the country for these regular checks and is sometimes called out for emergencies, which adds up to approximately five working days a month. Michael works for a combination of reasons: for social integration, especially to stay in contact with his old colleagues, to be active and to have some extra money. In the interview, it does not become completely clear how important the extra income is for him. He presents the financial aspect and the other reasons always in the same breath, as can be seen in this quote:

> Yes [money has] something to do [with it] as well, you know [the job] gets you of out of the house and I mean I'm only usually there for a day or two or three days on each particular site, but you know you sort of renew acquaintances with people and keep your hand in as they say.

The focus of his life is on playing music, and work is rearranged around it. Michael states that he has never done any pension planning, since he worked for a long period at a state-owned company with a generous occupational pension. Currently, his income for five days' work per month is as high as Henry's monthly income from full-time employment. Additionally, he lives in his own and paid-off home, and if necessary, he can fall back on his savings from the invested redundancy package. All of these aspects add up to the conclusion that Michael lives the typical life of a well-situated pensioner, with unpaid activities in the centre of his life and some additional paid work to stay in contact with old acquaintances and to supplement the household income. His continuous work career at one employer with an occupational pension made his very early retirement financially secure. Having a partner who is also retired and the hobby of playing music have reinforced the transition from being a worker to being a pensioner.

3.6.3 Gerd Schmitt

Turning to Germany, Gerd Schmitt's working career is very discontinuous and includes one major period of inactivity. He left school after finishing general elementary education, completed vocational training as a bricklayer and started further vocational training as a technician, which he never finished. His career and life were strongly affected by alcohol addiction. He mostly worked on short-term contracts and as a day labourer on construction sites. In his late 30s, his life was turned upside down by his addiction, and he was inactive for several years. After therapy and aged 43, he turned his life around, completed vocational training as an office clerk and found work after several unsuccessful attempts. He started working in the administration of a group of hospitals owned by the local authority which was later privatized. Just like Henry, he joined a big employer in a low-qualified job and continuously worked his way up to the administration of the maintenance of the buildings. Together with the architects in his department, Gerd is responsible for hiring and supervising construction firms for the upkeep of the buildings. When Gerd was approaching state pension age, his employer asked him to stay on a part-time basis because three new colleagues had just started in his department. He took this offer, because he had already planned to go on working because of his small pension due to his long period of inactivity. In Germany, retiring at pension age is still the norm, especially in the public service sector, so the hospitals' works council had to approve his individual case, which they did, acknowledging his financial situation. As can be seen in this quote, his reasons for working are manifold:

> First of all because it's fun. I can't imagine going to some pensioners club every morning to play cards, that would be boring, and here I spend time with young people every day, besides that it was the company that came to me [...] [and] if I had to live off the pension

I get from the state, then I would have to go to the social security office.

His work capacity is negatively affected by his health, but he is allowed to set his own working hours and that helps. He officially works ten hours a week for €8 an hour, but says that he often works more, because he wants to. As he has not paid into social insurance for longer periods, he receives only a small state pension. He also receives an occupational pension, which tops up his overall pension income to 73 per cent of his full-time working wage before retirement. Additionally, he financially supports his wife, who he is separated from. At first sight, Gerd's biography looks rather untypical because of the long period of inactivity due to his alcohol addiction. However, in our qualitative sample it is not uncommon to have a period of inactivity. This subgroup of working pensioners then had to find their way back into regular employment and now need to make up for those years away from the labour market.

3.6.4 Holger Weiss

Holger's work career is very typical for German men of his generation. He left school after finishing general elementary education, completed vocational training as a precision mechanic and then went back to school to obtain the university entrance qualification. He went to university and became an engineer and worked for a couple of years for two different employers, then started working at a big car manufacturer. From then on Holger's work career ran very smoothly. Similar to Michael, but a few years later, Holger went into early retirement aged 59. His employer encouraged early retirement, as many employers still did in the early 2000s in Germany. He did not really want to stop working, but he felt that his time at the company was ending and he wanted to be in control of his retirement passage. Here again it becomes apparent, as in the three other cases, that the employer plays an important role in the timing of retirement. Holger experienced the first years in retirement as a productive and satisfactory phase in his life. He planned and supervised the building of his new house, spent time with his family and took up several activities such as ballroom dancing and learning a foreign language. He also took up an honorary office in his sports club. However, after a couple of years he felt that he wanted to have something else to do.

> [...] a bit of fun, see something else again, because when you're a pensioner you can look after the lawn, and build mini houses and

put up the pool and fill it with water and look after the children or babysit the grandchildren, that's all very well, it's all very nice, but you also want to get out again.

He started to look for a job and applied for jobs but was not successful, until one of his friends told him about his own job. Holger started working for the same company that employs mostly pensioners to drive cars from one rental car location to another. Holger does approximately five trips a month and never earns more than €150 per month. He is very ambivalent about this job. On the one hand, it satisfies his urge to get out and to see new things, but on the other hand, he finds the pay and the working conditions bad. However, Holger does not have to work because of financial necessity: his household income is (more than) sufficient, because he receives a decent state pension and a generous occupational pension. Furthermore, he owns his house outright and the one next door where one of his sons lives, who also pays rent to him. Nevertheless, his state pension is lower than expected, because he went into early retirement and that is penalized by pension reductions in the German pension system. Looking at Holger's case as a whole, he is above all a pensioner whose life centres around his family and unpaid activities. In a way, work seems to be similar to these activities, because it gives him something to do and the pay is so low.

3.6.5 Summary of qualitative results

The qualitative results show that the socio-economic situation plays a decisive role in the subjective accounts of the two exemplary cases in poor socio-economic circumstances (Henry and Gerd), who emphasize the importance of their working wage in the context of their job, because they only receive small pensions. By contrast, the two exemplary cases in good socio-economic circumstances (Michael and Holger) more strongly articulate non-financial reasons (especially social integration and the desire for activity). However, such non-financial reasons are present in the subjective accounts of the two men in poor socio-economic circumstances as well.

The subjective accounts also give an idea of how strongly the country-specific institutional arrangements and conventions structure employment biographies and retirement decisions. In Germany, both interviewees had to change their work arrangements around state pension age. Holger even had to leave his main career job while Gerd needed to seek permission from the works council to go on working because of the strong norms around pension age in Germany. In contrast, in the UK Henry and Michael could go on working without such a change. Here, the importance of the employer in the decision to work after retirement becomes obvious and cannot be overestimated. Additionally, the country differences in the occupational system and the rigidity of the labour market are reflected by the interviewees. Gerd completed an entirely new course of vocational training in order to change his occupation and sector. Henry, who also changed both, reported no training before this change. The differences in the pension systems also appear in the individual cases: The need for more individual responsibility in the UK can be sensed when Henry says that even after working all his life the state pension is not enough to live on and he was only able to start paying into an occupational pension scheme in midlife. By contrast, other interviews of the sample clearly show that in Germany the state pension alone guarantees an adequate income in this generation after a continuous work career, it is perceived as such by the interviewees, and other motives for post-retirement employment take centre stage in their accounts.

3.7 Summary and conclusions

In this chapter, we have examined working beyond pension age in Germany and the UK with quantitative and qualitative methods. The quantitative analyses show that, in the UK, low-qualified work is more common among people working beyond pension age than for those aged 45 to 55. In Germany, there is no such difference. However, in Germany, there is a much higher share of self-employed without employees among those working beyond pension age compared to those aged 45 to 55. Furthermore, the gap in hours worked and in income between the two age groups is larger in Germany than in the UK.

These results speak for better employment opportunities but also for higher financial needs in the UK compared to Germany. This probably reflects institutional differences: Flexible labour markets and weak occupational boundaries in the UK provide higher chances of late employment for all who want to or need to work longer. At the same time, the British pension system contributes to a higher level of old-age income inequality and thus leads to higher economic pressure to work longer for some groups. In Germany, by contrast, the more rigid labour market creates obstacles for post-retirement work, while the pension system does not cause economic pressure to the same degree as in the UK. However, by comparing those working with those not working beyond pension age, we found two important groups working in both countries: one group with an equivalent income that would be very low without the working income and one (well educated) group that works despite an income that would be high even without extra earnings. This pattern seems to contradict other studies which find higher probabilities of post-retirement employment mostly for those in poor economic circumstances in Germany (Hochfellner and Burkert 2013) and mostly for those in good economic circumstances in the UK (Lain 2011 and in Chapter 2 of this book). However, it should be noted that all of these studies, including ours, use different indicators (different types of income, wealth, level of state pension benefits). None of these captures the economic situation at the transition from work to retirement in its full complexity.

The qualitative analyses indicate that, in both countries, exemplary respondents of the group in good socio-economic circumstances primarily work because of non-financial reasons such as social integration and enjoyment of the activity itself, while exemplary respondents of the group in poor circumstances primarily work for financial reasons. However, non-financial aspects are also important in the subjective accounts of this group. In addition, the qualitative findings also reflect the specific institutional structure in Germany and the UK and point to the importance of the family, as marital status affects both the financial motives and the non-financial motives for working beyond pension age.

Returning to the often-asked question 'Do pensioners work past state pension age because of financial need or for other motives?', we would answer as follows: both seem to be true in both countries. Some older people work because they like the work and its additional (non-financial) aspects; others work because they need the money. Furthermore, in the subjective accounts of these workers, there is often a mix of these reasons, with a varying relative importance of different kinds of reasons. Stressing just one group or aspect would obscure a considerable part of the complexity of the phenomenon. This is especially important for discussions about prolonging working lives which become one-sided and unjust if the focus is not broad enough.

Lastly, this chapter was also aimed at the integration of quantitative and qualitative data. Several findings have been confirmed by the other data source, respectively. Furthermore, qualitative and quantitative analyses have also complemented each other. However, the fact that both data sets were collected separately has made true mixing difficult. Additionally, it has only been possible to include four qualitative cases in this chapter, which limits the integration. Even though the cases are typical for the qualitative sample, they could only serve as illustrations. This has nevertheless proven to be fruitful and promising for future, more comprehensive attempts to combine quantitative and qualitative data.

Notes

1. The data was analysed (and in the qualitative case also collected) within the context of the Emmy Noether research group 'Paid work beyond retirement age in Germany and the UK', funded by the German Research Foundation and based at the Research Center on Inequality and Social Policy (SOCIUM) (University of Bremen).
2. The German welfare state also provides a modest means-tested basic flat-rate benefit (*Grundsicherung im Alter*) for those in pension age and without sufficient income. For those receiving this benefit, there is little incentive for working beyond retirement age as a large share of the income from such a job would be deduced from the pension.
3. Means-tested benefit for people beyond pension age in the UK (Pension Credit) is similar to that in Germany (*Grundsicherung im Alter*) with respect to the disincentives for working beyond retirement age through earnings limitations.
4. Recently, that is after the time our quantitative analyses relate to, old age inequality in the UK (defined as income poverty after taxes and transfers of those 65 or older) seems to have dropped considerably. According to OECD data (OECD 2013b: 165), old-age poverty in the UK was lower than in Germany in 2009/2010, and according to Eurostat (Eurostat 2014) data it was also so in 2012.
5. The country difference in unemployment rates of those aged 55 to 64 has shrunk after the years our quantitative analyses relate to. However, the share of long-term unemployed in this age group is still much higher in Germany than in the UK (OECD 2013a: 245, 257).
6. It should be noted that the results of the quantitative analyses do not change substantially if we exclude British women younger than 65.
7. By choosing people aged 45 to 55 as the reference group, we exclude those younger than 45 in order to reduce the bias from (cohort-related) sectoral and occupational change. We also exclude those aged over 55 but younger than retirement age in order to remove those who retired early and have a post-retirement job.
8. We did not include class as this variable would reduce the number of cases in the multivariate analysis.
9. Both the household income and the working income contain imputed values in Germany and the UK which are provided by the GSEOP and the BHPS. Because of this, we have also carried out additional regression analyses only for those without imputed incomes and those with only some imputations. The results show no important deviations between the different versions. In Germany, these imputed income versions provided by the GSOEP are measured on a yearly basis. Therefore, we divided these incomes by 12 to get comparable upper limits of the quintiles. Furthermore, we also carried out additional regression analyses on the basis of the current monthly incomes (which contain a lot of missing values). We did not find any important deviations in comparison to the other versions.
10. The upper limits of the other quintiles are £940 and €1287 for the second quintile, £1189 and €1568 for the third quintile and £1626 and €2110 for the fourth quintile.

11. In our reduced class scheme, the high and low service classes are merged into the 'service class' and the two classes of self-employed into the 'self-employed' following the reduced class scheme of Erikson and Goldthorpe (1993). Furthermore, the classes of low routine non-manuals and of semi- and unskilled manuals are merged as both classes share important aspects with respect to their employment relation (Goldthorpe 2007) and both classes comprise jobs which often – but not always – only need little vocational training.
12. We also analysed the difference in educational level between workers beyond retirement age and middle-aged workers (not shown) in order to get an idea of the social position of the older self-employed without employees in both countries. In Germany, the self-employed (without employees) post-retirement age are somewhat more often low and very low educated than their middle-aged counterparts, which is not the case in the UK. However, about 60 per cent of German self-employed (without employees) beyond retirement age have a middle or high education (77 per cent in the UK).
13. An alternative interpretation would be that self-employed in small businesses in Germany have a low income in old age and need to work because of poor finances. However, additional analyses (not shown) indicate that the reduced household equivalent income of the majority of the self-employed who work beyond retirement age is not particularly low.
14. We carried out these stepwise regressions (not shown) using the average marginal effects to account for rescaling differences, and the method proposed by Karlson, Holm and Breen (2012) to account for rescaling differences and differences in the shape of the error distribution between the different models.
15. It is also possible that some individuals working beyond retirement age defer receipt of an occupational or private pension while receiving a state pension (or also vice versa in Germany). However, further analyses of aspects which are usually correlated with the receipt of occupational and private pensions (working income and the educational level) do not indicate that a large share of such deferrers can be found in the lowest quintile (not shown).
16. We decided against the inclusion of women only because of space restrictions. However, we will take up this topic in further publications.
17. Due to the focus on a region with low unemployment rates in Germany and on a region with high unemployment rates in the UK, we find higher probabilities for the exemplary groups in Germany compared to the UK, although the share of working people beyond retirement age is generally higher in the UK than in Germany (see Figure 3.1).
18. This and all following names are pseudonyms.

References

Allmendinger, J. and Hinz, T. (1998), 'Occupational careers under different welfare regimes: West Germany, Great Britain and Sweden', in: L. Leisering and R. Walker (eds.), *The dynamics of modern society*, Bristol: Policy Press, 63–84.

BBC News (2013), *The pensioners who choose to work*, by K. Peachey, 9 March 2013, http://www.bbc.co.uk/news/business-21668880, date accessed 7 January 2014.

Brauns, H. and Steinmann, S. (1999), 'Educational reform in France, West-Germany and the United Kingdom: Updating the CASMIN educational classification', *ZUMA Nachrichten*, 40 (2), 7–44.

Brenke, K. (2013), 'Immer mehr Menschen im Rentenalter sind berufstätig', *DIW Wochenbericht*, 80 (6), 3–13.

Buchholz, S., Rinklake, A., Schilling, J., Kurz, K. and Blossfeld, H.-P. (2011), 'Ageing populations, globalization and the labour market: Comparing late working life and retirement in modern societies', in: H.-P. Blossfeld, S. Buchholz and K. Kurz (eds.), *Aging populations, globalization and the labour market. Comparing late working life and retirement in modern societies*, Cheltenham: Edward Elgar, 3–32.

Erikson, R. and Goldthorpe, J. H. (1993), *The constant flux: A study of class mobility in industrial societies*, Oxford: Clarendon Press.

Erzberger, C. and Kelle, U. (2003), 'Making inferences in mixed methods: The rules of integration', in: A. Tashakkori and C. Teddlie (eds.), *Handbook of mixed methods in social and behavioral research*, London: Sage Publications, 457–87.

Esping-Andersen, G. (1990), *The three worlds of welfare capitalism*. Princeton: Princeton Universtity Press.

Eurostat (2014), *Income and living conditions*, Luxembourg: Eurostat, http://ec.europa.eu/eurostat/web/income-and-living-conditions/data/database, date accessed 11 March 2015.

Goldthorpe, J. H. (2007), 'Social class and the differentiation of employment contracts', in: J. H. Goldthorpe (ed.), *On sociology. Volume II*, Standford: Standford University Press, 101–25.

Haisken-DeNew, J. P. and Frick, J. R. (2005), *DTC. Desktop companion to the German Socio-Economic Panel (SOEP). Version 8.0*, Berlin: Deutsches Institut für Wirtschaftsforschung.

Hall, P. A. and Soskice, D. (2001), 'An introduction to varieties of capitalism', in: P. A. Hall and D. Soskice (eds.), *Varieties of capitalism. The institutional foundations of comparative advantage*, Oxford: Oxford University Press, 1–68.

Hochfellner, D. and Burkert, C. (2013), 'Berufliche Aktivität im Ruhestand – Fortsetzung der Erwerbsbiographie oder notwendiger Zuverdienst?', *Zeitschrift für Gerontologie und Geriatrie*, 46 (3), 242–50.

Karlson, K. B., Holm, A. and Breen, R. (2012), 'Comparing regression coefficients between same-sample nested models using logit and probit. A new method', *Sociological Methodology*, 42 (1), 286–313.

Komp, K., van Tilburg, T. and Broese van Groenou, M. (2010), 'Paid work between age 60 and 70 years in Europe: A matter of socio-economic status?', *International Journal of Ageing and Later Life*, 5 (1), 45–75.

Lain, D. (2011), 'Helping the poorest help themselves? Encouraging employment past 65 in England and USA', *Journal for Social Policy*, 40 (3), 493–512.

Möhring, K. (2013), 'Altersarmut in Deutschland und Großbritannien: Die Auswirkungen der Rentenreformen seit Beginn der 1990er', in: C. Vogel and A. Motel-Klingebiel (eds.), *Altern im sozialen Wandel: Die Rückkehr der Altersarmut?*, Wiesbaden: VS Verlag für Sozialwissenschaften, 292–311.

OECD (Organisation for Economic Co-operation and Development) (2013a), *OECD employment outlook 2013*, Paris: OECD.

OECD (2013b), *Pensions at a glance 2013. Retirement-income systems in OECD Countries*, Paris: OECD.

OECD (2014), *Employment rates by age group*, Paris: OECD, http://stats.oecd.org/Index.aspx?DataSetCode=LFS_SEXAGE_I_R, date accessed 24 October 2014.

Scherger, S., Hagemann, S., Hokema, A. and Lux, T. (2012), *Between privilege and burden. Work past retirement age in Germany and the UK*, ZeS-Working Paper No. 4/2012, Bremen: Centre for Social Policy Research, http://www.zes.uni-bremen.de/veroeffentlichungen/arbeitspapiere/?publ=435&page=1, date accessed 25 March 2014.

Schulze, I. and Jochem, S. (2007), 'Germany: beyond policy gridlock', in: E. M. Immergut, K. M. Anderson and I. Schulze (eds.), *The handbook of West European pension politics*, Oxford: University Press, 660–710.

Schulze, I. and Moran, M. (2007), 'United Kingdom: pension politics in an adversarial system', in: E. M. Immergut, K. M. Anderson and I. Schulze (eds.), *The handbook of West European pension politics*, Oxford: University Press, 46–96.

Spiegel Online (2012), *Arbeiten trotz Rente. 'Ich bin verdammt zu putzen'*, 28 August 2012, http://www.spiegel.de/karriere/berufsleben/senioren-mit-job-warum-rentner-weiter-arbeiten-a-852613.html, date accessed 7 January 2014.

Taylor, M. F., Brice, J., Buck, N. and Prentice-Lane, E. (2010), *British household panel survey user manual. Volume A: Introduction, technical report and appendices*, Colchester: University of Essex.

Witzel, A. and Reiter, H. (2012), *The problem-centred interview*, London: Sage.

4
Characteristics of Working Pensioners in Italy: Between Early Retirement Tradition and Reforms to Extend Working Life

Andrea Principi, Pietro Checcucci, Mirko Di Rosa and Giovanni Lamura

4.1 Introduction: Pension receipt and paid employment in Italy

Across Europe, the call for developing and implementing more effective active ageing policies has increased in recent years, as these are considered to be effective tools for tackling current and future demographic and welfare challenges deriving from population ageing. Labour market participation is one crucial area in which active ageing can be pursued. One way to prolong working life is to join or to remain in the labour market after having started receiving an old-age pension. Work beyond retirement age is becoming more frequent across Europe (Dubois and Anderson 2012). This may also apply to Italy, where a considerable share of pensioners stay in or join the labour market: in 2010, about nine per cent of all pensioners (that is, people receiving a pension) aged 55 years and over were in employment (own calculations based on ISTAT-INPS 2012 and INPS-ISTAT 2012b).

The classical gerontological theory of disengagement suggests that as people age, they naturally withdraw from society (Cumming and Henry 1961). In contrast, the individual decision to work while receiving an old-age pension is in line with theoretical approaches claiming that older people strive to maintain their previous lifestyle and status (that is, continuity theory – Maddox 1968). On the one hand, this implies that retired older people decide to join (or remain in) the labour market because they need to or because they wish to (Dubois and Anderson 2012). On the other hand, working while receiving a pension is not only linked to an individual intention of the pensioner but also to the availability of labour market opportunities for pensioners. These are conditioned by factors on the meso (for example employers' behaviour) and macro levels (for example the pension and labour laws and regulations; see Scherger et al. 2012).

With a few exceptions (for example Principi et al. 2012), there is a substantial lack of studies about Italian working pensioners. Thus, the main purpose of this chapter is to contribute to a deeper understanding of work after the beginning of pension payments in Italy. Data from the National Institute of Social Security and the National Institute of Statistics (INPS-ISTAT) (Table 4.1) show an increasing trend in the number of working pensioners aged 55 years and over between 2007 (1,011,081) and 2010 (1,380,041) in Italy, including both genders. Compared to 2010, provisional[1] data for 2011 seem to indicate a decrease of working pensioners in terms of numbers, while their share decreased in particular in the age group 55 to 59 years, possibly because of the increased statutory retirement age. Table 4.1 also shows that while the bulk of working pensioners is in the age group 60 to 64 years for both genders, almost one out of two working pensioners (49 per cent) was over 65 in 2011.

To study the Italian case in depth is interesting for a number of reasons, including the apparent contradictions of some factors at the macro level which, on the one hand, promote work among pensioners and, on the other hand, seem to hinder it. Once an older person fulfils the requirements to receive a pension, he or she must fully withdraw from the labour market in order to receive this pension, with the exception of self-employed workers. In other words: no gradual retirement exists in Italy. This precondition to receive a pension does not facilitate work after retirement[2] since it is more difficult to access the labour market again once one has left it. Then again, since 2009 most pensioners have been permitted to fully combine income from pensions and from employment, and after having started to receive a pension, they are allowed to rejoin the labour market.

Another particularity of the Italian context is that it is characterized by many 'young pensioners', that is, people who have benefited from a wide-ranging early retirement scheme and started receiving a pension (quitting their work) before the statutory retirement age (of 60 years for women and 65 for men until December 2011). As a consequence,

Table 4.1 Working pensioners among pensioners by age group and gender, 2007–11, per cent (by column)

	2007[1]			2008[2]			2009[3]			2010[3]			2011[3,4]		
	M	F	All	M	F	All	M	F	All	M	F	All	M	F	All
55–59	19	27	21	18	25	20	16	25	19	19	26	21	15	23	17
60–64	32	36	33	33	37	34	34	37	35	34	39	35	33	36	34
65–69	27	22	25	26	22	25	25	22	24	24	19	22	26	22	25
70–74	14	10	13	15	10	13	16	11	14	14	10	13	17	12	15
75–79	5	4	5	6	4	5	6	4	6	6	4	5	7	4	6
80+	2	2	2	2	2	2	2	2	2	3	2	3	3	2	3
n[5]	736,566	274,515	1,011,081	759,003	290,130	1,049,133	797,871	301,370	1,099,241	969,611	410,430	1,380,041	788,854	319,966	1,108,820

Notes: [1] INPS-ISTAT 2010; [2] INPS-ISTAT 2012a; [3] INPS-ISTAT 2012b; [4] provisional data; [5] this row shows the total number of cases from which percentages by column are calculated.
Source: INPS-ISTAT 2010, 2012a, 2012b, own elaboration.

about one-quarter of all people receiving an old-age pension were aged between 40 and 64 years in 2011 (INPS-ISTAT 2013). It is therefore not surprising that the average effective retirement age, that is, the age at which people start to receive a pension, is low in Italy when compared with most other European countries.

Important factors to understand paid work after the beginning of pension payments in Italy are also individual motivations, individual characteristics of working pensioners and characteristics of the work they carry out. In the light of the lack of evidence on this topic in Italy, the main aim of this study is to understand how these individual and work characteristics help to describe and explain post-retirement work in Italy, and how they can mediate individual motivations of working pensioners.

The remainder of this chapter is structured as follows: Section 4.2 addresses the Italian institutional background characterizing paid work by pensioners, in terms of the pension system and regulations and programmes related to the inclusion of pensioners in the labour market. Section 4.3, based on quantitative data analyses from the ISTAT IT-SILC survey, will provide a profile of Italian working pensioners, their individual characteristics and the attributes of their work, to focus then on the explanation of post-retirement work in Italy by means of multivariate statistical modelling. Section 4.4 will complete the chapter by discussing the results and also relating them to possible individual reasons for post-retirement work and to consequences of the changed institutional framework.

4.2 Institutional background

4.2.1 The labour market

In the terms of the typology of varieties of capitalism by Hall and Soskice (2001), Italy can be classified as representing a regulatory regime between liberal market economies (like the USA or the UK) and coordinated market economies (with Germany, but also Sweden, as classical examples). Despite attempts to identify Italy as a case of 'Mediterranean capitalism', the typical characteristics of coordinated market economies seem to prevail. Among these characteristics, the pursuit of flexibility through controlled decentralization of industrial relations, along with a continued commitment to coordinate wage bargaining, is underlined as being important (Thelen 2001). Since the beginning of the 1990s, a recentralization of labour negotiations can be observed, which implies a clear mandate for national contracts to set the parameters for local negotiations and/or company-level bargaining, and was supported by employers (Thelen 2001).

After the progressive introduction of fiscal incentives to hire younger workers within training on the job programmes since 1984, the Italian labour market witnessed a massive 'young in, old out' pattern (Contini and Rapiti 1999). Although this pattern lost importance after the peak of the recession in 1993, the male employment rate of the age group 55 to 64 continuously dropped until 2000, while it gained only one percentage point among women. The employment rate of both men and women aged 55 to 64 years has then grown continuously from 2000 onwards. This trend was boosted by reforms of the pension system, which gradually raised retirement age and restricted previously available early retirement schemes and similar measures. As a consequence of this, the 'young in, old out' pattern changed into a sort of 'blocking strategy' – with these measures actually 'blocking' older workers from retiring (Thijssen and Rocco 2010, as cited in Checcucci 2013). The introduction of antidiscrimination legislation (including anti-age discrimination), which acknowledged the specific European Directive 2000/78/EC on employment equality, may also have played a positive role in this context.

The current situation of the very segmented Italian labour market shows that its main divides – in particular gender and age – have remained unaddressed during the last 15 years (Schindler 2009; Addabbo and Maccagnan 2011) and been partly aggravated by the recent economic crisis, resulting in a dramatic increase in the unemployment rate of younger people aged 15 to 24 (20.3 in 2007 and already 40 per cent in 2013). During the crisis, a number of measures were enforced by the central Italian government and the regions, which were addressed at widening the range of workers receiving unemployment benefits and better coordinating active and passive employment measures. In this context, older workers were often explicitly identified as one of the target groups. In 2012, the government and the social partners, as part of a comprehensive labour market reform, included the possibility of firm level agreements aimed at allowing early retirement for workers four years before pension age. The same reform also introduced, from 2013 onward, incentives to hire people over 50 who have been unemployed for 12 months or more, by subsidizing 50 per cent of employers' contributions. The new legislation has also confirmed the longer coverage of unemployment benefits for workers aged 55 and older in comparison to younger workers and has left traditional job protection schemes untouched. As an important exception, the 'mobility allowance' (*Mobilità*) will be abolished from 2017 onwards. This allowance offers benefits and a special fast track into public employment services for workers dismissed due to a crisis in their firm and/or the conclusion of the coverage of other job protection measures.

4.2.2 The pension system

In the 1990s, two historical issues entered the political agenda which had affected the Italian welfare system since the recovery after the Second World War: the marked polarization of public spending towards pensions (to the detriment of family, unemployment and social inclusion investments) and its unequal distribution among different social groups and generations (to the detriment of women, younger people and workers without full-time standard jobs) (Ferrera et al. 2012). The Italian pension system, as designed after the Second World War, was conceived as a pay-as-you-go scheme, guaranteeing a minimum pension level and giving the option of early retirement under certain conditions. It was funded through a pay-roll tax shared between employees and the employer who paid one-third and two-thirds of the contributions, respectively, and benefits were calculated based on average real earnings of the last five years, or, after the 1992 reform, on average real career earnings. Until 1992, men could retire at 60 years of age and women at 55, with a minimum contribution period of 15 years. Alternatively, everybody could retire at any age with 35 years of contributions in the private sector and 20 years in the public one ('seniority pension') (Aben 2011).

Questions about the financial sustainability of the system arose in the early 1990s, mainly due to demographic trends and the expensive fragmentation of the public system in sectorial pension funds. As a consequence, between 1992 and 2010, the government undertook a series of reforms. After the last pension reform in 2011 (Law 214/2011), the 'Fornero reform', the Italian pension system is organized in three pillars: a mandatory public pension scheme which is accompanied by voluntary occupational and/or private pension schemes. The first pillar consists of a compulsory pay-as-you-go insurance plan comprising, most importantly, the pension insurance for employees and for the self-employed as well as pensions for civil servants. This pillar is based on a newly introduced notional defined contribution scheme, which has completely replaced the previous scheme for all workers who started their employment career in 1996 or later. The rate of return is related to

GDP growth, and the accumulated notional capital is converted into an annuity after retirement.

Amongst other reactions to the European pension recommendations (Commission of the European Communities 2001), the adoption of the defined contribution scheme was aimed at strengthening the relationship between the number of contribution years and the level of individual pension payments. Employees enrolled after January 1996 will also receive less generous pensions: in 2060, they will be entitled to a pension equal to 63.4 per cent of their last wage received (2010: 74.1 per cent) (Ministero dell'Economia e delle Finanze 2013). At the same time, tax incentives to participate in the private pillar of pension provision have been introduced (Aben 2011). This also implies the need to raise the financial education of workers and the awareness of the amount of their future pension wealth. Occupational and private pension schemes are still not widespread. As regards this point, the Monitoring Commission on Private Pension Schemes (COVIP) reported in June 2012 that only 22 per cent of the labour force were contributing to a private pension scheme (ISFOL 2013).

The Fornero reform also introduced, from January 2012 onwards, a higher statutory retirement age of 66 years for men in the private and the public sector and for women working in the public sector. For women working in the private sector, the new requirement is 62 years (for more details, see Table 4.2). From January 2018 onwards, workers of both genders will retire at the same age, 66 years, with this limit gradually rising to 67 years from January 2021 onwards. The minimum contribution period has been set at 20 years (Ministero dell'Economia e delle Finanze 2012). From 2012, early retirement is only possible after a contribution period of 42 years and 1 month for men and 41 years and 1 month in the case of women (see Table 4.2); a gradual annual increase of the required contribution period started in 2013. There were, however, some exceptions to this general rule. For example, workers employed in physically demanding jobs retained the option to retire after 35 years of contributions. From 2013, statutory retirement age and contribution requirements are recalculated every three years in accordance with average life expectancy at 65 years. From 2021, the recalculation will occur every two years (Ministero dell'Economia e delle Finanze 2012). To discourage early retirement, the reform introduced a reduction in the pension income of one per cent for each year that early retirement is accessed before the age of 62 years and two per cent per year before 60 years.

Table 4.2 Main requirements for retirement before and after the 'Fornero reform'

		Before the 2011 reform		From 2012 onwards	
		Statutory retirement age	Early retirement[1]	Statutory retirement age[2]	Early retirement: years of contributions[3]
Private sector	Men	65	58 until June 2009; 60 from July 2009[4]	66	Men: 42 years and 1 month

Women: 41 years and 1 month[5] |
	Women	60		62	
Public sector	Men	65		66	
	Women	61		66	
Self-employed	Men	65	see above + 1 year	66	
	Women	60		63½	

Notes: [1] With at least 35 years of contributions and at any age with 40 years of contributions; [2] With a minimum of 20 years of contributions and a pension per month 2.8 times the minimum social benefit; statutory retirement age will be 66 years for all workers from 2018 onwards, gradually rising to 67 years from January 2021 onwards.
[3] Workers hired after 1996 can retire up to 3 years before the statutory retirement age with 20 years of contributions and 2.8 times the minimum social benefit.
[4] Then the age requirement gradually increased until 2011 (Ministero dell'Economia e delle Finanze 2012).
[5] From January 2014 to December 2015, 42 years and 6 months for men and 41 years and 6 months for women; after January 2016, this requirement will be revised taking life expectancy into account.
Source: Own compilation.

Against the background of continued sectorial fragmentation and the discontinuity of working careers, the difference between Italy and many other countries in the EU in terms of effective retirement age has remained substantial, despite the various legislative interventions described (OECD 2013). In 2012, the average duration of working life in Italy was still five and a half years shorter than the EU-27 average, also showing a relevant gender gap. The effects of the last reforms are, however, already visible in terms of older workers' labour supply. While 9 per cent of workers aged 55 to 67 ceased their activity on the labour market each year from 2008 to 2011, this percentage dropped to 4.9 per cent in 2012. The majority of these persons were previously employed rather than unemployed, thus generating an increase in the number of employed persons aged 55 and older by 6.8 per cent from 2011 to 2012 (CNEL 2013). The pension reform is expected to generate a further increase in the labour supply of people in the age group 57 to 66 until 2020 (CNEL 2012).

4.2.3 Policies for working pensioners

From January 2009, Italian legislation (DL n. 112/2008 as enforced by Law n. 133/2008) has allowed the full combination of paid (self-)employment and pension incomes from both previous and new schemes, thus facilitating regular work for older people receiving pensions. Limits to the combination and consequent reductions in individual pension payments remain in the cases of invalidity pensions and allowances (*assegno di invalidità*) and for employees who transform fulltime into part-time work. The latter is an exception to the general rule from 1996 that in order to be eligible for receiving a pension, an Italian employee must quit his or her employment relationship – an exception that is not very often realized. Furthermore, the recent labour market reform abolished the possibility for large employers to fire a worker who reaches the statutory retirement age, in order to allow workers the possibility to work until the age of 70. At this age, large companies then have the right to dismiss a worker.

Once retired, pensioners are often hired under one of the non-standard work contracts provided by Italian legislation, for example as freelance workers. These and other forms of non-standard contracts basically imply a lower tax burden for both employees and employers. However, restrictions to these flexible and temporary contracts were introduced in 2012 in order to avoid irregular work. This probably reduced some of the opportunities previously exploited by employers to re-employ older workers after their retirement from their earlier main career (ISFOL 2013).

4.3 Describing and explaining work among pensioners in Italy

4.3.1 Methods

Our analysis is based on the ISTAT IT-SILC database, from a representative national survey on Italian households' income and living conditions.[3] The main aim of this survey, which is part of the project European Statistics on Income and Living Conditions (EU-SILC), is to provide comparable data at the European level. Even though this survey has not been designed to specifically investigate post-retirement work, IT-SILC data allow for an analysis of this phenomenon in Italy. The Italian sample includes about 20,000 households and 50,000 individuals each year. In this study, we define working pensioners as people who simultaneously received income from work and from a pension based on their own contributions through paid work (that is, from an old-age or early retirement pension) in the year before the interview (for a different, more subjective definition, see Principi et al. 2012). To identify possible differences between the working pensioners who have (mostly) not yet reached the official retirement age and those who have, we differentiate between two age groups: people aged 55 to 64 years, and 65 years and over. As we analyse data from 2004 to (in particular) 2011, the increase of retirement age to 66 in 2012 is not yet covered.

In the following, bivariate analyses will be firstly carried out to investigate the trend among working pensioners, comparing the cross sections from 2004 to 2011. Then we will examine individual characteristics of working pensioners using the 2011 data (n = 16,919). For these analyses, the sample is grouped in four main categories: working pensioners as defined above (n = 1,297); workers, that is people who only received income from work and no pension (n = 2,879); pensioners, that is people who did not receive income from work, but (only) a pension based on their own contributions through paid work (n = 8,703); other people, that is, those who neither received income from work nor from a pension based on their own contributions through paid work (n = 4,040). People who received other kinds of pensions (for example disabled or survivors) make up 57 per cent of the latter category and people without any income from work or pensions 43 per cent (for example housewives and poor people, or people with income from properties). Then we compare the work characteristics of working pensioners and other workers. In all our descriptive analyses, we use weights to compensate for systematic non-response or to filter incomplete response. For responding households, usually all members received an individual questionnaire (Eurostat 2011).

The chapter ends by analysing post-retirement work using multivariate logistic regression analyses at first in the overall sample and then dividing it by age groups (55 to 64 years and 65 and older). This kind of model was chosen since it can be applied to dichotomous dependent variables (such as working and not working). Average marginal effects (AMEs) were estimated in order to make comparisons across the models for the two age groups (Mood 2010). The logistic regression included the following explanatory variables: gender, age (included as a continuous variable), educational level, subjective health status, marital status and payment of real estate property tax (used as a proxy for individual wealth). To avoid heteroskedasticity problems (which would make substantial testing on estimates impossible) and to take into account the possibility of respondents belonging to the same family, clustered standard errors were applied in the logistic estimation.

4.3.2 Prevalence of working pensioners

According to the ISTAT IT-SILC data, the share of working pensioners among the population aged 55 years and older is decreasing (Table 4.3): it was 9.6 per cent in 2004 and 6.8 per cent in 2011. However, the trend

Table 4.3 Prevalence of combinations of working and pension receipt among the population aged 55+, different years, per cent (by row)

		n	Working pensioners	Workers	Pensioners	Other
			2004			
55–64	Male	3,902	27.5	7.0	57.3	8.2
	Female	4,025	12.3	6.3	56.3	25.2
65+	Male	5,242	9.3	0.9	74.7	15.1
	Female	6,965	2.5	0.8	52.7	43.9
Total	Male	9,144	15.0	2.9	69.2	12.9
	Female	10,990	4.8	2.1	53.6	39.6
	All	20,134	9.6	2.5	60.8	27.2
			2007			
55–64	Male	3,374	24.5	6.7	60.6	8.1
	Female	3,514	14.8	6.5	57.1	21.6
65+	Male	4,789	10.0	0.8	77.7	11.5
	Female	6,387	2.5	0.4	58.2	38.9
Total	Male	8,163	14.0	2.4	73.0	10.5
	Female	9,901	4.9	1.6	58.0	35.5
	All	18,064	9.2	2.0	65.1	23.8
			2011			
55–64	Male	3,132	12.2	51.2	27.5	9.2
	Female	3,303	5.6	32.1	20.6	41.6
65+	Male	4,616	10.2	1.8	79.1	9.0
	Female	5,868	2.4	1.0	50.1	46.5
Total	Male	7,748	11.0	21.6	58.4	9.1
	Female	9,171	3.5	11.7	40.1	44.8
	All	16,919	6.8	16.1	48.3	28.8

Source: Own calculations with ISTAT, Indagine sulle condizioni di vita (UDB IT – SILC, 2011), weighted data.

among workers aged 55 and over without pensions is a sharp increase from around 2 per cent in 2007 and 2004 to 16.1 per cent in 2011. Correspondingly, the share of recipients of pensions (based on contributions through paid work) decreased drastically from between 60 and 65 per cent in 2004 and 2007 to around 48 per cent in 2011. The main reason for these changes in pension receipt and working are the recent pension reforms, more precisely the gradual increase in the pension age and the increasing difficulties to access early retirement. In accordance with this, the differentiation by age groups shows that all these trends can be observed almost exclusively and very strongly in the young-old group of people aged 55 to 64 years, rather than in the population aged 65 and older.

Among the population aged 55 and older, higher shares of males, compared to females, are working pensioners. Nonetheless, the previously mentioned trends concern both men and women, even if a gender differentiation can be seen. On the one hand, the proportion of working pensioners aged 55 to 64 dropped by half among men between 2007 and 2011, while it decreased by around two-thirds among women. On the other hand, even if the share of workers increased steeply for both genders among the young-old and between 2007 and 2011, the increase is noticeable especially among men (plus 44.5 percentage points vs. plus 25.6 percentage points for women). This indicates that women still face greater difficulties than men in the labour market and withdraw more quickly from the workforce, as also witnessed by the almost doubled percentage (41.6 per cent) of women aged 55 to 64 without an income from work or a pension based on their own contributions in 2011. Among those aged 65 and older, the 'pure' pensioners become a somewhat smaller group among women in 2011, and the share of women without a pension or an income grew.

4.3.3 Who tends to work after having retired from their main career?

Although there are differences between countries in the structure and characteristics of post-retirement work (see the other contributions in this book), the literature agrees in many points on what the main individual characteristics of this work are. A very important precondition for work after retirement seems to be good health (Crawford and Tetlow 2010; Lain 2011). Since health generally tends to worsen as people age, it is not surprising that working pensioners are younger than other pensioners (Crawford and Tetlow 2010; Dubois and Anderson 2012). Furthermore, it is mainly men, rather than women, who are still present in the labour market at retirement age or while receiving a pension. One reason for this, additional to the lower female employment rate earlier in the life course of these cohorts, is that older women more frequently deal with family commitments (for example care of older relatives) than men (Milazzo 2000; Dubois and Anderson 2012). However, divorced women, who are often in economic need, are over-represented among working pensioners (Smeaton and McKay 2003; Scherger et al. 2012).

Educational level and the economic situation are other influences of key importance. On the one hand, working pensioners have a higher educational level than other pensioners (Dubois and Anderson 2012; Scherger et al. 2012). This has been explained with reasons such as their particular interest in their work and better work opportunities offered by companies that are keen to retain highly qualified employees to avoid skill loss. In line with this, working pensioners have been found to be particularly represented among those in a better economic situation (Smeaton and McKay 2003; Crawford and Tetlow 2010) and among those who might have a desire to improve or maintain their previous lifestyle (Smeaton and McKay 2003; Dubois and Anderson 2012). On the other hand, working pensioners are also over-represented among those with debts and mortgages (Scherger et al. 2012). This indicates economic necessity as a main driver at least for some working pensioners.

In Table 4.4, we describe the distributions of some of these characteristics for Italy, considering two different age groups: 55 to 64 years and 65 and older. People aged 55 to 64 years are more likely to be working pensioners than people 65 and older (8.8 per cent vs. 5.7 per cent). Table 4.3 above shows that the distribution of working pensioners is highly disproportionate in terms of gender, since (for example in 2011) men aged 55 to 64 years work much more often (12.2 per cent vs. 5.6 per cent for women), while women are much more often in the category 'other'. These differences are even higher in the older age group (Table 4.3). Table 4.3 also shows the different shares of 'workers' among men and women without a pension, which demonstrates that, although the employment rate of older women is increasing in Italy (Eurostat 2014a), older women still face considerable barriers to joining or remaining in the labour market. Table 4.4 shows that in the younger age group, people in good health are working pensioners more than twice as frequently as those in poor health (10.3 per cent vs. 4.0 per cent). To a lesser extent, this relationship also applies to workers not receiving a pension. Among working pensioners in the older age

Table 4.4 Combinations of working and receiving a pension among the population aged 55+, by individual characteristics (2011, row per cent)

		Working pensioners	Workers	Pensioners	Other
Age	55–64	8.8	41.4	24.0	25.8
	65+	5.7	1.4	62.4	30.6
Subjective health					
Good	55–64	10.3	47.1	21.6	21.0
	65+	10.7	2.4	66.5	20.5
Neither poor nor good	55–64	7.7	35.5	30.3	26.5
	65+	5.4	1.0	67.1	26.5
Poor	55–64	4.0	28.6	20.9	46.4
	65+	1.9	0.9	53.1	44.1
Marital status					
Married/cohabiting	55–64	9.2	40.3	25.2	25.3
	65+	7.2	1.6	67.6	23.6
Divorced/single	55–64	7.6	51.5	18.2	22.7
	65+	8.9	1.8	68.0	21.3
Widowed	55–64	7.4	30.2	23.9	38.5
	65+	2.1	0.8	51.6	45.4
Educational level[1]					
Low	55–64	6.8	21.0	28.7	43.6
	65+	3.4	0.7	58.9	37.0
Medium	55–64	9.4	43.8	24.5	22.3
	65+	8.5	2.2	69.3	20.0
High	55–64	10.2	70.6	13.0	6.2
	65+	18.8	4.5	69.2	7.6
Tax on home ownership					
No	55–64	7.1	39.8	21.8	31.2
	65+	3.5	1.2	60.1	35.3
Yes	55–64	12.0	44.4	27.9	15.8
	65+	10.3	1.8	67.2	20.8
Yearly income[2]	55–64	35,602	23,960	16,388	2,308
in € (mean)	65+	35,310	27,542	14,419	7,769

Notes: Chi square test for association between individual characteristics and combinations of work/pension receipt: all differences are significant at the level p < 0.001; [1]Low, ISCED 0–2; intermediate, ISCED 3–4; high, ISCED 5–6; [2]Individual income from work and/or from pensions after taxes and deductions, mean by column.
Source: Own calculations with ISTAT, Indagine sulle condizioni di vita (UDB IT-SILC, 2011), weighted data.

group, this relationship between health and being a working pensioner is even more evident (10.7 per cent in good health vs. 1.9 per cent in poor health). Table 4.4 also shows that among those aged 55 to 64 who are married or cohabiting there are more working pensioners (9.2 per cent) than among the divorced (7.6 per cent) or widowed (7.4 per cent) of this age group. Interestingly, younger-old workers without pensions are over-represented among divorced people, while people in 'other' situations are more common among the widowed of this age. People aged 65 and older, by contrast, have a greater chance to be working pensioners when they are divorced (8.9 per cent), in comparison to the married/cohabiting and the widowed.

Concerning the educational level of younger-old people, people with high education are more likely to work whilst receiving a pension, and this applies even more to working without receiving a pension, which is more than three times more likely among those with high education than among the low educated (70.6 per cent vs. 21.0 per cent). This relationship is broadly similar for those aged 65 and older, with the qualification gradient being stronger for older working pensioners than among the younger. We also measured wealth, using as a proxy a dichotomous variable on whether someone had to pay the annual tax on home ownership, which implies being a homeowner.[4] We can observe that, among younger-old people, working pensioners, workers and pensioners are over-represented among homeowners, with the greatest difference for working pensioners (almost double, 12 per cent vs. 7.1 per cent of non-owners). This difference becomes threefold when those aged 65 and older are considered. Correspondingly, young-old people in 'other' situations are more represented among those not owning a home. For both age groups, working pensioners have the highest mean individual income, because they receive both a pension and income from work.

4.3.4 Job characteristics and working conditions

In order to learn more about the consequences of post-retirement work, it is also crucial to analyse the characteristics of the jobs that working pensioners do. For example, it is known that self-employment is quite common among working pensioners (Dubois and Anderson 2012) and the self-employed often keep on working after having started to receive pension payments (Menning et al. 2007). In terms of hours worked, part-time work is much more common among retirees than both among the population as a whole and among older non-retired workers (Dubois and Anderson 2012). Regarding the occupational position, in the UK, pensioners are particularly recruited into low-paid part-time elementary occupations requiring few qualifications (Smeaton and McKay 2003; Lain 2012). This might also be the case in Italy since people employed in high-qualified jobs, such as managers and supervisors, tend to postpone retirement rather than working as pensioners (Villosio 2008). This kind of positive selection into postponing retirement might also be the reason for the relatively high incomes of workers without pensions aged 65 and older, as shown in Table 4.4. In many countries, working pensioners are over-represented in the agricultural sector (Smeaton and McKay 2003; Dubois and Anderson 2012) and under-represented in manufacturing and among civil servants (Smeaton and McKay 2003; Dubois and Anderson 2012). At the same time, working pensioners seem very often to be professionals (Dubois and Anderson 2012).

Table 4.5 shows the job characteristics of Italian working pensioners compared with those older workers who do not receive a pension. As to the employment status, working pensioners are much more often self-employed (with or without employees) than workers without pensions, especially after the age of 64 years. With regard to the number of hours worked and independently of their employment status, working pensioners work part-time to a greater extent than workers not receiving a pension, this difference again being much more marked in the older age group. For employees, we also have information on their occupational position, which allows some speculation on the occupational class older people work in. It can be noted that younger Italian working pensioners work less often than workers without pensions in managerial and white-collar positions. However, those aged 65 years and older are more often in white-collar jobs than older non-retired workers. Correspondingly, working pensioners aged 55 to 64 years are in blue-collar jobs more often than workers without a pension of the same age. Examples of sectors in which Italian working pensioners (of both age groups) work more often than workers without pensions are agriculture, hunting and forestry, fishing (12.3 per cent vs. 3.2 per cent) and wholesale and retail trade (25 per cent vs. 11.2 per cent). By contrast, they are found less often than workers without pensions in the public sector (1.8 per cent vs. 10 per cent), in education (3.8 per cent vs. 14.2 per cent) and in the health and social work sector (5.1 per cent vs. 11.4 per cent), and this in particular after 65 years of age. Among those aged 65 and older, more than 20 per cent of older workers without pensions are in professional, scientific and technical activities, compared to 12.5 per cent among working pensioners.

Table 4.5 Work characteristics of working pensioners and non-retired workers in 2011, per cent (by column)

	Working pensioners			Workers		
	55–64	65+	Total	55–64	65+	Total
Employment status[1]						
Employee	60.5	33.7	46.5	76.6	56.7	75.5
Self-employed without employees	29.1	46.4	38.2	17.6	31.3	18.4
Self-employed with employees	10.4	19.9	15.4	5.8	12.0	6.1
Employment status and hours worked						
Full-time employee	30.7	8.2	18.6	71.0	52.0	70.4
Full-time self-employed	6.3	(3.4)	4.7	7.5	(10.0)	7.6
Part-time employee	54.1	64.8	59.9	19.0	28.3	19.2
Part-time self-employed	9.0	23.7	16.9	2.6	(9.7)	2.8
Occupational position (only employees)						
Manager/director	(10.1)	(12.8)	10.8	18.0	(20.2)	18.1
White collar	34.5	(27.5)	32.6	44.0	20.3	43.5
Blue collar	55.4	59.8	56.6	38.0	59.5	38.4
Sector						
Agriculture, hunting and forestry, fishing	10.9	13.5	12.3	3.2	(4.3)	3.2
Mining, manufacturing, energy	15.7	19.2	17.6	16.4	21.4	16.6
Construction	8.0	7.1	7.5	6.3	(7.3)	6.3
Wholesale and retail trade	22.8	27.0	25.0	11.2	(10.6)	11.2
Hotels and restaurants	6.6	(3.3)	4.8	6.3	(2.9)	6.2
Transport, storage communication	4.5	3.7	4.1	2.7	(0.5)	2.6
Financial intermediation	(0.9)	(0.8)	0.8	1.7	(1.7)	1.7
Real estate, renting and business activities	(2.4)	(0.7)	1.5	3.3	(1.2)	3.2
Professional, scientific and technical activities	4.9	12.5	9.0	7.6	21.2	8.0
Public administration and defence	(3.9)	(0.1)	1.8	10.2	(5.0)	10.0
Education	5.5	(2.3)	3.8	14.3	(9.3)	14.2
Health and social work	6.8	3.7	5.1	11.6	(6.2)	11.4
Other community, social and personal service activities	7.1	6.1	6.6	5.3	(8.3)	5.4

Notes: Chi square test for association between being a working pensioner or worker and work characteristics: all differences between workers and pensioners are significant at the level $p < 0.001$; numbers in parentheses indicate that case numbers are below ten.
[1] Asked to all individuals who indicated that they have worked for at least one hour in the week before the interview.
Source: Own calculations with ISTAT, Indagine sulle condizioni di vita (UDB IT-SILC, 2011), weighted data.

This points to the prolonged working lives in these (usually well-paid) jobs already mentioned above, probably due to relatively good labour market opportunities. In all other sectors considered in Table 4.5, differences between working pensioners and non-retired older workers are less pronounced.

4.3.5 Work whilst receiving a pension in Italy: What matters?

In order to make a first step towards explaining work despite receiving a pension, we investigate potential influences on being a working pensioner in a multivariate way. We include all recipients of pensions based on their own contributions through paid work into a logistic regression model, both working and nonworking ones. Accordingly, the main question would be: once retired from the main career and receiving a pension, what is important for the decision to rejoin the labour market? As observed above, work despite receiving a pension can be driven by individual socio-economic characteristics, family-related circumstances and individual motivation. At the same time, it is connected to labour market opportunities which are shaped at the macro (that is, institutional) and meso (that is, organizational) levels. We explore the role of individual characteristics in three models: for all pensioners, only for those aged 55 to 64 years and only for those aged 65 and older. Characteristics of the work carried out before retirement could not be included in the analysis, since information about this was only asked to nonworking pensioners.

Table 4.6 displays the results of the multivariate analysis in the form of average marginal effects which express the percentage change in the probability of working due to a certain characteristic, compared to the reference group; in the case of age, the effect shows the change in probability due to each one-year increase in age. Considering the pooled sample, pensioners are more likely to work if they are male, younger, highly educated, in good health, married/cohabiting (both in comparison to divorced or single people and to widowed people) and homeowners. Things do not change much when we consider the different models for the young-old and the oldest people. However, the effect concerning gender, age, education and health is a bit stronger for those aged 55 to 64, while the influence of marital status loses its significance for the younger age group. When considering different models for males and females (results not shown), the negative influence (on working whilst receiving a pension) of being divorced or single when compared with being married/cohabiting is significant only for women.

Table 4.6 Logistic regression analysis on working (1 = yes, 0 = no) among people aged 55 to 64 and 65+ receiving a pension (2011, average marginal effects)

	Total n = 9,480		55–64 n = 2,160		65+ n = 7,320	
	AME (dy/dx)	Standard error	AME (dy/dx)	Standard error	AME (dy/dx)	Standard error
Gender (ref. male)						
Female	−0.07***	(0.01)	−0.10***	(0.02)	−0.06***	(0.00)
Age	−0.01***	(0.00)	−0.02***	(0.00)	−0.01***	(0.00)
Educational level (ref. low)						
Medium	0.02**	(0.01)	0.02	(0.02)	0.02*	(0.01)
High	0.12***	(0.01)	0.17***	(0.04)	0.10***	(0.01)
Subjective health (ref. good)						
Neither poor nor good	−0.04***	(0.01)	−0.09***	(0.02)	−0.03***	(0.01)
Poor	−0.08***	(0.01)	−0.13***	(0.03)	−0.06***	(0.01)
Marital status (ref. married/ cohabiting)						
Divorced/single	−0.03**	(0.01)	−0.04	(0.03)	−0.03*	(0.01)
Widowed	−0.05**	(0.01)	−0.05	(0.05)	−0.04**	(0.01)
Property tax paid (ref. no)						
Yes	0.06***	(0.01)	0.06**	(0.02)	0.06***	(0.01)
r^2	0.16		0.05		0.15	

Notes: Average marginal effect (dy/dx) for factor levels is the discrete change from the base level; standard errors adjusted for 7,188 family clusters in the total sample, 1,935 family clusters in the sample 55–46 years and 5,731 family clusters in the sample 65+; *$p < 0.05$, **$p < 0.01$, ***$p < 0.001$.
Source: Own calculations with ISTAT, Indagine sulle condizioni di vita (UDB IT-SILC, 2011).

4.4 Discussion

The main aim of this chapter was to contribute to a deeper understanding of post-retirement work in Italy by examining individual and work characteristics of working pensioners. Consistent with previous studies carried out in other countries and with slightly stronger effects for those aged 55 to 64 than those aged 65 and older, we found that working after having retired from one's main career in Italy is associated with being male and having more resources: being younger, in good health, having a higher education and being well-off. An explanation for this is that people with these characteristics probably belong to a medium or high occupational class and have better labour market opportunities. A surprising result (when compared to findings in many other countries, see, for example, Smeaton and McKay 2003; Scherger et al. 2012) is that older married Italians tend to work, once retired, more frequently than divorced and single ones (similarly, for Denmark, see Larsen and Pedersen 2012). The reason probably is that the divorced, single and widowed people are often among the poorer pensioners who avoid retiring (early), because they often live alone and cannot pool their economic resources with a partner (for the higher poverty risk of people living alone, see also ISTAT 2013b). As it is difficult to find a post-retirement job, poorer people postpone retirement as long as possible, possibly even beyond pension age – and those among the divorced, single and widowed people who do retire are probably the few ones who are well-off. This selection effect is in line with our result that marital status is not significant when only the age group 55 to 64 is analysed – before retirement age, people tend to work while receiving a pension regardless of their marital status. In general, this also explains stronger effects of having more resources in terms of health and education in the young-old group: people with better health and education (and probably in higher occupational classes) can afford to withdraw from the labour market by retiring before the statutory retirement and then work more often. Furthermore, since in the models carried out by gender we found single or divorced women less likely to work after retirement (while this difference was not found for men), this could imply that the above-mentioned selection effect is much greater for women. This means that the few very well-off single and divorced women are over-represented among pensioners, since the poor ones postpone receiving a pension as long as possible for the simple reason that for women it is more difficult to find a post-retirement job, this mirroring the better position of and opportunities for men in the labour market in all age groups; indeed, the employment rate of men is higher than that of women in all age groups. This situation is mainly due to the fact that despite recent changes towards a de-familiarization, the Italian welfare model is still anchored in the family model of the male breadwinner and female carer (Principi et al. 2014).

Even if our results demonstrate that especially people with more resources join the labour market after having retired, working pensioners face inequalities in working conditions compared to workers without pensions of the same age. For example, they face barriers to accessing the public sector or similar sectors like education or health. Only (few) highly qualified older people in managerial and professional positions are hired in the public sector after retirement, through particular direct assignments as consultants (Principi et al. 2012). Compared to workers without pensions of the same age, working pensioners are overrepresented in sectors such as wholesale and retail trade and in agriculture, and seem to carry out more low-qualified jobs, as indicated by the higher share of blue-collar workers, especially among the younger-old. This suggests that employers sometimes consider them as a cheap employment reserve. Pensioners may often accept quite low wages compared to standard wages, also because they mainly work part-time and in a more flexible way than before retirement. Specifically regulated contracts, in principle available to all people of working age, allow pensioners to be considered as self-employed workers. In this way, both the company and the working pensioners benefit economically, the former since it saves money (in wages, taxation and social security contributions), and the latter since their overall income exceeds a standard wage when the pension is topped up by a wage (Principi et al. 2012).

In Italy, arrangements for working pensioners are mainly negotiated at the individual and company level. There is no public discussion about work after retirement, and this topic is not on the Italian political agenda. Rather, the labour market-related discussions centre on how to fight the high youth unemployment (Eurostat 2014b) and unemployment in middle and old age (Associazione Lavoro Over 40 2014), and how to retain more older workers in employment *before* they retire. Thus, even if policy makers are dealing with policies of active ageing in employment because of the future prospect of labour shortages and problems to finance pensions, the official policy agenda is to manage this by postponing retirement, rather than to facilitate work while receiving a pension. Only by postponing retirement, that is, pension payments, will the state save money. However, in terms of working conditions, this often implies a too rigid transition to retirement for older workers. For example, older workers in the final stage of their professional career tend to have the same working time and to carry out the same tasks as in previous life stages, although they may need different working conditions than workers of younger age. By contrast, those pensioners who are able to re-enter the labour market can negotiate better working conditions with employers, because of either their particular skills or a low(er) wage.

The (pension) policies to extend working lives, especially the reform from 2012, will have implications for the future of work after retirement in Italy. The increased retirement age and the discouragement of early retirement will result in a decreased number of pensioners, and consequently of working pensioners, in the young-old age group. This trend is already visible in the results above and will become increasingly evident in the future. Therefore, future working pensioners will be older than current ones. Due to changes in pension calculation and an increasing trend of career fragmentation (Principi et al. 2012), future pensions will be even lower than current ones. Already today and taking into account that some pensioners simultaneously receive different kinds of pensions, the share of people with less than €1,000 per month from (one or more) pensions is, with 44.1 per cent, very large (INPS-ISTAT 2013). Consequently, future pensioners will have worked longer than current ones and still be poorer. Thus the number of (older) working pensioners will probably rise, and a higher share among them will be driven by economic need. Being a working pensioner will be less closely connected to having above-average resources than shown above for the year 2011, and the less well educated, less healthy and those with less income (who are often divorced or single) might be pushed to work. All in all, the factors leading to work while receiving a pension and the resulting composition of working pensioners will become more diverse. In the light of the current strict rule that one must fully withdraw from the labour market in order to start receiving a pension, future retired people looking for a job due to economic necessity may face considerable barriers in terms of work opportunities available. Thus, perhaps Italian policy makers should consider revising this rule to prevent marginalization and social exclusion of older people, who then might be able to continue working in their old job. Future studies will have to shed light on the effects of the recent policy changes, which have only just begun, including the role of further new policies supporting work among pensioners.

This study is a first step towards a deeper understanding of work after retirement in Italy. In our analyses, working while receiving a pension in the age group 55 to 64 years seems to be less structured by the variables included in the model, as there is a lower proportion of explained variance for them. This suggests that post-retirement work also depends on a number of factors we could not consider in this study. For example, even if work off the books is widespread among pensioners (Principi et al. 2012), it is difficult to analyse its role based on the available data. Furthermore, future studies should more deeply examine the reasons for post-retirement work (including economic reasons by using longitudinal data) and consider the role of previous work histories, household dynamics including the provision of informal care to family members, class and regional differences.

Notes

1. However, in this data set provisional data very often change considerably when they become definitive.
2. Even if to retire literally means to fully withdraw from work (so that a working individual receiving a pension might not be defined as 'retired'), according to the Italian culture and context, to retire can be more easily associated with 'starting to receive a pension' based on one's own contributions through paid work. Actually, the two definitions in Italy are quite commonly coincident, given that people must quit work to start receiving this pension. Thus in this chapter, 'work after retirement' is defined as work of people who rejoin the labour market after having quit their work and started to receive a pension (of whatever kind) based on their own contributions.
3. Full name: ISTAT, Indagine sulle condizioni di vita (UDB IT-SILC). The responsibility for the data calculations lies with the authors of this chapter, not with ISTAT.
4. The tax on home ownership in Italy is paid by everyone owning a home, including people who still have to pay a mortgage for a property they are buying. Therefore, in the case of a high share of mortgages being paid by homeowners, this variable may be interpreted as a proxy 'opposite' to wealth, that is, debt. However, here, home ownership is actually a valid proxy for wealth for the following reasons: in Italy, to be homeowner is a goal, and generally only those people who cannot economically afford to buy live in rented homes or in other situations; in 2012, in Italy about 72 per cent of families owned their own home, and only 16.5 per cent of them still had to pay off a mortgage for this (ISTAT 2013a); in our sample, 81 per cent were homeowners and only 7 per cent of them were still paying off a mortgage. The latter means that most of the homeowners in our 'older' sample have completed the payment of the possible mortgage in the past. For this reason, homeownership in this study can be considered as a proxy for wealth rather than for debt.

References

Aben, M. (2011), *Overview of the Italian pension system*, Trento: APG World of Pensions Scholarship.

Addabbo, T. and Maccagnan, A. (2011), *The Italian labour market and the crisis*, Università di Modena e Reggio Emilia: Dipartimento di Economia Politica.

Associazione Lavoro Over 40 (2014), *Periscopio 16 Marzo 2014*, Milano: Associazione Lavoro Over 40, http://www.lavoro-over40.it/public/not/324/05%29%20Notizie%20in%20breve%2016%20marzo%202014%20LT.pdf, date accessed 11 January 2015.

Checcucci, P. (2013), *Actively ageing: Italian policy perspectives in light of the new programming period of ESF*, Brussels: 2013 Demography Forum.

CNEL (Consiglio Nazionale Economia e Lavoro) (2012), *Rapporto. Il mercato del lavoro in Italia 2011–2012*, Roma: CNEL.

CNEL (2013), *Rapporto. Il mercato del lavoro 2012–2013*, Roma: CNEL.

Commission of the European Communities (2001), *Communication from the Commission to the Council, the European Parliament and the Economic and*

Social Committee. *Supporting national strategies for safe and sustainable pensions through an integrated approach (COM (2001) 362 final)*, Brussels: European Commission.
Contini, B. and Rapiti, F. M. (1999), ' "Young in, old out" revisited: New patterns of employment replacement in the Italian economy', *International Review of Applied Economics*, 13 (3), 395–415.
Crawford, R. and Tetlow, G. (2010), 'Employment, retirement and pensions', in: J. Banks, C. Lessof, J. Nazroo, N. Rogers, M. Stafford and A. Steptoe (eds.), *Financial circumstances, health and well-being of the older population in England. The 2008 English Longitudinal Study of Ageing (Wave 4)*, London: Institute for Fiscal Studies, 11–75.
Cumming, E. and Henry, W. E. (1961), *Growing old*, New York: Basic.
Dubois, H. and Anderson, R. (2012), *Income from work after retirement in the EU*, Dublin: European Foundation for the Improvement of Living and Working Conditions.
Eurostat (2011), *Description of target variables: Cross-sectional and longitudinal*, EU-SILC 065: 2011 operation, Luxembourg: Eurostat, https://circabc.europa.eu/sd/a/1ad4dc02-7695-4765-b6db-c609acb1a162 /SILC065%20operation%202011%20VERSION%20MAY%202011.pdf, date accessed 11 January 2015.
Eurostat (2014a), *Employment rate of older workers*, Luxembourg: Eurostat, http://epp.eurostat.ec.europa.eu/tgm/table.do?tab=table&init=1&plugin=1& language=en&pcode=tsdde100, date accessed 11 January 2015.
Eurostat (2014b), *Unemployment rate by sex and age groups (Data Explorer, table une_rt_a)*, Luxembourg: Eurostat, http://appsso.eurostat.ec.europa.eu/nui /show.do?dataset=une_rt_a&lang=en, date accessed 11 January 2015.
Ferrera, M., Fargion, V. and Jessoula, M. (2012), *Alle radici del welfare all'italiana. Origini e futuro di un modello sociale squilibrato*, Venezia: Marsilio.
Hall, P. A. and Soskice, D. (2001), *Varieties of capitalism*, New York: Oxford University Press.
INPS-ISTAT (Istituto Nazionale Previdenza Sociale/Istituto Nazionale di Statistica) (2010), *Rapporto sulla coesione sociale. Anno 2010*, Roma: INPS-ISTAT, http://www.inps.it/portale/default.aspx?sID=%3b0%3b7730%3b7208%3b& lastMenu=7208&iMenu=1&p4=2, date accessed 11 January 2015.
INPS-ISTAT (2012a), *Rapporto sulla coesione sociale. Anno 2011*, Roma: INPS-ISTAT, http://www.inps.it/portale/default.aspx?sID=%3b0%3b7730%3b7679%3b& lastMenu=7679&iMenu=1&p4=2, date accessed 11 January 2015.
INPS-ISTAT (2012b), *Rapporto sulla coesione sociale. Anno 2012*, Roma: INPS-ISTAT, http://www.inps.it/portale/default.aspx?sID=%3b0%3b7730%3b7679% 3b8036%3b&lastMenu=8036&iMenu=1&p4=2, date accessed 11 January 2015.
INPS-ISTAT (2013), *Trattamenti pensionistici e beneficiari. Anno 2011*, Roma: INPS-ISTAT, http://www.istat.it/it/archivio/87850, date accessed 11 January 2015.
ISFOL (Istituto per lo sviluppo della formazione professionale dei lavoratori) (2013), *Rapporto di monitoraggio del mercato del lavoro 2012, I libri del Fondo sociale europeo 180*, Roma: ISFOL.
ISTAT (Istituto Nazionale di Statistica) (2013a), *Annuario Statistico Italiano 2013*, Roma: ISTAT.
ISTAT (2013b), *Reddito e condizioni di vita – Anno 2012*, Roma: Istituto nazionale di statistica.

ISTAT-INPS (Istituto Nazionale di Statistica/Istituto Nazionale Previdenza Sociale) (2012), *I beneficiari delle prestazioni pensionistiche. Anno 2010*, Roma: ISTAT-INPS, http://www.istat.it/it/archivio/68870, date accessed 11 January 2015.

Lain, D. (2011), 'Helping the poorest help themselves? Encouraging employment past 65 in England and the USA', *Journal of Social Policy*, 40 (3), 493–512.

Lain, D. (2012), 'Working past 65 in the UK and USA: Segregation into "Lopaq" occupations?', *Work, Employment and Society*, 26 (1), 78–94.

Larsen, M. and Pedersen, P. J. (2012), *Paid work after retirement: Recent trends in Denmark, IZA DP No. 6537*, Bonn: Institute for Labor Studies.

Maddox, G. L. (1968), 'Persistence of life style among the elderly: A longitudinal study of patterns of social activity in relation to life satisfaction', in: B. L. Neugarten (ed), *Middle age and aging: A reader in Social Psychology*, Chicago: University of Chicago Press, 181–83.

Menning, S., Hoffmann, E. and Engstler, H. (2007), *Erwerbsbeteiligung älterer Menschen und Übergang in den Ruhestand, GeroStat Report Altersdaten 1/2007*, Berlin: Deutsches Zentrum für Altersfragen.

Milazzo, R. (2000), 'Le dinamiche del lavoro in età anziana in Italia', in: G. Geroldi (ed.), *Lavorare da anziani e da pensionati*, Milano: Franco Angeli, 85–110.

Ministero dell'Economia e delle Finanze (2012), *Le tendenze di medio-lungo periodo del sistema pensionistico e socio-sanitario – Aggiornamento 2012, Rapporto n. 13*, Roma: Ministero dell'Economia e delle Finanze.

Ministero dell'Economia e delle Finanze (2013), *Le tendenze di medio-lungo periodo del sistema pensionistico e socio-sanitario, Rapporto n. 14*, Roma: Ministero dell'Economia e delle Finanze.

Mood, C. (2010), 'Logistic regression: Why we cannot do what we think we can do, and what we can do about it', *European Sociological Review*, 26 (1), 67–82.

OECD (Organisation for Economic Co-Operation and Development) (2013), *Pensions at a glance 2013*, Paris: OECD.

Principi, A., Checcucci, P. and Di Rosa, M. (2012), *Income from work after retirement in Italy*, Ancona (Italy): National Institute of Health and Science of Aging (INRCA), http://www.inrca.it/inrca/files/focuson/Income%20from%20work%20after%20retirement%20in%20Italy.pdf, date accessed 11 January 2015.

Principi, A., Chiatti, C. and Lamura, G. (2014), 'Older volunteers in Italy: An underestimated phenomenon?', in: A. Principi, P. H. Jensen and G. Lamura (eds.), *Active ageing: Voluntary work by older people in Europe*, Bristol: The Policy Press, 47–70.

Scherger, S., Hagemann, S., Hokema, A. and Lux, T. (2012), *Between privilege and burden: Work past retirement age in Germany and the UK, ZeS Working Paper No. 04/2012*, Bremen: Centre for Social Policy Research, http://www.zes.uni-bremen.de/veroeffentlichungen/arbeitspapiere/?publ=435&page=1, date accessed 11 January 2015.

Schindler, M. (2009), *The Italian labor market: Recent trends, institutions and reform options, IMF Working Paper WP/09/47*, Washington, DC: International Monetary Fund.

Smeaton, D. and McKay, S. (2003), *Working after state pension age: Quantitative analysis, DWP Research Report 182*, Leeds: Department for Work and Pensions.

Thelen, K. (2001), 'Varieties of labor politics in the developed democracies', in: P. A. Hall and D. Soskice (eds.), *Varieties of capitalism*, New York: Oxford University Press, 73–85.

Thijssen, J. and Rocco, T. (2010), 'Development of older workers: Revisiting policies', in: European Centre for the Development of Vocational Training (ed.), *Working and ageing – Emerging theories and empirical perspectives*, Luxembourg: Publications Office of the European Union, 13–27.

Villosio, C. (2008), 'Il mercato del lavoro dei "giovani anziani"', in: P. Brivio and M. Quarta (eds.), *Game over? Percorsi professionali per gli over 40*, Milano: Franco Angeli, 63–76.

5
Work Beyond Pension Age in Sweden: Does a Prolonged Work Life Lead to Increasing Class Inequalities Among Older People?

Björn Halleröd

5.1 Introduction

This chapter investigates work beyond pension age in Sweden. The focus is on class inequalities among older workers and whether or not class-related inequalities have increased during the first decade of the 21st century. For many reasons, Sweden is an interesting case when it comes to employment in old age. Compared to other European countries, the employment rate among older people has for a long time been high in Sweden, not least among women, and is continuing to increase. Sweden is also a forerunner when it comes to reforms of the state pension system and during the past decades, reforms that increase incentives and possibilities to work longer have been introduced. Using data from the years 2002–03 and 2010–11, three basic issues are addressed in this contribution. First, is it predominantly white-collar workers who are able to respond to and take advantage of legislative changes and economic incentives, and keep on working after retirement? Consequently, is the long-existing difference in work beyond pension age between white-collar workers and blue-collar workers growing? Second, does work after retirement affect the income distribution among older workers and, in that case, does it increase or decrease class-related income inequalities? Third, does an increase of work beyond retirement lead to a situation where larger shares of post-retirement age workers suffer from health problems?

5.2 Institutional background and basic assumptions

With regard to the USA, Ekerdt (2010: 73) states: 'The trend toward early retirement is history. People claim that they want to work longer, they are advised (if only for financial reasons) to work longer, policy directions point toward working longer, and population structure suggests a workforce that may need to retain older workers. And people are working longer.' In spite of considerable institutional differences, this observation fits just as well as a description of today's Sweden. When older workers (55 years or older) in 2002–03 were asked about their preferred retirement age, about 9 per cent of the men and 5 per cent of the women answered that they wanted to keep on working after turning 65. Eight years later, the corresponding figures were 17 and 10 per cent (Stattin 2013). The withering away of age 65 as a normative standard of retirement age is an effect of a long-term discourse emphasizing the need to work longer in order to counterbalance the costs generated by increasing longevity. In practice, the rhetoric has been combined with legislative changes and strong economic incentives to continue working in old age in Sweden (SOU 2013) – and indeed, just as in the USA and in a majority of EU-countries, Swedes do now work longer.

The dotted lines in Figure 5.1 show the employment rate between 1976 and 2013 among men and women aged 55 to 64. The employment rate of men decreased from the mid-1970s to the mid-1980s when it stabilized, only to drop dramatically in the early 1990s as Sweden was hit by a deep economic recession. After that, the male employment rate has increased more or less continuously, and in 2013 it was back at the same level as it was at the beginning of the 1980s.

For women, the period from the mid-1970s to the early 1990s was markedly different. The employment rate increased rapidly throughout this period. Women were also hit by the 1990 crisis; and just as among men, the employment rate among women has been increasing since then and is now at an all-time high. From the year 2005 onwards, the labour force survey also has included the population above the age of 64, which in itself is an indication of the transformed view on older workers. During the eight-year period for which data are available, the employment rate in this age group has increased among both men and women.

5.2.1 The Swedish pension system

The current Swedish pensions system has four main layers. The guaranteed pension does, as the name indicates, guarantee a basic pension

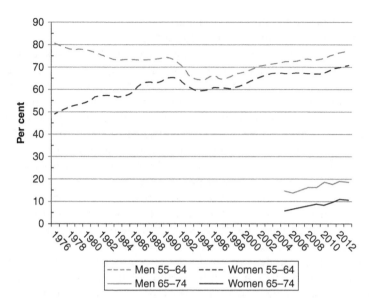

Figure 5.1 Employment rate of men and women aged 55 to 64 (1976–2013) and 65 to 74 (2005–2013)
Source: Statistics Sweden – Labour Force Survey.

for people with no or very low pension claims accumulated by labour market participation. The second layer is a mandatory earnings-related pension strictly linked to earnings during labour market participation. The third layer consists of collective occupational pensions that are negotiated between the unions and the employers' federation as a part of the collective bargaining process, which means that they differ by sector and branch of the labour market. The fourth layer consists of private pensions and savings. In the discussion of changes in the pension system in relation to work beyond pension age, it is mainly the second layer that, at least so far, is relevant. This is not only the case because the earnings-related pension for most people forms the largest share of their pension. The first layer, the guaranteed pension, is a system that mainly covers those who for different reasons are not working. Both the third and fourth layers are of course also based on incomes and thus mirror inequalities in labour participation and incomes, but they are heterogeneous and have so far not been systematically adapted to promote a prolonged working life.

If not before, then the deep crisis in the early 1990s convinced politicians that the old pay-as-you-go income pension system, which provided defined benefits and guaranteed the retired a price-indexed income, was unsustainable. The reformed earnings-related pension system, gradually introduced since the mid-1990s, has two central features. First, instead of guaranteeing pensioners a certain income, the new system has a defined contribution rate which is a safeguard against increasing pension costs. Contributions are fixed as a percentage of wages and pensions are supposed to follow real wages, not real prices. As a consequence, pensions are automatically lowered in the case of an economic downturn, which in fact happened as a result of the recent economic crisis which hit Sweden in 2008 and 2009. The costs for increasing longevity have also been transferred to the retired, as the level of pensions is actuarially adjusted to changes in life expectancy each year and in an automatic way. In this way, the risk has been shifted from the wage earners who finance the system to the pensioners. Second, in the previous second layer, pensions were based on the 15 best income years in the pensioner's career. In the reformed system, pensions are based on life-time income, and it is estimated that being employed an extra year beyond pension age increases the annual pension income by 10 per cent (SOU 2012). Thus, in contrast to the former system, working longer has a direct impact on the expected income pension.

In tandem with these changes, a flexible pension age has been introduced. The second-layer earnings-related pension can be claimed from age 61, and there is no longer a formal pension age at 65. However, there are strong economic incentives to work longer. In order to raise awareness about the economic gains of working longer, the Swedish Pensions Agency annually distributes the 'orange envelope' informing everyone about their estimated pension, given their income situation. The agency presents four different estimates depending on the age at which the individual decides to retire. A typical example would be: if you retire at age 61, you will receive a monthly state pension of €1,144; if you wait until 65, it will increase to €1,456; and if you keep on working until 70, you will receive a pension of €2,056 per month. The example refers to a situation in which a person chooses either full-time retirement or carries on working as before, although the system is actually flexible and allows the individual to combine retirement and part-time work or to retire but defer pension receipt. Income from work will always increase the expected earnings-related pension as will any postponement of payments from this second layer. The economic incentives have been further strengthened by changes in the taxation system. In Sweden pensions are, like most other benefits, taxed, and until 2007 the tax rate was the same for pensions as for income from work. In 2007, a so-called job tax deduction was introduced and thereafter gradually extended that lowered taxation of income from work but not from pensions and other social benefits such as sickness or unemployment benefit. This means that pensions are taxed higher than income from work.

The substantial changes to the early retirement system are also important. The system is now strictly health related and is no longer a part of the pension system but of the sickness benefit scheme, and has correspondingly been renamed 'sickness and activation benefit'. The related benefit is no longer permanent, and the ability to return to the labour market is tested regularly. The possibility to combine early pension receipt with other benefits, for example with payments from unemployment insurance, which used to facilitate a 'flexible' labour market exit, no longer exists in the current system. As a consequence, early retirement has decreased dramatically (Halleröd et al. 2013).

Even though 65 has been abolished as formal pension age, this is only true for the second layer of the pension system, that is, the earnings-related pension. The first layer, the guaranteed pension, is still tied to reaching age 65. Within the third layer, occupational pensions, different rules apply, which in some cases mean that pensions can be claimed well before age 61. As pointed out by a recent public investigation (SOU 2013), there is also a complicated web of legislation and negotiated collective agreements that are based on the assumption of a set pension age. The Employment Protection Act that used to cover employees until they reached the age of 65 years was extended to cover employees until the age of 67 in 2001. Hence, once an employee has turned 67, it is up to the employer to decide about further employment. Other parts of the income protection system are still tied to age 65. For example, unemployment insurance does not cover people over the age of 65. To increase the incentives for employers to employ older workers, there is a reduction of pay-roll tax from 31 per cent of the salary to 10 per cent that kicks in after the employee has turned 65. These and many other (very complex) regulations indicate that even though in the 'income pension' 65 has been abolished as a fixed age for receiving an old-age pension, other parts of the pension system, labour market legislation, negotiated collective agreements and welfare policies still apply the age limit of 65 implicitly or explicitly. And, even though it is possible that age 65 is slowly withering away as the normative retirement age, 65 is still the most common answer when Swedes are asked about preferred retirement age (Stattin 2013).

5.2.2 Why should we expect increasing class inequalities?

Socio-economic differences in mortality, health and general well-being are well documented, universal, persistent and seemingly unaffected by general improvements in living conditions (Marmot 2004; Erikson and Torssander 2008; Halleröd and Gustafsson 2011; Socialstyrelsen 2012). These differences also affect retirement patterns, and a clear social gradient exists when it comes to retirement timing (Halleröd et al. 2013; Scherger et al. 2012) and retirement preferences (Wahrendorf et al. 2013). These patterns of social stratification can be related to the notion of the accumulation of advantage or disadvantage, with health and well-being in old age primarily the results of not only enduring but also increasing inequalities across the entire life course (Blane 2006; Halleröd and Gustafsson 2011; Moen 2011 – see also Matthews and Nazroo in Chapter 12 of this book). In support of this perspective, a number of studies have shown that people in general feel, at least when discounting for the initial happiness bliss (Kim and Moen 2001), neither better nor worse than before when they retire. This implies that retirement per se has no significant universal effect on the subsequent well-being of pensioners (Kim and Moen 2001; Brockmann et al. 2009; Hult et al. 2010; Halleröd et al. 2013). As a consequence of this research so far, we would expect the increasing employment rate among older workers to go together with a selection by good health. This implies that white-collar workers are probably more able to respond to the economic incentives built into the reformed pension and taxation system by working longer and that these workers make up an increasing share of those who keep on working after pension age. *Thus, first, it is assumed that the difference in the employment rates above age 64 between white-collar workers and blue-collar workers has increased between 2002–03 and 2010–11.*

The economic situation among older people is largely determined by past actions, in particular employment participation. A central feature of the reformed Swedish pension system is that it is clearly profitable to work longer, as the pension gets larger during the extended work life. Hence, economically there is a double benefit from work after pension age and/or deferring pension receipt. Since wages vary depending on class position and bearing in mind the first assumption above, a prolonged working life will most likely increase income inequality among older people. *So, secondly, it is assumed that class-related income inequalities among people aged over 64 have increased between 2002–03 and 2010–11 and that the increase is mainly driven by income from work among white-collar workers.*

If the class-related employment gap between white- and blue-collar workers does *not* increase, larger or rising health and well-being inequalities among the older workforce can be expected. The reason is that blue-collar workers who previously left the labour market due to health and well-being problems now possibly remain in the workforce in old age. Thus, the economic incentives to keep on working might be so strong that people stay in paid work even though they suffer from health problems. This is further underpinned by the fact that incentives might be particularly strong for blue-collar workers who can only expect moderate pensions, which would make even a modest income from work relatively important. Furthermore, Westerlund et al. (2009) show that retirement in fact has a general positive impact on health, but especially so among low-skilled occupations. This gives support to the idea that a prolonged working life will increase health differences between blue- and white-collar workers because health improvements of blue-collar workers due to retirement are not realized or delayed. *So, thirdly, it can be assumed that an increase in working beyond pension age among blue-collar workers will lead to increasing class-related health inequalities among people aged over 64 between 2002–03 and 2010–11.*

5.3 Data and overview of sample

The following analyses are based on data from the first and second wave of the Panel Survey of Ageing and the Elderly (PSAE). PSAE is an integrated part of the Swedish Survey of Living Conditions (Undersökningen av levnadsförhållanden – ULF) (Vogel and Häll 2006), which has been conducted annually since 1975. What makes PSAE special in relation to ULF is that the sample size has been increased in the group of those aged 65 and older, the upper age limit previously used in ULF (age 84) has been removed, and the questionnaire has been extended by questions about living conditions in old age. Data were collected in two waves: 2002–03 and 2010–11. PSAE is partially designed as a panel, and about half of the sample is included in both waves. However, in this analysis we treat PSAE as repeated cross-sectional survey with a representative population sample. We focus on the age group 65 to 74 (n = 3,374), that is, an age span in which most people are retired and receive a pension.

5.3.1 Who is retired and who works after reaching pension age?

The notion of retirement is 'famously ambiguous' (Ekerdt 2010: 70) and increasingly difficult to pin down. Using information from the PSAE we can identify individuals who classify themselves as retired. Using this information we can conclude that work after pension age has become more common, in fact much more common. In 2002–03, 17 per cent of those aged 65 to 74 were still working. Eight years later, in 2010–11, this figure had increased to 44 per cent. If we are to believe what people say, there has been a massive increase in work beyond pension age. However, PSAE also provides information from the income register that makes it possible to identify those who have registered income from paid work. Using this information we can still observe an increase in work beyond pension age, but it is less dramatic. In 2002–03, 27 per cent had a registered income from work, a figure that in 2010–11 had increased to 35 per cent. A possibly more severe problem is that there is a substantial mismatch between the two measures. Among those who stated in 2002–03 that they were not retired, only 61 per cent actually had an income from work. In 2010–11, this figure had gone down to 52 per cent. One possible explanation for this mismatch is that older workers continue to work but are paid off the books, or that they do not work but still do not consider themselves as retired, for example because they do not receive any age-related pension, only a very small one, or continue to have an 'active' role as a homemaker in the case of women. However, in Sweden it is very uncommon not to receive a pension after age 65: data from the income register show that in 2008 in the age group 65 to 74 almost everyone, 98 per cent, received either the guaranteed pension or the 'income pension' and that most of the very few who did not receive either of these nevertheless received some other kind of age-related pension. Hence, when talking about work in the age group 65 to 74, it is almost always safe to conclude that we are studying work despite receiving a pension and/or after retirement from the main career. Nonetheless, it still has to be decided how to operationalize work.

At this stage we can only speculate, but it seems more reliable to use income data, not the survey data. Using income data, however, still leaves another problem unsolved. If people have income from work, we know that they do some kind of paid work. A large share of those who keep on working despite receiving a pension work very little. Among those who worked after retirement in 2002–03, 53 per cent had an income from work that made up less than a tenth of their total pre-tax individual income. A third had an income from work that was between 10 and 50 per cent of their total individual income, and only 14 per cent had an income from work that was above 50 per cent of their total income. In 2010–11, these figures had markedly changed, and 38 per cent had an income from work that was less than ten per cent of their total income, just as many had an income between 10 and 50 per cent, and 24 per cent (far more than in 2002–03) had an income from work that made up more than 50 per cent of their total income. Thus, many people only earn a little extra and seem to combine work income with other incomes, in particular from pensions; while for others, the earned income constitutes a substantial part of their income, be it because of low pensions or because they at least partly defer pension receipt. Since the aim is to study work beyond pension age in general, any income from work is used here as an indicator of this work.

5.3.2 Health and income

The subjective health question, in which respondents are asked if they consider their health to be very good, good, fair, bad or very bad, is used as an indicator of individual health. Those who deem their health to be fair, bad or very bad are classified into the group of those with health problems. The subjective health question has been thoroughly evaluated and has proved to be a good predictor of both future morbidity and mortality (Idler and Benyamini 1997; Idler et al. 2004). Recent studies have also shown that the question is a very good indicator of multidimensional well-being among older people (Halleröd 2009; Halleröd and Seldén 2013).

Three measures of income are used: the individual's total pre-tax income, the pre-tax income from work and the equivalent disposable household income. Using pre-tax income makes it possible to clearly distinguish income from work, which is, because of progressive marginal taxes, not possible using post-tax income. Individual income is used because it is related to the individual's class position and whether or not the individual continues to work beyond pension age. However, when looking at the consequences of this work, it is also important to look at the household's disposable income, which has a more direct impact on the individual's living conditions. The distributions of income measures are typically very skewed with a long right-hand-side tail, implying a small number of very high incomes. In order to adjust for this, the logarithm of income is used. By doing so, we get a model that is less sensitive to what is going on at the top of the income distribution. All the income measures are adjusted to 2011 prices and recalculated into Euro using the exchange rate 9 Swedish crowns for €1.

5.3.3 Class

Our measure of class position is the Swedish Socio-Economic Code (SEI), which provides a categorization of occupations based on formal qualification and educational requirements. We use this information to form four occupational groups: The 'blue-collar class' consists of both unskilled and skilled blue-collar occupations. The lower and middle-range white-collar classes have been merged into one class that henceforth will be labelled 'lower white-collar class'. Jobs in the 'upper white-collar class' consist of highly skilled employees and professionals. All self-employed including farmers are grouped into one category. The SEI and hence also our class categories closely resemble the well-known Erikson, Goldthorpe and Portocarero (EGP) schema (Erikson and Goldthorpe 1993; Bihagen 2000). The variable relates to the respondent's main occupation, and for those who have already reached pension age, it refers to the respondent's previous main occupation. A potential problem is that it is not clear what is meant by previous main occupation, especially not among those who continue working beyond pension age. To check the robustness of the results, all the analyses have been recalculated using panel data on occupation. Occupational data from 2002–03 have been added to the 2010–11 subsample included in the PSAE panel, and occupational data from 1994–95 have been added to the 2002–03 panel sample. The results from these analyses (not shown here but available from the author) do not deviate in substance from the results presented below, confirming the reliability of the measure of class as a measure of main pre-retirement occupation.

5.3.4 Method

Even though two of the outcomes are binary, ordinary least squares (OLS) regression is used in all models. There are two reasons to choose a linear model instead of a logistic model. First, the linear model allows a straightforward interpretation of a binary outcome in terms of proportions. It has also been shown that a linear model basically provides equal estimates of significance as the logistic model (Hellevik 2007). Second, the linear model also produces estimates that are comparable across models, which is not the case for the logistic model (Winship and Mare 1984; Mood 2010). Because some individuals are included in both the 2002–03 and 2010–11 samples, clustered standard errors are estimated.

5.4 Results

Descriptive statistics are presented in Table 5.1. Percentages are given separately for the total sample (aged 65 to 74) and separately for both nonworking and working persons. The latter refers to those who have a registered income from paid work. The percentages in the first row show the increase in work after pension age from 27 per cent to 35 per cent between 2002–03 and 2010–11. This development is also shown in

Table 5.1 Descriptive statistics for wave 1 and 2 – not working and working in the age group 65 to 74 (column per cent and averages)

	2002–03			2010–11		
	All	Not working	Working	All	Not working	Working
Working	27.2	0.0	100.0	34.7	0.0	100.0
Health: less than good	34.5	38.8	23.1	30.4	35.6	20.7
Class						
Blue-collar	43.2	47.6	31.4	39.3	43.6	31.3
Low/middle white-collar	35.7	35.5	36.3	36.2	36.2	36.2
Upper white-collar/professional	9.3	7.3	14.7	11.4	9.5	14.9
Self-employed/other	11.9	9.7	17.6	13.1	10.7	17.7
Gender: women	52.9	58.3	38.3	50.5	54.8	42.5
Age: mean	69.2	69.6	68.2	69.0	69.6	68.1
(standard deviation)	(2.9)	(2.8)	(2.9)	(2.9)	(2.8)	(2.8)
Median incomes (€)						
Income from work	–	0.0	1,991	–	0.0	6,133
Total pre-tax income	20,835	18,938	27,399	23,900	20,811	34,045
Equivalent disposable income	15,317	14,459	18,216	18,822	15,545	24,078
Log income from work	–	0.0	5.2	–	0.0	5.9
(standard deviation)			(1.9)			(1.9)
Log total pre-tax income	7.5	7.4	7.9	7.7	7.6	8.1
(standard deviation)	(0.7)	(0.7)	(0.6)	(0.7)	(0.7)	(0.6)
Log equiv. disposable income	7.3	7.2	7.4	7.5	7.4	7.7
(standard deviation)	(0.4)	(0.4)	(0.4)	(0.6)	(0.6)	(0.5)

Source: Panel Survey of Ageing and the Elderly (PSAE), 2002–3 and 2010–11, own calculations.

Figure 5.2. The graph shows that rather than a retirement age there is a retirement period in which employment rates steadily decline and that starts around age 60 and levels out after age 67.

Table 5.1 also shows that subjective health has somewhat improved over time among both the working and the nonworking respondents. At the same time, the share of those with less than good health is persistently different between the two groups, and those who keep on working after retirement report fewer health problems. Moreover, as seen, it is not only that a larger share of older people work after pension age in 2010–11, but they also have higher incomes from work. The median income from work has tripled from €2,000 per year to slightly more than €6,000, and because of the skewed distribution of income from

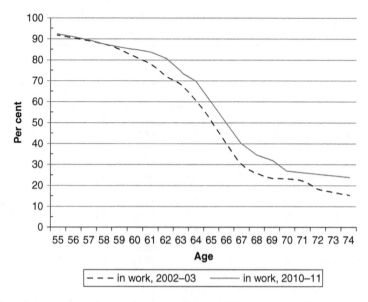

Figure 5.2 Per cent of PSAE sample in paid work by age in 2002–03 and 2010–11
Source: Panel Survey of Ageing and the Elderly (PSAE), 2002–03 and 2010–11, own calculations.

work, the mean has increased substantially more from about €7,700 to around €14,200 (the latter figures are not shown in the table). Partly as a consequence of this, total pre-tax incomes and equivalent household incomes have increased among those still working, although they are aged 65 years or older. However, total pre-tax income and disposable household income have increased for both the working and the nonworking respondents between 2002–03 and 2010–11. So, we can conclude that in 2010–11 people in this particular age group were more often in work, were more often in good health and generally had a somewhat higher income. Table 5.1 also shows the class composition of those who were working and those who were not. The share of the blue-collar class is decreasing both in the overall sample and among the inactive while the upper white-collar class and the self-employed are growing. Among those working, however, the share of those from the blue-collar class remains the same.

Table 5.2 shows the probabilities of being in paid work among 65- to 74-year-old Swedes with different characteristics. The model is aimed at describing class-related changes in the probability of working over time and includes class, time, interaction effects for class and time, and controls for gender and age. In this model, the class indicator gives the class-related probability of working after pension age in 2002–03. The constant refers to those in blue-collar occupations at the observation time of 2002–03. Because age is centred to the mean (of 69) and gender is coded 0 for men and 1 for women, the constant in the regression models refers to male blue-collar workers who are 69 years old. The estimates for 2002–03 show how the other classes deviate from the blue-collar class. The interaction effect estimating the changes for 2010–11 should be interpreted in the following way: The 2010–11 estimate for the blue-collar class shows the change over time for the blue-collar class; plus it serves as a baseline estimate for the interaction. Hence, the interaction estimates for 2010–11 show if the other classes have changed more or less than the blue-collar class between 2002–03 and 2010–11.

Looking at Table 5.2, the estimates for class in 2002–03 show the expected pattern. Work after pension age is significantly more common among those who previously worked in white-collar, especially upper white-collar occupations, than among the blue-collar class. Regarding the comparison between 2002–03 and 2010–11 (interaction effects for 2010–11, shown in the second horizontal panel of the table), the expectation was that the biggest increase of work after pension age is to be found in the white-collar class and, hence, that class differences should increase. However, the data show almost the opposite result. There is a significant employment increase, six per cent, among those from the blue-collar class between the two observation points. As pointed out above, this increase constitutes the baseline to which the other 2010–11 estimates are to be compared. As can be seen, none of the other estimates deviates significantly from the blue-collar estimate; hence there is no indication of a widening post-retirement age employment gap between those from blue-collar and those from white-collar occupations. In fact, the estimate for those from the upper white-collar class is negative and about the same size as the blue-collar estimate for 2010–11, which indicates no change over time in this group and a narrowing class-related employment gap. As this group is small, the confidence interval is wide and the estimate is not significant. Anyway, *there is no indication of a widening class difference when it comes to work after pension age.*

The additional analyses presented in Table 5.2 are restricted to those who work after pension age. The dependent variables are the estimated pre-tax income from work and the share of the income from work in total pre-tax income. Looking at 2002–03, incomes from work are substantially higher within the upper white-collar class, the estimated income from work for a blue-collar worker is €1,097 (exponent of

Table 5.2 The probability of working after pension age among 65 to 74-year-old Swedes (estimates of OLS regressions and confidence intervals[1])

	Working or not (all respondents)		Log income from work (only working pensioners)		Income from work as a percentage of total income (only working pensioners)	
	Estimate (confidence intervals in brackets, p<0.05)					
Estimates 2002/03						
Blue-collar (reference category)	0		0		0	
Low/middle white-collar	0.07*	(0.02, 0.12)	0.39	(−0.02, 0.78)	−1.13	(−6.39, 4.13)
Upper white-collar/professional	0.19*	(0.11, 0.28)	1.13*	(0.57, 1.66)	7.58*	(0.15, 15.01)
Self-employed/other	0.17*	(0.09, 0.24)	0.65*	(0.14, 1.14)	3.05	(−3.76, 9.85)
Estimates 2010/11 (interaction class/2010–11, divergence from 2002/03)						
Blue-collar	0.06*	(0.02, 0.10)	1.15*	(0.76, 1.52)	10.70*	(5.19, 16.20)
Low/middle white-collar	0.02	(−0.05, 0.08)	−0.43	(−0.95, 0.09)	−2.50	(−9.68, 4.67)
Upper white-collar/professional	−0.05	(−0.15, 0.06)	−1.01*	(−1.71, −0.30)	−10.10*	(−20.08, −0.13)
Self-employed/other	0.00	(−0.10, 0.10)	−0.21	(−0.87, 0.45)	−0.75	(−9.76, 8.26)
Age	−0.04*	(−0.04, −0.03)	−0.24*	(−0.28, −0.20)	−3.68*	(−4.19, −3.16)
Gender	−0.11*	(−0.14, −0.08)	−0.32*	(−0.55, −0.09)	0.01	(−3.35, 3.37)
Constant (male, 69 yrs old, blue-collar)	0.27*	(0.24, 0.31)	7.00*	(6.67, 7.33)	14.95*	(10.86, 19.04)
r^2	0.10		0.19		0.18	
n	3,374		1,078		1,078	

Note: [1] Clustered standard errors; *significance: $p < 0.05$.
Source: Panel Survey of Ageing and the Elderly (PSAE), 2002–3 and 2010–11, own calculations.

7.00) and the estimated income from work for a white-collar worker is €3,395 (exponent of 7.00 + 1.13). Incomes from work also make up a substantially larger share of white-collar workers' total pre-tax income, around 23 per cent compared to 15 per cent among blue-collar workers. The self-employed also have a significantly higher income from work compared to the blue-collar class, while there is no difference between those from lower/middle white-collar and blue-collar occupations. Neither for those from low and middle white-collar occupations nor for the self-employed do work incomes make up a larger share of the total pre-tax income compared to the case of blue-collar occupations. Looking at the estimates for 2010–11 it is clear that estimated incomes from work among the blue-collar class have more than tripled from €1,097 to €3,475 (exponent of 7.00 + 1.15) and that incomes from work in this class make up a larger share of the total pre-tax income than in 2002–3, increasing from around 15 per cent (see constant of model) to around 25 per cent (14.95 + 10.70). Again, the 2010–11 estimates for the blue-collar class make up the baseline for 2010–11 to which the other estimates are compared. All 2010–11 estimates for the other classes are negative; in the case of the higher white-collar class, this negative effect is statistically significant. Thus, from a class perspective, *income from work has become more equal, and it is mainly within the blue-collar class that the relative contribution of income from work to the total income has increased.* In fact, within the higher white-collar class it seems that the income from work beyond pension age and its share in total income have not increased at all.

In Table 5.3 estimates of OLS regressions with the dependent variables total pre-tax individual income, equivalent disposable household income, and self-reported health problems are presented. The analyses are organized in the same way as in Table 5.2, with the difference that classes are subdivided into those who work beyond pension age and those who do not. Starting with total individual pre-tax income in 2002–03, we can see that those from the blue-collar class who work after pension age have a higher income than those blue-collar retirees who do not work, and that they have approximately the same income as lower white-collar retirees who do not work. Lower white-collar respondents who work have about the same income as those from the upper white-collar class who do not work, but substantially less than workers in upper white-collar occupations who work after pension age. The self-employed who do not work have about the same income as the blue-collar class, but the income of those who continue to work is clearly higher. By and large, these class-related income differences correspond

Table 5.3 Estimated incomes and probability of self-reported health problems among 65 to 74-year-old Swedes (estimates of OLS regressions and confidence intervals[1])

	Log pre-tax individual income		Equivalent disposable hh income		Self-reported health problems	
	Estimate (confidence intervals in brackets, p<0.05)					
Estimates 2002–03						
Blue-collar, not working (reference category)	0		0		0	
Blue-collar, working	0.23*	(0.16, 0.30)	0.10*	(0.04, 0.16)	−0.16*	(−0.24, −0.08)
Low/middle white-collar, not working	0.23*	(0.18, 0.28)	0.21*	(0.17, 0.25)	−0.12*	(−0.18, −0.07)
Low/middle white-collar, working	0.47*	(0.40, 0.53)	0.34*	(0.28, 0.39)	−0.25*	(−0.32, −0.17)
Upper white-collar/professionals, not working	0.52*	(0.39, 0.65)	0.44*	(0.35, 0.52)	−0.10	(−0.20, 0.00)
Upper white-collar/professionals, working	0.92*	(0.81, 1.03)	0.63*	(0.55, 0.72)	−0.28*	(−0.39, −0.16)
Self-employed, not working	−0.02	(−0.15, 0.12)	0.05	(−0.05, 0.15)	−0.08	(−0.17, 0.01)
Self-employed, working	0.30*	(0.14, 0.45)	0.24*	(0.11, 0.37)	−0.15*	(−0.25, −0.04)
Estimates 2010–11 (interaction, divergence from 2002/03)						
Blue-collar, not working	0.10*	(0.06, 0.15)	0.16*	(0.12, 0.19)	−0.01	(−0.07, 0.04)
Blue-collar, working	0.12*	(0.03, 0.21)	0.13*	(0.05, 0.20)	−0.02	(−0.14, 0.09)
Low/middle white-collar, not working	−0.02	(−0.09, 0.05)	−0.05*	(−0.11, 0.01)	−0.04	(−0.12, 0.05)
Low/middle white-collar, working	0.11*	(0.02, 0.20)	0.07	(0.00, 0.14)	−0.03	(−0.14, 0.08)
Upper white-collar/professionals, not working	−0.06	(−0.24, 0.13)	−0.03	(−0.17, 0.11)	−0.06	(−0.20, 0.08)
Upper white-collar/professionals, working	−0.02	(−0.17, 0.13)	0.05	(−0.07, 0.17)	0.07	(−0.09, 0.22)
Self-employed, not working	0.01	(−0.16, 0.19)	0.00	(−0.13, 0.13)	0.03	(−0.10, 0.16)
Self-employed, working	0.18	(−0.01, 0.38)	0.27*	(0.09, 0.46)	−0.03	(−0.18, 0.11)
Age (69 years)	−0.02*	(−0.02, −0.01)	−0.02*	(−0.02, −0.02)	0.01	(0.00, 0.01)
Gender	−0.34*	(−0.37, −0.31)	−0.07*	(−0.10, −0.04)	0.01	(−0.02, 0.04)
Constant (male, 69 years old, blue collar)	9.94*	(9.90, 9.98)	9.56*	(7.12, 7.18)	0.34*	(0.23, 0.45)
r^2	0.36		0.28		0.04	
n	3,359		3,365		3,372	

Notes: [1]Clustered standard errors; *significance: p < 0.05.
Source: Panel Survey of Ageing and the Elderly (PSAE), 2002–03 and 2010–11, own calculations.

to what was expected, while at the same time they reveal the income benefit from working after pension age.

Now, looking at the interaction estimates for 2010–11, individual pre-tax incomes of blue-collar pensioners who are not working have increased significantly in comparison to the reference group of nonworking blue-collar retirees in 2002–03. Just as in the analyses in Table 5.2, the blue-collar-related estimate for 2010–11 can be seen as the baseline estimate for this observation period, but this time it is the estimates for those blue-collar workers who are not working. We can see that in this group there has been a significant albeit small increase. The estimate for blue-collar workers above age 64 who keep on working in 2010–11 is 0.12 and significant, which shows that the income increase has been somewhat bigger for them. The same goes for those from lower white-collar classes who work. The remaining interaction estimates for 2010–11 are insignificant (although the positive estimate for the working self-employed is relatively large and almost significant).

In the next step of the income analysis, equivalent disposable household income is used as the outcome. Looking at equalized total post-tax household income does not change but replicates the story. According to the interaction estimates for 2010–11, it is not the higher (working or nonworking) white-collar pensioners who have increased their income the most. There has been a general income increase, but it is working pensioners from blue-collar classes who have experienced an income increase on top of that, as well as the working self-employed. In contrast, there has been a significantly lower increase for nonworking lower white-collar classes. Thus, *the analysis does not support the assumption that the increase in work beyond pension age has generally caused larger class-related* differences in disposable income. However, the increase in work beyond pension age does seem to generate increasing income differences within the lower white-collar and blue-collar classes. This means that the overall income distribution might become increasingly uneven not because of rising class inequalities but because of increasing differences between those who are fully retired and the growing group of people that carry on working. Looking at Gini coefficients indicates that this in fact is the case (no tables). The Gini for pre-tax individual income for those who did not work after retirement was 0.29 in 2002–03 and almost the same (0.30) in 2010–11. Looking at the total population, that is, adding those who continue to work in old age, the Gini figures increase to 0.33 in 2002–03 and 0.37 in 2010–11. The Gini coefficients for disposable income reveal the same pattern, although the difference between the two measurement points is larger. Hence, income inequality among retired people has grown and does so not only because a larger share of pensioners keep on working, but also because incomes from work within this group are at the same time increasing more quickly than other (in particular pension) incomes.

The final model, also presented in Table 5.3, shows linear probability estimates of the risk of reporting bad health. Given that we know that there is a class-related health gradient, with blue-collar workers being more exposed to bad health, and that we also know that mainly those from blue-collar occupations drive the increase of work beyond pension age, it is plausible, as stated above, that a larger share of the blue-collar class keeps on working even though they suffer from health problems. The estimates, again using a model including 2002–03 as baseline and 2010–11 as interaction, reproduce the expected picture, that is, substantial class differences and within each class, significantly better health among those who keep on working after pension age compared to those who do not. However, there are no significant changes over time. Even though the trend is towards a general improvement of health (so fewer health problems), there is neither anything that indicates increasing class-related inequalities nor any support for the assumption that those from the blue-collar classes keep on working to a larger extent even though they have health problems. *The class-related health gradient seems stable (and has not become steeper), despite the increase in work beyond pension age.*

5.5 Conclusion

As pointed out by Ekerdt (2010), people in many industrial countries want to work longer because they are healthier, more active and are urged to work longer by less secure and decreasing pension incomes, and they in fact do work longer. Sweden is no exception from this trend; in fact, Sweden is a forerunner when it comes to reforms of the pension system, making it possible and economically profitable to work longer. Thus labour market participation among older people in Sweden is high, in comparison with similar countries. Today's pension system, in combination with taxation, is based on strong economic incentives to carry on working and offers few alternative routes to exit the labour market. At the same time, we know that social stratification affects pension age, incomes and health status: typically, blue-collar workers leave the labour market earlier, have lower incomes and have worse health (for example Marmot 2004; Erikson and Torssander 2008; Halleröd and Gustafsson 2011; Halleröd et al. 2013; Wahrendorf et al. 2013). Therefore changes to the pension system that increase the economic incentives for individuals to carry on working beyond pension age are likely to increase class inequalities.

The well-documented class-related health differences suggest that it is first and foremost upper white-collar workers who can take advantage of the system and work longer. As a consequence, it was assumed above that the class-related retirement age gap would increase over time. This assumption was not confirmed in the analyses presented here; if anything, the opposite was the case, and class differences in working longer have decreased, not increased. Even though class-driven differences in retirement age in fact seem to be decreasing, it could still be the case that upper white-collar retirees benefit most from working after pension age in economic terms. They might earn more, and their income gains from working after pension age should be larger, especially compared to the blue-collar class. Again, the analyses showed another picture. Among the 65- to 74-year-olds, there was a general income increase between 2002–03 and 2010–11. Incomes increased particularly among blue-collar and lower white-collar workers who continued to work after pension age. Thus, between-class inequalities did not increase, but for blue-collar and lower white-collar classes the within-class inequality did increase. Does this mean that first and foremost blue-collar workers carry on working even though they suffer from health problems and that class-related health inequalities among older workers increase? Again the answer seems to be no. It appears as if the health situation has improved generally, but without any significant changes in class-related health inequalities.

We are in the midst of a historical shift; the trend towards earlier retirement has been halted and reversed. From what we know about social stratification, it seems very likely that this trend towards a prolonged working life will lead to increasing class-related inequalities among older people. This initial study could not confirm such a development in Sweden for the recent decade investigated. The comparison of older Swedish people aged 65 to 74 in 2002–03 and 2010–11 showed that a larger share of them carry on working after retirement. However, this increase took place without any considerable increase in class-related inequalities when it comes to the chances of working after pension age, incomes or health among all older people of that age group. One reason for these unexpected results might be that they reflect a trend towards improved health among older workers in conjunction with improved working conditions. Since the potential for improvements in both health and working conditions is largest among blue-collar workers, it might be that the increase in the number of individuals who are willing *and able* to work beyond retirement is also largest among blue-collar workers. Or, to put it in another way, if in 2002–03 most white-collar workers who were willing and able to work beyond pension age did already work, the potential for an increase is perhaps low. Thus it might be among blue-collar workers that the bulk of *potential* pension-age workers who do not yet realize this potential might be the largest. Time and repeated studies will tell if these results are consistent and persistent within the Swedish context, and comparative studies will tell if the results also hold true for other countries. Extending the analysis to other countries is important as differences and similarities between countries can reveal the role of public policies and the importance of general differences in living conditions (see also the other contributions in this book).

Finally, even though class inequalities seem hardly affected or seem even to be counterbalanced by the increase in older people who continue to work beyond pension age, income inequality among the retired has increased. It has done so because more people who have reached pension age continue to work and have higher incomes from work. So, even though the increase in work beyond pension age does not lead to larger class differences, it seems to lead to larger overall income inequality among the retired – new and growing cleavages among those who work longer and those who do not might be the consequence.

References

Bihagen, E. (2000), *The significance of class*, Umeå: Department of Sociology, Umeå University.
Blane, D. (2006), 'The life course, the social gradient, and health', in: Michael G. Marmot and R. Wilkinson (eds.), *Social determinants of health*, Oxford: Oxford University Press, 64–80.
Brockmann, H., Müller, R. and Helmert, U. (2009), 'Time to retire – Time to die? A prospective cohort study of the effects of early retirement on long-term survival', *Social Science and Medicine*, 62 (2), 160–4.
Ekerdt, D. J. (2010), 'Frontiers of research on work and retirement', *Journals of Gerontology Series B-Psychological Sciences and Social Sciences*, 65 (1), 69–80.
Erikson, R. and Torssander, J. (2008), 'Social class and cause of death', *European Journal of Public Health*, 18 (5), 473–8.
Erikson, R. and Goldthorpe, J. H. (1993), *The constant flux: A study of class mobility in industrial Societies*, Oxford: Clarendon Press.

Halleröd, B. and Seldén, D. (2013), 'The Multi-dimensional characteristics of wellbeing: How different aspects of wellbeing interact and not interact with each other', *Social Indicator Research*, 113 (3), 807–25.
Halleröd, B., Örestig, J. and Stattin, M. (2013), 'Leaving the labour market: The impact of exit routes from employment to retirement on health and wellbeing in old age', *European Journal of Ageing*, 10 (1), 25–35.
Halleröd, B. (2009), 'Ill, worried or worried sick? Inter-relationships among indicators of wellbeing among older people in Sweden', *Ageing & Society*, 29 (4), 563–84.
Halleröd, B. and Gustafsson, J.-E. (2011), 'A longitudinal analysis of the relationship between changes in socio-economic status and changes in health', *Social Science & Medicine*, 72 (1), 116–23.
Hellevik, O. (2007), 'Linear versus logistic regression when the dependent variable is a dichotomy', *Quality and Quantity*, 43 (1), 59–74.
Hult, C., Stattin, M., Janlert, U. and Jarvholm, B. (2010), 'Timing of retirement and mortality – A cohort study of Swedish construction workers', *Social Science & Medicine*, 70 (10), 1480–6.
Idler, E., Leventhal, H., McLaughlin, J. and Leventhal, E. (2004), 'In sickness but not in health: Self-ratings, identity, and mortality', *Journal of Health and Social Behavior*, 45 (3), 336–56.
Idler, E. L. and Benyamini, Y. (1997), 'Self-rated health and mortality: A review of twenty-seven community studies', *Journal of Health and Social Behavior*, 38 (1), 21–37.
Kim, J. E. and Moen, P. (2001), 'Is retirement good or bad for subjective wellbeing?', *Current Directions in Psychological Science*, 10 (3), 83–6.
Marmot, M. (2004), *Status syndrome: How your social standing directly affects your health and life expectancy*, London: Bloomsbury Publishing.
Moen, P. (2011), 'A life course approach to the third age', in: D. C. Carr and K. Komp (eds.), *Gerontology in the area of the third age. Implications and next steps*, New York: Springer Publishing Company LLC, 13–32.
Mood, C. (2010), 'Logistic regression: Why we cannot do what we think we can do, and what we can do about it', *European Sociological Review*, 26 (1), 67–82.
Scherger, S., Hagemann, S., Hokema, A. and Lux, T. (2012), *Between privilige and burden. Work past retirement age in Germany and the UK*, ZeS-Working Paper No. 4/2012, Bremen: Centre for Social Policy Research.
Socialstyrelsen (2012), *Folkhälsorapport*, Stockholm: Socialstyrelsen.
SOU (2012), *Längre liv, längre arbetsliv: Förutsättningar och hinder för äldre att arbeta längre*, Statens offentliga utredningar 2012:28, Stockholm.
SOU (2013), *Åtgärder för ett längre arbetsliv*, Statens offentliga utredningar 2013:25, Stockholm.
Stattin, M. (2013), *Svenska befolkningens inställning till sin pensionsålder 2002/2003 och 2010/2011*, Working paper 2/2013, Umeå: Department of Sociology, Umeå University.
Vogel, J. and Häll, L. (eds.) (2006), *Living conditions of the elderly: Work, economy and social networks 1980–2003*, Stockholm: Statistics Sweden.
Wahrendorf, M., Dragano, N. and Siegrist, J. (2013), 'Social position, work stress, and retirement intentions: A study with older employees from 11 European countries', *European Sociological Review*, 29 (4), 792–802.

Westerlund, H., Kivimaeki, M., Singh-Manoux, A., Melchior, M., Ferrie, J. E., Pentti, J., Jokela, M., Leineweber, C., Goldberg, M., Zins, M. and Vahtera, J. (2009), 'Self-rated health before and after retirement in France (GAZEL): a cohort study', *Lancet*, 374 (9705), 1889–96.

Winship, C. and Mare, R. D. (1984), 'Regression models with ordinal variables', *American Sociological Review*, 49 (4), 512–25.

6
Work Beyond Pension Age in Russia: Labour Market Dynamics and Job Stability in a Turbulent Economy

Jonas Radl and Theodore P. Gerber

6.1 Introduction

This chapter explores the socio-economic characteristics, job mobility and retirement attitudes of working pensioners in the Russian Federation. Working beyond pension age is a relatively common phenomenon in Russia where pensionable ages are low, old-age benefits are slim and pension benefits are fully compatible with labour earnings. These institutional features, together with the country's tumultuous economic transition and peculiar demographics, make Russia an interesting case for the study of working pensioners.

Conceptually, the literature on work past retirement age is closely related to the (much larger) literature on older workers and early retirement (Blossfeld et al. 2011; Radl 2013a), which deals with a broader spectrum of the workforce. Almost by definition, working pensioners and persons working[1] past the statutory retirement age are a minority in affluent societies, where, since at least half a century, the great majority of people withdraw from paid work by age 65. However, improved health and fitness among older people, changing social perceptions around the 'third age', as well as pension reforms incentivizing a later exit from work have made this minority grow over the last years. In view of the accumulating financial pressures on the welfare state induced by population ageing, there is considerable policy interest in understanding the factors contributing to continued employment beyond conventional age thresholds. Participation in the labour market allows older people to continue contributing economically and in many cases can be a central component of 'successful ageing'.

By studying work beyond retirement age, we focus on a specific span of the work life course: the very last. This focus allows us to overcome some of the limitations of most retirement research, which concentrates on the transition from work to labour market inactivity within a wider age range (usually approximately between ages 50 and 70). This strand of research typically employs techniques of event history analysis which define retirement as an 'absorbing' event, neglecting the possibility of a later return to employment and often adopting a simplified dichotomous conceptualization of work versus retirement that struggles to map the roles of part-time work and partial retirement. While this approach is legitimate when the aim is to understand the dominant trends driving the retirement process of the entire workforce – because the direct shift from full-time employment to non-employment is still the overwhelming norm in most advanced societies – it has clear methodological limitations if our aim is to understand the smaller (and potentially select) group of workers who remain active past the normal pension age.

The first part of the chapter will present the institutional background of work exit decisions in Russia. It will touch upon regulations and reforms in the fields of the pension system and pensionable ages as well as describe existing 'active ageing' policies at the company and sectoral levels. In the second part of the chapter, we present empirical evidence on working pensioners in Russia. We trace the trend of the proportion of people working beyond retirement age in Russia during the last two decades and analyse the socio-economic profile of working pensioners in terms of characteristics such as gender, education and social class. Two sources of data are employed. Using data from the European Social Survey (ESS), we examine subjective attitudes towards post-retirement work. Exploiting the panel structure of the Russia Longitudinal Monitoring Survey (RLMS), the chapter furthermore explores the dynamics of Russian retirement processes. By contrasting the jobs held by Russians post pension entrance with those that they had before, we analyse the dynamics of labour market mobility at the tail end of the working career. In addition, the study compares workers who continue in their old employment relations with those who start a new employment. One focal point in this context is to find out what share of working pensioners accepts downward mobility in order to remain in gainful employment.

6.2 Demographics

The Russian demographic situation has undergone a period of notorious deterioration during the 1990s, with both life expectancy and fertility rates falling significantly below prior levels. However, since the mid-2000s, both parameters of population growth have picked up notably (Forbes 2013). Partly thanks to a dramatic increase in non-marital childbearing (Perelli-Harris and Gerber 2011), in 2013 Russia reached a total fertility rate of 1.7, which is among the highest in Eastern Europe. Moreover, the country has long had a positive net migration rate, making immigration 'the major compensatory factor in buoying Russia's population' (Migration Policy Centre 2013: 3) since the economic transition. Russia's peculiar mortality problem – according to the 2012 Human Development Report, the country ranks 12th in the world in terms of average years of schooling and 56th in terms of Gross National Income per capita, but only 134th according to life expectancy – has been attributed chiefly to unhealthy cultural practices, and in particular to high levels of alcohol consumption (Guillot et al. 2011; Zaridze et al. 2014). Life expectancy at birth is still much lower among men (63 years) than among women (75 years) (World Bank 2013). As a consequence of high mortality among working-age men, the Russian labour market is characterized by considerable mismatch between labour supply and labour demand with a shortage of skilled labour and pronounced regional disparities (Korovkin 2011).

6.3 Institutional framework

6.3.1 Pension system and the situation of older people

For current retirees, the Russian pension system has been essentially a conventional single-pillar pay-as-you-go scheme. It was established in 1991 preserving many characteristics of the former Soviet system. With more than 40 million current recipients, the most common pension benefit is called old-age labour pension, featuring a standard pension age of 60 for men and 55 for women.[2] Disabled people as well as persons who do not qualify for a labour pension are entitled to the lower social security pension (three million recipients), with pensionable ages of 65 and 60 years for men and women, respectively.[3] In addition, a relatively large share of older Russians qualifies for one of the various types of early-retirement pensions, often claiming a pension as soon as age 50 or even earlier. Workers in a myriad of hazardous (or otherwise special) occupations are eligible, as are workers in the peripheral far Northern regions of the country (Turner and Guenther 2005; OECD 2011). Unlike in many Western countries, where labour earnings and pension income are not fully compatible until age 65 or even later (because, historically, the raison d'être of old-age pensions is to replace income from wages), Russian pensioners are not penalized when engaging in gainful employment.[4]

A mandatory supplementary scheme of private pensions with a strong default option – administered by the state financial agency – has only been introduced in 2002 (Rudolph and Holtzer 2010). A recent reform law approved by the government has fuelled fears concerning a possible nationalization of these pension savings to alleviate the state's general budget.

In 1997, a national framework programme targeted at the social integration of the older population called 'Older Generation' was set up (Fortuny et al. 2003), and it has been renewed several times thereafter. However, this policy agenda has largely remained lip service. In fact, some claim that the programme did not have any practical effects on the living situation of older people because it does not contain any specific measures (Kopyrina 2007). Moreover, 'financial resources set aside for the actualization of the program either were never allotted or were pocketed by municipal authorities' (Kopyrina 2007: 31). As a consequence, many issues, such as precarious access to health care for large proportions of the senior population, have not been effectively addressed.

6.3.2 Labour market policy

The Soviet system was characterized by extremely rigid labour markets structures and low turnover rates (Clarke and Donova 1999). Persistent labour shortages made enterprises retain internal reserves of labour to meet fluctuating demands of production. During the reform era of the 1990s, job mobility rose substantially, but internal labour markets remained important as employers increasingly employed strategies based on multitasking and multi-skilling (Clarke and Donova 1999). The substantial insider-outsider divide on the Russian dualistic labour market particularly undermines employment opportunities for younger people. A distinctive feature that capitalist Russia has inherited from the Soviet labour market is low overall returns to education (Gerber and Hout 1998), although there is evidence of increasing benefits of education among older workers (Gerber and Radl 2014).

Collective bargaining in Russia is preponderantly carried out via agreements reached between trade unions and the government at the national level without involving employers. The coverage of collective agreements is 62 per cent. However, regional-level agreements may depart from these standards, and company-level agreements are often not effectively enforced (Venn 2009), although according to enterprise surveys the pervasiveness of labour inspections is comparable to Western European standards (OECD 2008: 304).

National employment legislation in the Russian Federation specifically mentions older workers in pre-retirement age as a vulnerable group potentially eligible for job quotas (Fortuny et al. 2003). However, each region has to decide about the actual necessity and level of such a quota for older workers, taking into account the specific labour market situation at the local level. In 1998, the proportion of registered unemployed persons participating in active labour market programmes was 35.4 per cent. Among those, about half were involved in training measures, whereas the other half carried out public works (Fortuny et al. 2003: 40). Older jobseekers are usually offered participation in public works rather than training measures.

The economic crisis of the 1990s following the shock therapy reforms led to rapid shrinkage of employment, especially in manufacturing and construction (Gerber 2012). People who lost their jobs were often left in dire straits as unemployment benefits were below the subsistence level, and many unemployed Russians did not even receive those due to defunct administration (Standing 1994).[5] 'Informal and secondary employment mushroomed in the mid-1990s, providing some workers with the opportunity to offset declining real wages in their primary employment' (Tchetvernina et al. 2001: 2). The recession culminated in the 'Rouble' crisis of 1998. Pensioners suffered mightily from the deterioration of purchasing power due to rampant inflation (Fornero and Ferraresi 2007). As a consequence of the state's inability to sustain the senior population, many older Russians looked for other ways to generate income for their households, such as by hunting, fishing or growing livestock, with produce destined for either self-consumption or sale to neighbours or on the streets (Tchernina and Tchernin 2002). In fact, informality remains a salient feature of the Russian labour market. A recent paper estimates that between 2003 and 2011 the share of informal employment amounted to between 7 and 20 per cent of dependent employment (depending on measurement and definitions) and between 45 and 72 per cent of self-employment (Lehmann and Zaiceva 2013). However, the study also shows that the incidence of informal employment is significantly lower among older than among younger workers.

6.4 Working pensioners in Russia

In retirement research, much of the theoretical debate has centred around the question of whether economic incentives or structural constraints are the main driving force of (early) retirement processes (see Radl 2013a for an extensive discussion of the theoretical ramifications). Ultimately, however, both perspectives are retirement-centred in that they are chiefly concerned with the reconstruction of the reasons for completed exit transitions of retirees. As mentioned above, our focus here is narrower, as we are centrally interested in those who work past statutory pension age; it is also more work-centred, as we deal with older workers who are still active in the labour market. In this vein, to understand why Russian pensioners do or do not continue to work, in an earlier study we proposed three alternative mechanisms (Gerber and Radl 2014): accordingly, older workers may be (1) pushed into gainful employment by economic necessity, (2) pulled into the labour market by the opportunities it offers or (3) find themselves blocked from the job market entirely due to lack of demand.[6]

Our empirical analysis of 20 labour force surveys spanning the period between 1991 and 2007 led to the conclusion that all three mechanisms operate to some extent in post-Soviet Russia, albeit with different saliency among different segments of the workforce (Gerber and Radl 2014). Drawing on the same conceptual framework, in this analysis we examine further evidence to contrast these three perspectives and find out which mechanism dominates empirically for different groups of workers. We focus on the 2000s, because employment beyond pension age in Russia during the 1990s has already been extensively studied (see Gerber and Radl 2014), and the 2000s represent a qualitatively different phase in Russia's post-socialist transition, characterized by long-term stable growth, punctuated only by the global financial crisis of 2008.

6.5 Data

Our data come from two sources. The Russia Longitudinal Monitoring Survey (RLMS) is a household panel data set conducted almost annually in Russia since 1992. The current panel was started in 1994 and includes 18 waves (rounds 5 to 22), the most recent of which was carried out in 2013 (and was not available when we conducted the analyses for this chapter). For the most part, the study is a household survey fielded in the same dwellings from one year to the next: when families move to new dwellings, they are replaced with those new families that have moved into the previously sampled dwellings. However, in some waves families and some individuals who moved were followed up in their new locations. Household response rates have consistently exceeded 80 per cent. Typically, all adults in a household are interviewed, yielding annual sample sizes from about 8,000 to 17,000 adults per wave. The survey covers a wide range of socio-economic topics, health, nutrition, consumption and attitudes. It has been widely analysed by economists, sociologists and public health scholars. The survey has been implemented by a collaborative team including two Russian sociologists and researchers at the Carolina Population Center (at the University of North Carolina, Chapel Hill, USA), who have worked with different institutions and partners in Russia for different waves. We use a harmonized version of the complete current panel that was prepared by investigators at the Higher School of Economics in Moscow.[7]

Although it is not a 'pure' panel study (because dwellings are the fundamental unit of analysis, not individuals or households), a sizable number of adult RLMS respondents are interviewed in multiple rounds, some in all 15. This feature and the relatively large total sample size make the data set especially useful for analysing whether and how reaching pension age affects the labour market participation of older Russians, because we are able to observe a large number of respondents just before and just after they turn 55 (statutory retirement age for women) or 60 (statutory retirement age for men). The survey includes the full date of birth of respondents, which we use to ascertain each respondent's exact age on the day they complete the interview. Accordingly, for those respondents who reach pension age between two successive rounds, we can measure their employment activity (including hours worked) and occupation (for those working) immediately prior to and immediately following statutory retirement age. We produce cross-classifications of these two variables in order to show whether and how reaching pension age affects labour supply and also occupational mobility/stability for older Russians. For this purpose, we restrict analyses to those individuals who newly reached pension age in the year prior to a particular survey round. Accordingly, our units of analysis are respondents who reach pension age between successive surveys that they completed. We use data from the survey they took immediately prior to their reaching statutory retirement age to characterize their employment status and occupation at the moment they turn 55 (for women) or 60 (for men), and we use data from the subsequent wave they completed to characterize their employment status, hours worked and occupation immediately after reaching pension age. We also present several analyses based on the entire sample of pension-age respondents included in the data during particular years.

In addition, we use the European Social Survey (ESS) for evidence on subjective attitudes towards work among Russian retirees. The ESS was initiated by the European Science Foundation with the aim of gathering rigorous and comparable empirical evidence on attitudes, beliefs and behaviours in European societies. The survey has been conducted on a biannual basis since the year 2002, and more than 30 countries have participated in it since its creation. The ESS was recently granted the status of European Research Infrastructure, securing its funding in the long term. The Russian Federation entered the pool of participating countries in round three, carried out in 2006. We use data from round five, carried out in Russia between December 2010 and May 2011, because it includes specific questions about retirement as part of a special module on 'Family, Work and Wellbeing'. Analogous to the analyses based on the RLMS, we examine the subsample of pension-aged Russians, that is, men older than 60 years of age and women older than 55.

6.6 Empirical evidence

6.6.1 Labour market outcomes

According to official statistics, the proportion of employed people among the 60- to 72-year-olds was 4.2 per cent in 2011, and the rate of registered unemployment was 2.2 per cent. Considering the prevalence of informal employment and long-existing issues with the reporting of unemployment (Standing 1994), both figures may underestimate the respective true numbers. Re-evaluating these questions using the RLMS data (albeit with a slightly different sample population), we first present a time series on the share of pension-aged persons (men aged 60 and older, women aged 55 and older) in paid employment and also the unemployment rate since the turn of the century (Figure 6.1).

At the outset in 2000, men were substantially more likely than women to be gainfully employed after being entitled to an old-age pension (17 per cent vs. 10 per cent). This is consistent with existing evidence that, despite historically high female employment rates, there is still strong support for the traditional male breadwinner model in Russia (Motiejunaite and Kravchenko 2008). Older men's employment rates initially stayed relatively stable. Parallel to economic growth picking up,

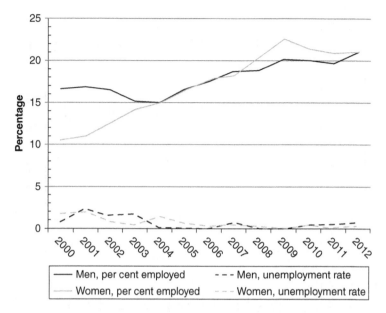

Figure 6.1 Percentage of pension-age RLMS respondents employed and unemployed, 2000–12, by gender
Source: Russia Longitudinal Monitoring Survey, rounds 9–21 (41,978 observations based on 9,428 unique respondents), own elaboration.

employment rates then increased gradually up to 21 per cent in 2012. However, women's employment rates increased much more swiftly, first catching up with men during the first years of observation and even temporarily overtaking them during the late 2000s before eventually reaching the same level as men in 2012.

As for joblessness, the unemployment rate among older men and women was hovering around two per cent during the first years of the 2000s and later fell to practically zero. In sum, while the official statistics seem to underreport the amount of labour market activity among older Russians (probably due to informal employment), our data regarding unemployment are well in line with the official figures.

Table 6.1 shows changes in employment status around the statutory pension age. We construct this variable using information on whether the respondent worked at the time of the survey, how many hours they normally work (coding anything below 40 hours per week as 'part-time'), and an additional standard question asking respondents to indicate their primary activity, which included the response categories

Table 6.1 Employment status of older Russians, before and after reaching state pension age, by gender

	After state pension age (destination)						
Before state pension age (origin)	Work full-time	Work part-time	Retired, no work	Unemployed	Other	Total N	Total % origin
Men and women							
Work full-time	71	8	20	1	0	864	50
Work part-time	25	56	19	0	0	236	14
Retired, no work	4	3	89	0	4	319	19
Unemployed	8	4	74	8	5	74	4
Other	3	1	80	0	16	228	13
Total N	*695*	*216*	*740*	*14*	*56*	*1721*	
Total % Destination	40	13	43	1	3		100
Men							
Work full-time	68	4	27	1	1	305	53
Work part-time	28	45	28	0	0	29	5
Retired, no work	5	1	91	0	3	139	24
Unemployed	10	3	74	10	3	31	5
Other	1	0	83	0	16	75	13
Total N	*227*	*28*	*300*	*5*	*29*	*579*	
Total % destination	39	5	52	1	3		100
Women							
Work full-time	72	11	17	1	0	559	49
Work part-time	25	57	17	0	0	207	18
Retired, no work	3	3	88	1	5	180	16
Unemployed	7	5	74	7	7	43	4
Other	3	2	78	1	16	153	13
Total N	*468*	*188*	*440*	*9*	*37*	*1142*	
Total % destination	41	16	39	1	3		100

Notes: Cell entries are row percentages; row variable is pre-pension age work status (measured less than one year before the respondent reaches pension age); column variable is post-pension age work status (less than one year after respondent reaches pension age).
Source: Russia Longitudinal Monitoring Survey, rounds 9–21, own calculations.

'retired and not working' (that is, completely retired), 'unemployed and looking for a job' (that is, unemployed by the standard ILO definition), standard categories denoting work for an organization, self-employment and entrepreneurial activity and additional categories (attending school full-time, on maternity leave and so on) that we combine into a single 'other' category because they capture distinctions that are not relevant for our current purposes.[8] Like conventional mobility tables, it shows the combined distribution of origins (employment status immediately before reaching pension age: 59 years for men and 54 years for women) and destinations (employment status immediately after reaching statutory pension age at ages 60 and 55, respectively).

As we can see in the marginal distributions in Table 6.1, every second pre-pension-age Russian works full-time and 14 per cent work part-time (preponderantly women, not shown in the table), whereas 19 per cent are already retired. Initial inactivity (retired without work) is more prevalent among men, which is obviously related to the fact that the men under analysis here are five years older than the women. After reaching pension age, only 40 per cent remain employed full-time, while the share of retired (inactive) persons rises to 43 per cent. Interestingly, the overall share of part-time workers hardly changes around the statutory pension age.

Looking specifically at the transitions originating in full-time employment, we observe that seven out of ten workers remain full-time employed when reaching pension age. In turn, only one in five retires. This retirement trajectory is more common among men than among women. Overall, eight per cent of former full-time workers reduce their labour supply to become part-time employed, but this percentage is larger among women than among men. One might expect that part-time work is more sustainable at an advanced age than full-time employment, yet the retirement rate of part-time workers is virtually identical to that of full-time workers.

Notably, there are also cases of increasing employment participation, even at this late point in the work life course: every fourth former part-time worker ups his or her work hours to full-time. Moreover, seven per cent of people who were already retired return to work (either full- or part-time) upon reaching pension age. Finally, a total of 12 per cent find a job after having been unemployed. Although these numbers are not very large, they illustrate that there still exists a labour market for pension-aged Russians.

Table 6.2 examines intragenerational class mobility around pension age among Russian workers, using a modification of the well-known

Table 6.2 Occupational class pre- and post-retirement (per cent, rounded), RLMS respondents who reached retirement age 2000–12

Men

EGP pre-retirement	Post-retirement EGP class								
	Ia/IIa.	Ib.	IIb.	IIIa.	IIIb.	IV.	V/VI	VIIa.	VIIb.	Total
Ia/IIa. Managers	**54**	11	24	5	0	0	3	3	0	6
Ib. Upper professional	4	**83**	4	0	4	4	0	0	0	4
IIb. Lower professional	8	5	**68**	3	0	2	10	4	0	18
IIIa. Upper routine NM	6	0	35	**41**	0	0	0	18	0	3
IIIb. Lower routine NM	0	0	0	0	**100**	0	0	0	0	1
IV. Proprietors	11	2	6	2	0	**32**	32	15	0	7
V/VI. Skilled manual	1	1	2	1	0	1	**79**	14	0	27
VIIa. Semi/unskilled manual, industry	1	0	3	0	1	0	8	**85**	2	35
VIIb. Semi/unskilled manual, agriculture	0	0	0	0	0	0	0	0	**100**	0
Total	6	5	16	2	1	3	29	36	1	

Women

EGP pre-retirement	Post-retirement EGP class								
	Ia/IIa.	Ib.	IIb.	IIIa.	IIIb.	IV.	V/VI	VIIa.	VIIb.	Total
Ia/IIa. Managers	**53**	9	12	7	13	2	1	3	0	6
Ib. Upper professional	1	**80**	4	7	2	3	0	4	0	10
IIb. Lower professional	5	3	**75**	6	3	1	1	7	0	24
IIIa. Upper routine NM	1	5	7	**75**	2	1	1	7	1	19
IIIb. Lower routine NM	4	0	4	3	**79**	0	0	9	0	10
IV. Proprietors	1	2	7	7	13	**38**	8	21	3	7
V/VI. Skilled manual	0	0	3	2	1	0	**78**	16	0	7
VIIa. Semi/unskilled manual, industry	0	0	0	1	3	2	2	**91**	1	16
VIIb. Semi/unskilled manual, agriculture	0	0	0	6	0	0	0	61	**33**	1
Total	5	11	22	18	11	4	7	22	1	

Note: Basis of percentage in main part of table: EGP class pre-retirement (row per cent).
Source: Russia Longitudinal Monitoring Survey, rounds 9–21, own calculations.

Erikson-Goldthorpe class schema (EGP) developed by Gerber and Hout (2004).[9] Analogous to Table 6.1, it cross-classifies class positions held before and after reaching statutory pension age. More specifically, for each class of origin the table shows the percentage of workers with a given origin that ends up in a given destination, along with the attendant marginal distributions. Following the convention in mobility research, we present separate tables for men and women.

The most striking finding from these cross-classifications is the considerable stability in class positions among workers who remain in paid employment upon reaching pension age. Among both men and women, for most origin categories the large majority of workers remain in the same class (that is, on the diagonal). In other words, there are many more stayers than movers. With respect to the latter, although the class schema does not follow a strictly hierarchical logic, one can say that the bottom-left triangle (below the diagonal) largely represents upward mobility, whereas the top-right triangle (above the diagonal) largely represents downward mobility. Using this simplified criterion, there is clearly more downward mobility than there is upward mobility, although the numerical difference is relatively small. For example, there are substantial movements from the managerial class (Ia/IIa) to lower-ranking white-collar occupations such as lower professionals (IIb). Similarly, the largest single movement originating in the upper professional class (Ib) (where more women are found than men) takes place among women moving towards upper routine non-manual employees (IIIa), while only very few make an upward move to become managers. Individuals who are skilled manual workers (V/VI) before reaching pension age (mostly men) with some frequency move down to semi- or unskilled (VIIa) occupations in industry. A very interesting case is the class of self-employed and small proprietors (IV), because there is more mobility here than among the other classes,[10] although the pattern is highly gender specific. Almost a third of male proprietors become skilled manual workers (V/VI) upon reaching the statutory retirement age, whereas among women the most common changes are observed towards unskilled manual work in manufacturing (VIIa, 21 per cent) as well as towards routine non-manual jobs (in particular IIIb, 13 per cent). These movements may well reflect the passing over of small family businesses to the next generation, with senior heads of the company taking on a secondary role (as employees or unpaid advisors) when becoming pensioners. The most notable instance of upward mobility takes place from upper routine non-manual occupations (IIIa) into the lower professional class (IIb) for men, but only three per cent of pre-retirement men started out in IIIa, so we should not make too much of this tendency.

A further aspect that helps us understand the labour market trajectories of older Russians refers to changes between employers. This dimension of job mobility is related to class mobility, yet it also taps into other questions, such as whether employers are willing to employ people beyond retirement age. The RLMS contains a question that asks employed respondents whether they worked in the same job the prior survey round, or whether they changed job title (occupation), employers or both. A worker can change workplaces but remain in the same class, and vice versa. Table 6.3 only includes those who continued working beyond pension age and displays the proportion of employed older workers who stayed in the same workplace, changed their workplace or occupation or both immediately after reaching pension age.

As Table 6.3 demonstrates, first, the vast majority of older workers remain in the same job – same workplace and same occupation – after reaching retirement age. In fact, this pattern holds when we examine all older workers (not shown, available upon request): changes

Table 6.3 Employed RLMS respondents' job stability and changes within one year of reaching pension age (row per cent, rounded; only 'stayers')

	Same job (same workplace, same occupation)	Same workplace, new occupation	Same occupation, new workplace	New workplace and occupation	Did not work prior year
2000	84	2	6	9	0
2001	92	1	3	3	2
2002	86	1	3	4	7
2003	84	3	4	3	6
2004	90	1	2	3	4
2005	88	1	4	3	4
2006	86	1	4	5	4
2007	87	2	4	4	4
2008	84	1	4	4	7
2009	86	1	4	5	4
2010	91	1	2	4	2
2011	93	1	3	3	0
2012	91	1	3	2	2
Total	89	1	3	4	3

Note: Only includes persons who work before and after reaching the statutory pension age.
Source: Russia Longitudinal Monitoring Survey, rounds 9–21, own calculations.

in workplace and/or occupation are rare for older Russian workers. A plausible explanation is that the concept of a career job still has considerable validity in the Russian labour market and that this continuity in employment relations even persists beyond the statutory retirement age. Moreover, by the time they reach pension age, workers may be in jobs they consider optimal and where, by virtue of firm-specific skills acquired during their careers, their productivity is highest. Across time, the average share of pension-aged individuals working in the same job as the year before is just below 90 per cent, and since there is no clear time trend emerging from the data, it seems safe to assume that this regularity is largely insensitive to macroeconomic context conditions. Second, the lion's share of mobility occurs between workplaces, and not within the same firm. This implies that the downward mobility we observed in Table 6.2 does not usually take place in the form of demotions with the same employer. When employment contracts end because the worker reaches the statutory pension age (for example, due to regulations in collective bargaining agreements), it is more likely that he or she takes a lower-ranking job someplace else rather than being re-employed in his or her former company in a lesser position. Actually, older workers are more likely to be hired in their customary occupation with a new employer than to be demoted with the previous employer. Finally, we note that job mobility is substantially higher among working-age Russians (not shown), which further suggests that the signature experience of older Russians who remain in work is job stability, not mobility.

6.6.2 Work orientations and attitudes towards retirement

Focusing on the subjective dimension of work motivations, a previous study by Linz (2004) finds that different factors motivate Russians to work. Although the amount (and regularity) of pay emerges from this survey analysis as one of the most important job motivators among older workers, other factors such as the respect received by and the friendliness of co-workers are equally important (Linz 2004).

To further elucidate the subjective dimension of work past pension age, we explore the Russian subsample of the European Social Survey. In round five of the ESS, (nonworking) retired respondents were asked whether they had wanted to retire at the moment they did or would instead have preferred to continue working. This item appears to be an adequate indicator of involuntary retirement.[11] We select all respondents who have reached the standard pension age. Among those retired (pension-age) Russians, 36 per cent claimed that they would rather have

Table 6.4 Percentage of retired older persons who would have preferred to continue in paid work at the time they retired

Level of Education	Men older than 60	Women older than 55
Less than lower secondary education	31.8	30.4
Lower secondary education	35.8	36.4
Upper secondary education	34.1	31.0
Tertiary education	42.4	40.0
Total	37.1	35.5

Note: n = 630.
Source: European Social Survey, round 5, 2010–11.

remained employed at the time they left their last job. This percentage is slightly larger among men, but the gender difference is not statistically significant. Table 6.4 shows the distribution of perceived involuntary retirement by gender and highest level of educational attainment.

Although the pattern is not perfectly linear, the perception of involuntary retirement is most widespread among persons with tertiary education. Older people with less than lower secondary education, by contrast, less often express that they would have preferred to work longer. This finding is unexpected considering that previous research (on Western Europe) has shown that the risk of being blocked from the labour market is higher for low-skilled workers (Radl 2013b). However, as additional analyses reveal (not shown), while workers with tertiary education have a higher average retirement age than poorly educated persons when work withdrawal is voluntary, there are no noteworthy differences with respect to the average age at which these different educational groups leave the labour market *in*voluntarily in Russia. Based on these explorations, the most plausible explanation of the pattern observed in Table 6.1 is that better-educated workers might generally aspire to later retirement because of greater work attachment or better health and thus more frequently perceive their retirement as involuntary if it occurs early. Individuals who subsequently regret having left work were not necessarily forced to do so at the time (Ebbinghaus and Radl 2015, forthcoming). Factors other than structural barriers to employment can lead individuals to subjectively lament the decision to retire after the fact: the retirement experience itself may simply not be what these individuals hoped it would be; yet, they lack the energy or initiative to seek a new job, having already left the jobs they earlier held. Of course, it is also possible that some who did not regret having left their job at the time they entered retirement now wish to work. But altogether it seems most reasonable to deduce from the ESS data that the majority of Russian pensioners who are not working do not perceive any particular barriers to employment. Consistent with this interpretation, further explorations (not shown) based on the ESS suggest that better-educated older Russians have a more intrinsic work orientation than their less well-educated peers.

Another question included in the survey worth exploring taps into the degree of age discrimination perceived by respondents. Interestingly, the percentage of pension-aged respondents feeling discriminated against based on their age is only three per cent among workers, but twice as much (six per cent) among nonworkers. Still, these percentages are overall at such a low level that ageism does not appear to be perceived as a salient issue in Russian society.

6.7 Discussion

According to the evidence presented in this chapter, in 2012 more than one in five pension-aged Russians were in paid employment. Consistent with earlier research (Gerber and Radl 2014), the labour market situation of older workers has profited in no small degree from the prolonged period of economic growth in Russia since the turn of the century, which was only briefly interrupted by the financial crisis of 2008.[12] Employment rates have been increasing among both older men and women, and unemployment (in the conventional sense) is practically non-existent.

To further our understanding of the situation of working pensioners in Russia, in this chapter we have examined the patterns of labour market mobility occurring around the statutory pension ages, which are set to 60 years for men and 55 years for women. The observed 11 percentage point reduction in employment rates occurring at the statutory retirement ages is smaller than in other developed countries, where we usually see greater drops at the principal age thresholds established by the public pension system. Thus, the pension system seems to have a rather limited effect on the labour market. This Russian idiosyncrasy can be attributed to the country's unusual combination of low pensionable ages, low replacement rates and full compatibility between pension incomes and labour earnings, including early retirement pensions.

Furthermore, our analysis of social class mobility occurring at pensionable ages demonstrates that there is limited movement in terms of the occupational profile of older workers. Most pension-aged Russians maintain their previous class position. This being said, there is more downward mobility than there is upward mobility during this advanced stage in the work life course. The overwhelming majority of pension-aged Russians (almost 90 per cent) stay employed in the same job as before, suggesting that internal labour markets remain crucial in post-Soviet Russia. Nevertheless, there is still some mobility between workplaces. Thus, if workers are downwardly mobile, this is mostly related to a change of employer as very few people are demoted within their accustomed work environment. Since the relationship between education and employment is positive (Gerber and Radl 2014), the majority of working pensioners may be driven by opportunity rather than necessity. Nevertheless, given the meagre level of pension benefits, there are certainly also older Russians who are pushed into employment because of material hardship.

A limitation of the present analysis is that we only considered changes in employment status and job mobility within one year of the statutory pension age. It could be worthwhile for future research to examine life course transitions within a broader age window. Moreover, our study did not take account of the potential role of early-retirement pensions, which may in part explain the limited movement observed at statutory pension ages. Exploring these issues further would be an interesting task for a more encompassing study of older Russians' labour market behaviour.

Our results do paint a fairly positive picture of employment among pension-aged Russians. Participation rates remained stable for older men and grew for older women during the 2000s. Unemployment, understood as lacking a job and actively seeking one, was negligible throughout the period, implying that those older workers who wish to work are usually able to find jobs. Yet, we are unable to distinguish between formal and informal employment, which represents a considerable segment of the Russian labour market. We find no evidence at all of a tendency for workers to have their hours cut upon reaching retirement age or to be forced into less desirable occupations by their employers. Even during the aftermath of the 1998 financial collapse in Russia (whose effects were still being felt at the outset of our observation period) and the global recession of 2008, employment rates among older persons remained steady and robust. Thus, we find little evidence of systematic barriers or obstacles to employment for senior Russians during the 2000s. The one possible exception is the somewhat high percentages of retirees who regret halting work according to the 2011 ESS data. However, even by this subjective measure the majority of retirees are content to be retired, and perceived involuntariness was higher among the better educated, who generally face fewer labour market constraints.

Several decades of low birth rates and high working-age mortality have put Russia in a position where it now faces rapid population ageing and growing old age dependency ratios, which will surely pose a challenge to Russia's economy. As a result, the state has an interest in bolstering employment beyond pension age, as do authorities in many developed nations facing similar demographic issues. Russian policy makers have discussed increasing pensionable ages, but the prospect of them doing so soon is unlikely, given widespread public opposition. The fact that there are no earnings limits whatsoever in the state pension may be a next-best mechanism for encouraging older Russians to continue working well into their retirement years, although this policy may at the same time put a strain on the federal budget, with people claiming a pension as early as they can. Moreover, although many working seniors no doubt are drawn into the labour market by opportunities to make good wages in their chosen occupations, others are probably driven to seek employment by low pension levels. From the perspective of the latter group, it would surely be preferable to raise pensions, even though this might discourage some from continuing to work in the labour market. This option, however, is unlikely to have much appeal to policy makers confronting looming labour shortages and growing budget deficits. Flexible arrangements involving partial and gradual retirement schemes deserve some consideration as middle ground between fiscal discipline and labour market activation.

Notes

1. Note that in this chapter the terms 'pension age' and 'statutory retirement age' are used as interchangeable synonyms that refer to the minimum age to be eligible for a standard old-age pension.
2. On grounds of preventing gender discrimination and ageism, some observers have proposed to raise pensionable ages in Russia and equalize them between men and women (Kamasheva et al. 2013).
3. Numbers according to the official site of the Russian pension scheme: http://2014.pfrf.ru/ot_en/system/
4. For a brief period between 1997 and 2001, 'pensioners were penalized for continuous work by reducing the pension only to the basic part (which cut on average the level of pension by some 50 per cent)' (Fortuny et al. 2003: 31).
5. By 1998, the share of registered unemployed persons receiving benefits amounted to 89.5 per cent, but in relation to the total number of (registered

and unregistered) unemployed the numbers are much lower (Fortuny et al. 2003: 40).
6. Some people may be pushed and pulled at the same time.
7. 'Russia Longitudinal Monitoring Survey, RLMS-HSE', conducted by Higher School of Economics and ZAO 'Demoscope' together with Carolina Population Center and the Institute of Sociology RAS. For more information on the RLMS, see Carolina Population Center (2015) and Higher School of Economics (2015).
8. The 'other' category includes housekeeping, not working and not looking for work, studying full-time, not working due to disability, and infrequent write-in 'other' responses. It is possible that some retirees engaged in these activities may have chosen them rather than the 'retired' category to indicate their 'main' activity. Because we used a separate variable to measure whether respondents worked at the time of the survey, we are confident that respondents who combined work and another activity are consistently coded as working. That is, although there may be slightly more 'retired' respondents in a given wave than our measure captures (because they may be coded as 'other'), we doubt any respondents coded as 'retired' or 'other' were in fact working full-time or part-time.
9. In contrast to the original Erikson-Goldthorpe classification, our classification distinguishes managers from other professionals, combines skilled manual workers with technicians and foremen and aggregates employers with and without employees into a single 'proprietor' class (reflecting the minuscule number of employers with employees in contemporary Russia).
10. Another exception are semi/unskilled manual workers in agriculture (VIIb) who seem to frequently make a shift towards semi/unskilled manual work in industry (VIIa), but on closer look this group of workers is too small to make reliable inferences.
11. More precisely, it provides a *subjective* measure of involuntary exit from work, which focuses on the perception of individual actors. See Ebbinghaus and Radl (2015, forthcoming) for a comparative analysis of objective and subjective measures of constrained retirement.
12. This chapter was written before the geopolitical crisis involving Ukraine and thus does not take into account its recent economic implications.

References

Blossfeld, H.-P., Buchholz, K. and Kurz, K. (eds.) (2011), *Aging populations, globalization and the labour market. Comparing late working life and retirement in modern societies*, Cheltenham: Edward Elgar.

Carolina Population Center (2015), *The Russia Longitudinal Monitoring Survey of HSE (RLMS-HSE)*, Chapel Hill: University of North Carolina, http://www.cpc.unc.edu/projects/rlms-hse, date accessed 5 January 2014.

Clarke, S. and Donova, I. (1999), 'Internal mobility and labour market flexibility in Russia', *Europe-Asia Studies*, 51 (2), 213–43.

Ebbinghaus, B. and Radl, J. (2015, forthcoming), Pushed out prematurely? Comparing objectively forced exits and subjective assessments of involuntary retirement across Europe, *Research in Social Stratification and Mobility*, 40.

Forbes (2013), *11 things everyone should know about Russian demography*, by M. Adomanis, 24 October 2013, http://www.forbes.com/sites/markadomanis/2013/10/24/11-things-everyone-should-know-about-russian-demography/2/, date accessed 17 January 2014.
Fornero, E. and Ferraresi, P. M. (2007), *Pension reform and the development of pension systems: An evaluation of World Bank assistance. Russia country study, IEG Background Paper*, Washington, DC: World Bank.
Fortuny, M., Nesporova, A. and Popova, N. (2003), *Employment promotion policies for older workers in the EU accession countries, the Russian Federation and Ukraine, Employment Paper 2003/50*, Geneva: International Labour Organization.
Gerber, T. P. (2012), 'The structural perspective on postsocialist inequality: Job loss in Russia', *Research in Social Stratification and Mobility*, 30 (1), 17–31.
Gerber, T. P. and Hout, M. (1998): 'More shock than therapy: Market transition, employment, and income in Russia, 1991–1995', *American Journal of Sociology*, 104 (1), 1–50.
Gerber, T. P. and Hout, M. (2004), 'Tightening up: Declining class mobility during Russia's market transition', *American Sociological Review*, 69 (5), 677–703.
Gerber, T. P. and Radl, J. (2014), 'Pushed, pulled, or blocked? The elderly and the labor market in Post-Soviet Russia', *Social Science Research*, 45, 152–69.
Guillot, M., Gavrilova, N. and Pudrovska, T. (2011), 'Understanding the "Russian Mortality Paradox" in Central Asia: Evidence from Kyrgyzstan', *Demography*, 48 (3), 1081–104.
Higher School of Economics (2015): *Russia Longitudinal Monitoring Survey – HSE*, Moscow/Chapel Hill: National Research University Higher School of Economics and Carolina Population Center, University of North Carolina, http://www.hse.ru/en/rlms, date accessed 5 January 2014.
Kamasheva, A., Kolesnikova, J., Karasik, E. and Salyakhov, E. (2013), 'Discrimination and inequality in the labor market', *Procedia Economics and Finance*, 5 (1), 386–92.
Kopyrina, I. (2007), 'The elderly in the Russian Federation: Implications for socioeconomic development', *BOLD*, 17 (4), 21–31.
Korovkin, A. G. (2011), 'The problems of labor supply and labor demand adjustment on the Russian labor market', *Studies on Russian Economic Development*, 22 (2), 177–90.
Lehmann, H. and Zaiceva, A. (2013), *Informal employment in Russia: Incidence, determinants and labor market segmentation, Quaderni – Working Paper DSE No. 903*, Bologna: University of Bologna.
Linz, S. J. (2004), 'Motivating Russian workers: Analysis of age and gender differences', *Journal of Socio-Economics*, 33 (3), 261–89.
Migration Policy Center (2013), *MPC – Migration profile: Russia*, Florence: European University Institute.
Motiejunaite, A. and Kravchenko, Z. (2008), 'Family policy, employment and gender-role attitudes: A comparative analysis of Russia and Sweden', *Journal of European Social Policy*, 18 (1), 38–49.
OECD (Organisation for Economic Co-operation and Development) (2008), *Employment outlook 2008*, Paris: OECD.
OECD (2011), *OECD reviews of labour market and social policies: Russian Federation*, Paris: OECD.

Perelli-Harris, B. and Gerber, T. P. (2011), 'Nonmarital childbearing in Russia: Second demographic transition or pattern of disadvantage?', *Demography*, 48 (1), 317–42.

Radl, J. (2013a), *Retirement timing and social stratification: A comparative study of labor market exit and age norms in Western Europe*, London: Versita (De Gruyter Open).

Radl, J. (2013b), 'Labour market exit and social stratification in Western Europe: The effects of social class and gender on the timing of retirement', *European Sociological Review*, 29 (3), 654–68.

Rudolph, H. and Holtzer, P. (2010), *Challenges of the mandatory funded pension system in the Russian Federation*, Policy Research Working Paper Series No. 5514, Washington, DC: World Bank.

Standing, G. (1994), 'Why measured unemployment in Russia is so low: The net with many holes', *Journal of European Social Policy*, 4 (1), 34–49.

Tchernina, N. V. and Tchernin, E. A. (2002), 'Older people in Russia's transitional society: Multiple deprivation and coping responses', *Ageing & Society*, 22 (5), 543–62.

Tchetvernina, T., Moscovskaya, A., Soboleva, I. and Stepantchikova, N. (2001), 'Labour market flexibility and employment security: Russian Federation', *Employment Paper 2001/31*, Geneva: International Labour Office.

Turner, J. and Guenther, R. (2005), 'A comparison of early retirement pensions in the United States and Russia: The pensions of musicians', *Journal of Aging & Social Policy*, 17 (4), 61–74.

Venn, D. (2009), *Legislation, collective bargaining and enforcement: Updating the OECD employment protection indicators*, OECD Social, Employment and Migration Working Papers No. 89, Paris: OECD.

World Bank (2013), *Gender at a Glance: Russia*, Washington, DC: World Bank/Europe and Central Asia Poverty Reduction and Economic Management Network.

Zaridze, D., Lewington, S., Boroda, A., Scélo, G., Karpov, R., Lazarev, A., Konobeevskaya, I., Igitov, V., Terechova, T., Boffetta, P., Sherliker, P., Kong, X., Whitlock, G., Boreham, J., Brennan, P. and Peto, R. (2014), 'Alcohol and mortality in Russia: Prospective observational study of 151,000 adults', *The Lancet*, 383 (9927), 1465–73.

7
Working Pensioners in China: Financial Necessity or Luxury of Choice?

Ge Yu and Klaus Schömann

7.1 Introduction

Retirement refers to a socially defined and usually irrevocable event in the life course of every worker in modern societies with an established pension system (Kohli 1986) and involves the cessation of work in the labour market. Increasingly, however, retirement is no longer an irrevocable life event, since many older workers do not withdraw from the labour market completely, but continue working past pension age (Scherger et al. 2012). It is important to clarify whether continued work means greater choice or is the consequence of a necessity to finance old age, especially in a rapidly transforming society such as China where the pension system does not yet cover the whole population. Working pensioners and their motivations and reasons for working past retirement are, therefore, of interest to researchers and policy makers interested in social inequalities.

Socio-economic transformations in China have had a profound impact on individual life chances in the past three decades. The opening of China, since 1978, from a centrally planned economy to a market economy brought about not only opportunities but also risks for both young and senior workers. For example, the relaxation of the *Hukou* system allowed rural residents to live and work in urban cities, and the introduction of the New Rural Social Pension Scheme (NRSPS) in 2009 also provided some rural workers with the opportunity to be covered by a small pension at the age of 60. The new market economy has also provided opportunities for both younger and older workers to work in the various non-state sectors. This is particularly evident in self-employment, a sector which, however, did not really re-emerge until the market-orientated reforms were introduced in the late 1970s. Meanwhile, economic restructuring and reforms of enterprises have negatively impacted many workers since the mid-1990s, particularly urban middle-aged and older workers, forcing them to leave the labour market earlier than the official pension age.

It is not surprising that, under a planned economy, the state took care of labour planning, which made working after the mandatory pension age impossible. Therefore, in China, the phenomenon of work beyond pension age has been observed only in recent decades. Nonetheless, there are a considerable number of pensioners re-entering the workforce (Jiang and Du 2009). At the same time, China is experiencing a rapidly ageing population (Giles et al. 2012), motivating warnings about the lack of sustainability of the current pension system (Zeng 2011). To reduce the cost of older dependents and relieve the financial burden on the pension system, recent and planned pension policies have aimed at raising the pension age of (urban) workers, which is currently 60 for men, 55 for female white-collar workers and 50 for female blue-collar workers (China State Council 1978). These policies have raised concerns about whether the prolonged working lives will deepen social inequalities. It is therefore imperative for researchers to investigate the determinants of working beyond pension age, that is, whether older workers remain in the workforce out of economic necessity or their own choice. In this article, we examine the factors that lead pensioners to work or not to work past pension age, by which we mean only paid employment. In particular, we focus on individual and family characteristics of working pensioners and on the relevant institutional setting, such as retirement and pension policies. We proceed as follows: Section 7.2 presents findings regarding the determinants of post-retirement work from previous studies. Section 7.3 discusses the relevant institutional setting, that is, retirement policies and pension schemes. Section 7.4 shows results from our empirical analysis, starting with a description of the China Health and Retirement Longitudinal Study (CHARLS) data and methodology, followed by the empirical results and their interpretation. The article ends with a summary of the results and a discussion of policy implications.

7.2 Findings of previous studies

Although researchers have devoted some attention to paid work after pension age in the Chinese context, their studies have been limited to employment participation of adults aged 60 or over (for example Zhang 1999; Zhang and Li 2000; Wang 2001; Qian and Jiang 2006; Lin and Chi 2008; Zhang 2010; Chou 2010), neglecting younger working pensioners under 60 years. Findings from these studies include push and pull factors that influence the decision to work after pension age or despite receiving a pension. Push factors include low pension payments (Xiao 2004) and connected financial needs of the workers themselves and financial support of others, for example family members (Lin and Chi 2008). Pull factors are, amongst others, self-fulfilment and the desire for economic independence (Zhang 1999). Individual and family characteristics are also discussed as being systematically related to the propensity to work among pensioners, including age, gender, health, education and marital status. Findings suggest that men, individuals who are healthy and those with high educational attainment are more likely to remain in the workforce despite reaching pension age (Zhang 1999; Qian and Jiang 2006; Lin and Chi 2008).

Evidence from other countries complements our understanding of the determinants of paid work beyond pension age in China. Parker and Rougier (2007) find that, in Britain, the older long-term self-employed who have been running a business for more than six years are significantly less likely to withdraw from the labour market than those who are short-term self-employed. Similarly, in Germany, the older self-employed have a stronger motivation to work after pension age (Deller and Maxin 2009a, 2009b). Additionally, results from a study by Blau and Riphahn (1999) suggest that among older German married couples, one spouse's employment status affects the other spouse's work decisions in later life. Lastly, traditional family cultures in continental European countries (especially Southern Europe) lead to a strong involvement of grandmothers in the care of their grandchildren, which in turn reduces the attachment of (these) older women to the labour market (Hank and Buber 2007).

Although there are fundamental differences in social and economic backgrounds between China and other countries, at least some of these factors described may be transferrable to China, especially those which are purely based on the micro level. This will be examined in more detail in the empirical analysis of this article. Beyond investigating the effect of individual and family characteristics on work choices of pensioners, in the following section, we discuss the role of the institutional setting on the propensity to work in terms of the pension system and retirement policies in China.

7.3 The institutional setting: Retirement and pension policies in China

The Chinese pension system differentiates between urban and rural workers, with completely different rules for these two groups. The urban-rural divide accounts for differences in many domains of social policy in China and is based on the household registration system, known as the *Hukou* system.[1] Additionally, there is a separate pension system for the civil service. Essentially then, the current Chinese pension system consists of three main parts: the urban workers' pension system, the civil service pension system and the recently introduced pension schemes for non-employed urban residents and rural workers.

The old pension system was first established in 1952 on a pay-as-you-go basis, with a defined benefit plan (Zeng 2011). It was initially limited to urban employees who worked in the state formal sectors, including government agencies, public institutions and state-owned enterprises (SOEs). When China made the transition from a planned economy to a market economy, however, the old pension system became difficult to sustain for a number of reasons. First, as the number of pensioners increased due to downsizing and bankruptcy of enterprises, it became extremely difficult for these enterprises to finance pensions without government subsidies. Second, as the old pension system only covers a small proportion of the urban workforce (state formal sectors), it has been unable to meet the demands of China's structurally changing economy with its increasingly large number of employees from the rapidly growing non-state sectors, such as foreign-owned companies, joint ventures and collective as well as private enterprises. These employees also need to be included into the pension system. Against this background, the pension system has been undergoing various reforms on a trial basis in several cities since the mid-1980s. In 1997, a contributory pension system was established. Since this reform, all urban workers have been covered, and the self-employed may voluntarily join and make full contributions. This reformed system combines characteristics of both defined benefit and defined contribution retirement plans. It comprises three pillars. The first pillar is the basic public pension insurance. The employer is required to contribute 20 per cent of the employee's total wages to the social pooling account, while employees contribute 8 per cent of their wage to their individual account (Salditt et al. 2008: 54). To receive a pension, workers must have contributed for at least 15 years by the time they reach the state retirement age (Salditt et al. 2008).

The second pillar relates to the firm level and is known as 'enterprise supplementary retirement insurance' or 'enterprise annuity plan' based on voluntary contributions. A survey report indicates that only some multinational companies and almost no local enterprises are able to participate in this programme, because it is very hard to meet the prerequisites of an 'enterprise annuity plan' for local enterprises, and approval is very strict (Watson Wyatt Worldwide 2006).

Voluntary private pension insurance makes up the third pillar of the new pension system. However, it does not yet play an important role in the pension system. The private insurance market does not offer many related benefit plans, because the replacement rate from the basic public pension system is relatively high and there are few tax incentives for private retirement savings plans (Salditt et al. 2008).

More recently, the establishment of the New Rural Social Pension Scheme (NRSPS) in 2009, followed by the Urban Social Pension Scheme (USPS) in 2011, has enabled some rural residents and all non-employed urban residents to make voluntary contributions to individual accounts, which are subsidized by local and central governments. Combined with personal savings, these schemes should provide people with some financial security starting at the age of 60 when the official pension age is reached. Still, in 2012, only an estimated 55 per cent of China's population was covered by a pension scheme of some kind (HelpAge International 2013).

As previously mentioned, civil servants, military officers and most employees of public institutions have a separate pension system, which still follows the old approach of defined benefits (Salditt et al. 2008). This implies more generous pension payments on the basis of seniority and the wage level before retirement (Cai 2004), but without contributions to the state pension system by either the individual or the employer.[2]

Following the lines of the fragmented Chinese pension system, Chinese retirement policies differentiate between rural and urban workers. For rural workers, there is no mandatory retirement. As described above, however, some rural workers are now eligible to receive a basic pension at the age of 60 with the introduction of the New Rural Social Pension Scheme in recent years. For urban workers, there is a relatively low mandatory retirement age: it is 60 for men to withdraw from their main career, while for women it is 55 in white-collar jobs and 50 in blue-collar jobs (China State Council 1978). This retirement policy has been in effect since the 1950s. The mandatory retirement age is also the official pension age. Due to specific circumstances, pensions may be deferred, but the pension benefits will not increase if claimed at a later date (OECD 2013: 232). Additionally, men can retire at the age of 55, and women at 45, if they are in occupations classified as dangerous, harmful to health or extremely arduous, such as working underground, in high altitudes, under high temperatures, or doing heavy labour and similar jobs. Furthermore, workers with health problems and work-related disabilities can retire from their positions early (men at 50 years and women at 45) if either medical proof or disability evaluation is provided.

'Internal retirement' and early retirement have also been in practice in recent decades. The market reforms between the mid-1990s and early 2000s brought about a movement of privatization of enterprises and restructuring of industries. Reforms of enterprises frequently removed older workers without skills from the workforce, as well as persons unable to adapt to new technologies. Jobs also disappeared as a result of bankruptcy of the industries in which older workers worked. Due to financial difficulties of these enterprises, the state allowed the employers to make their workers redundant while still paying them reduced wages. Reaching normal retirement age, these 'internal' early retirees were finally eligible to the usual pension benefits. Meanwhile, early retirement policies have been widespread among urban employees. Economic incentives such as early retirement schemes or redundancy settlements have often forced older workers to withdraw from the labour market before the normal pension age, and a considerable number of people have exited early from their career employment at the age of 40, 45 or 50.

With regard to earning extra income after retirement, there is no differentiation between the 'normal' statutory pension age and the early pension age in the existing state pension scheme. All retirees, including early retirees, are allowed to combine their pensions with income from paid work. Evidently then, such a pension scheme actually encourages workers to retire earlier than the official pension age and then work, if possible, after retirement.

To investigate later whether pensioners need to or want to re-enter the workforce, we now turn our focus to pension benefits. Regarding current pension payments, The Epoch Times (2012), citing from the Chinese National Audit Office, reports that China's overall pension replacement rate (or percentage of a worker's pre-retirement income) was 42.9 per cent in 2011, below the international warning level of 50 per cent. Government agencies and related government bodies replace their retirees' incomes to 100 per cent, or sometimes pay them more than they made on the job (The Epoch Times 2012).

Altogether, the current Chinese pension system continues to maintain the urban-rural divide. In contrast to urban residents, rural residents are often not covered by the system. By the end of 2012, approximately 460 million out of 656 million rural individuals were covered by the New Rural Social Pension Scheme (Citylab 2012; Pozen 2013). Family-related old-age provision is the main support for older adults in these settings. Additionally, to make a living, 'ceaseless toil' characterizes the lives of the elderly in rural China (Benjamin et al. 2003: 1; see also Pang et al. 2004). Among urban workers, civil servants and most employees of public institutions are entitled to receive government-funded pensions substantially higher than employees of firms, without making any contributions to the system (Wang et al. 2009). Looking at these unequally designed pension schemes and benefits for different groups of people, we expect that such structures will have an effect on work decisions for pensioners, in addition to individual and family-related factors.

7.4 Data and methods

The empirical analysis is based on the 2011–12 national baseline survey data of the China Health and Retirement Longitudinal Study (CHARLS).[3] This is a biennial survey in China, conducted by the National School of Development at Peking University. CHARLS aims to be representative of residents of China who are aged 45 and older, with no upper age limit. The baseline national wave of CHARLS was conducted in 2011–12 and includes approximately 10,000 households and 17,500 individuals in 150 counties or districts and 450 villages. CHARLS is similar to the ageing surveys from the USA, Korea, Japan, England and other countries in continental Europe. It covers a wide range of topics around old age, health, work and retirement that allow for multidisciplinary analyses. The CHARLS data provide detailed information on workers before, in the transition to and after retirement, for example personal characteristics (amongst others age, gender, health, marital status and education), family structure (such as number of grandchildren younger than six years), labour market status and occupation and economic characteristics (for example income from work or pensions).

Our analysis focuses on working pensioners. These are defined based on two criteria. The first one relates to workers who have started to receive pension payments, including early pension receipt. Second, working pensioners are defined as engaging in paid employment. People can continue their job or be re-employed in the same or another job after they have started receiving a pension, but we cannot differentiate between the different groups. They are allowed to receive both their pension and their income from (post-retirement) employment, without any deductions. To examine the determinants of work beyond pension age in China, a binary logistic regression model was calculated (Agresti 1996) which models the probability of becoming a working pensioner. This model includes the following *independent variables: Pension age* is defined as the age when the individual starts receiving a pension. We use pension age to examine the role of age when receiving a pension for the first time for re-employment decisions in later life. It is used as a categorical variable and summarized into five groups (aged 45–49, 50–54, 55–59, 60–64, 65+). Furthermore, age in 2011 is included, as well as its squared value. *Gender* is used as a dummy variable (male = 1) to examine gender differences in entering or continuing paid work after retirement. *Health* relates to the respondent's self-reported health at the time of his/her first pension receipt. It is a dummy variable with at least fair health = 1, including people who reported their health as excellent, very good, good and fair. Those with poor health are the reference group. Subjective health in 2011 is also included, with the same variable description as the self-reported health at the time of first pension receipt. The level of *education* is divided into three categories: no and little education (the reference group), middle education (middle school and high school) and high education (vocational school, college and university degree and above). *Hukou (residence registration)* is used as a dummy variable (with an urban *Hukou* = 1) to test whether the respondent's *hukou* status has an effect on work decisions of pensioners. Rural *Hukou* is used as the reference group. The *last workplace/employer before retirement* refers to the type of the respondent's last workplace at the time of first pension receipt, including government agencies, public institutions, non-government organizations (NGO), firms (state, collective and private), self-employment and other types. It is used as a proxy of pre-retirement job. Work in government agencies at retirement is the reference group.

To examine the effect of pension payments on individual pensioners' work decisions, we use the *logged*[4] *monthly pension in 2010* as a continuous variable. The monthly pension in 2010 is a proxy of the monthly pension at the beginning of pension receipt. We do not use the latter as this information is retrospective and may be missing or inaccurate due to the unreliability of recipients' memories. The monthly pension in 2010 which we use instead is the most recent information available.

The existence of grandchildren younger than six years is captured by a dummy variable, in which a value of 1 is for an individual pensioner who has grandchildren and 0 is for those without grandchildren who are also the reference group. This variable is included to examine how the potential obligation to care for grandchildren is linked with the individual decision to work beyond pension age, and how this impact varies with gender.

Spouse's health is controlled in the model to examine its effect on the respondent's likelihood of working after retirement. Respondents reported their *spouse's health* at the time of the respondent's retirement. It is a dummy variable (at least fair health = 1), including people who self-reported their spouse's health as excellent, very good, good and fair. Poor health is the reference group. We also look at the effect of *spouse's employment status* on the respondent's decision to work. In the original data, this variable is defined as the kinds of economic activities that the respondent's spouse is engaged in at the time of the respondent's retirement. It is a categorical variable consisting of several groups: (1) spouse in paid employment (either self-employed or as an employee) or looking for a job, (2) not in paid work and not looking for a job or (3) spouse engaged in farming.

7.5 Descriptive results

Figure 7.1 shows the pattern of retirement timing for men and women, which means the age when respondents received a pension for the first time. Most women do so for the first time around the age of 50, with around 36 per cent exactly at the age of 50, and 10 per cent at the age of 55. Men most frequently receive a pension for the first time at the age of 60 (about 42 per cent), and the second largest share is at the age of 55 (about 12 per cent). These variations by gender in the age of receiving a pension for the first time are attributable to retirement and pension policies, with normal pension age being 60 for men and 55 or 50 for women.

Figure 7.2 displays how labour force participation is distributed among pensioners and non-pensioners aged 45 and over and differentiates by age group and by gender. Generally, there are more men than women who work despite receiving a pension in 2011, and there is a similar gender difference for working non-pensioners aged 45 and older. Importantly, the bar chart indicates that for women after state pension age, labour participation among pensioners (darkest grey in the bar chart) is most frequent among those aged 50 to 54 compared to

Figure 7.1 Age of first pension receipt by gender
Note: n/men = 699, n/women = 657 (unweighted).
Source: CHARLS data 2011/2012.

older age groups, and the second largest share is at the age of 55 to 59. However, a few women keep working after the age of 60. For men, the rates of working after state pension age follow a similar line. It is highest among pensioners aged 60 to 64, declines somewhat in older age groups, and then falls rapidly after the age of 70. This implies that a number of pensioners do not withdraw from the workforce completely after state pension age (50 and 55 for women, and 60 for men), but continue working for a certain time until retiring completely.

Among younger pensioners who retired before the state pension age (women aged 49 and below, men aged 45 to 54), many work as well; among men aged 45 to 49, there are even slightly more working pensioners than nonworking pensioners. This is despite the fact that their withdrawal from the labour market at a young age may be caused by health problems. For those not continuing to work, this might be connected to being made redundant and not finding a new job, for example due to low qualifications. Furthermore, the figure shows that the majority of older workers (probably from rural areas) do not receive a pension (two lighter shades of grey) and a large share of them continue working, especially men. At the same time, the employment

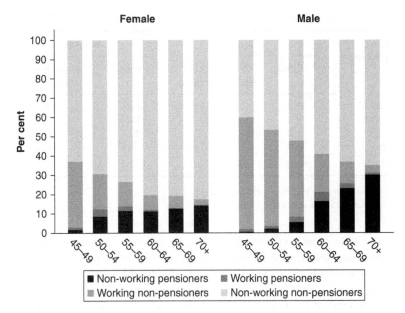

Figure 7.2 Labour force participation of pensioners and non-pensioners aged 45 and over in 2011, by age group and gender
Note: n/men = 8,451, n/women = 8,850 (unweighted).
Source: CHARLS data 2011/12.

participation rate for both male and female non-pensioners decreases steadily with age among the population aged 45 and over.

The following descriptive analyses only include those receiving a pension. They focus on the characteristics of post-retirement jobs and the characteristics of post-retirement workers in comparison to nonworkers. On average, both male and female working pensioners work around 46.1 hours every week. This indicates that working pensioners generally work full-time. Working hours do not differ a lot by gender (see Table 7.1).

Table 7.2 shows the distribution of economic sectors in which working pensioners and non-pensioners aged 45 and over are employed. Jobs of working pensioners in manufacturing represent 31.3 per cent, in wholesale and retail 14.4 per cent, in public administration 9.5 per cent and in education 6.9 per cent. Further, working pensioners aged 45 and over share a similar distribution pattern across these sectors as non-pensioners.

Table 7.1 Average working hours per week of working pensioners by gender

Hours per week	Men		Women	
	Per cent	Mean hours within category	Per cent	Mean hours within category
1–30	13.2	12	13.5	13
31–50	59.4	45	58.6	45
51–70	21.4	60	23.0	60
70+	6.1	84	4.9	84
All	100.0	46	100.0	46

Note: n/men = 283, n/women = 180 (missing = 278).
Source: CHARLS data 2011–12.

The only exception is construction jobs, where pensioners are under-represented but non-pensioners are over-represented. Notably, construction labourers in China are predominantly migrant workers who moved from rural contexts into the cities and the unemployed/laid-off who previously worked in state-owned enterprises (Cai et al. 2009). One of the characteristics of construction jobs is that they involve heavy physical labour, so workers must be robust and healthy enough. Thus, construction jobs would not be the best fit for older pensioners. The opposite might be true of the wholesale and retail sector, which offers jobs that can also be done by (less well-educated) older persons and importantly also in self-employment.

Table 7.3 presents characteristics of working and nonworking pensioners. There is no distinctive difference between working pensioners and nonworking pensioners regarding their average monthly pension. Regarding the average age of first pension receipt of workers and nonworkers, it is not markedly different, with around 53 to 54 years[5] for both groups. However, among working pensioners, more received their first pension relatively late (aged 60 to 64) compared to their nonworking counterparts. Among working pensioners, male workers account for 60 per cent, and high-educated workers make up 24 per cent; by contrast, 46 per cent of the nonworking pensioners are men, and only 19 per cent of them have a higher education. Among those working, 93 per cent report they were healthy at first pension receipt, which is only slightly higher, by about 3 per cent, than the nonworkers (90 per cent).

Regarding health status in 2011, a similar pattern is also seen between working and nonworking pensioners (87 per cent and 83 per cent). The

Table 7.2 Distribution of working pensioners and workers across economic sectors

Sector	Working pensioners (45+)	Workers (45+) without pension
	Column per cent	
Agriculture, forestry, livestock farming, fishing	3.03	3.91
Mining	2.74	1.79
Manufacturing	31.27	27.01
Energy sector (including electricity, gas and water)	1.59	1.11
Construction	6.48	20.30
Transportation, warehousing and postal services	6.48	5.56
Information transmission, computer services and software industry	0.72	0.22
Wholesale and retail	14.41	15.28
Hotels and catering	2.74	3.42
Finance	1.15	0.68
Real estate	1.59	1.09
Leasing and business services	0.58	0.98
Scientific research, technical services, geological prospecting	1.30	0.43
Water conservation, environmental and public facilities management	0.58	1.11
Personal services and other service industries	3.89	3.99
Education	6.92	3.77
Public healthcare, social security and public welfare	4.03	2.06
Culture, sports and entertainment	1.01	1.03
Public administration and social organizations	9.51	6.24
Total per cent	100	100
Total number of observations	*694*	*3,684*
Missing	*47*	*1,080*

Source: CHARLS data 2011–12.

health status of the respondent's spouse at the time of the respondent's first pension receipt does not differ between working and nonworking pensioners. 94 per cent of either workers or nonworkers among the pensioners hold an urban *Hukou*, and 6 per cent a rural *Hukou*, which reflects that the current pension system in China covers only a small

Table 7.3 Characteristics of working pensioners and nonworking pensioners (aged 45 and older)

	Working pensioners	Nonworking pensioners
Monthly pension (mean)	1751.1 Yuan (282 US$)	1721.8 Yuan (277 US$)
Age at first pension receipt (mean)	53.8	53.3
Age 2011 (mean)	65.0	65.1

	Per cent among working pensioners	Per cent among nonworking pensioners
Age when a pension is first received		
Up to 49	19.1	19.3
50–54	28.5	36.1
55–59	22.6	24.9
60–64	29.2	17.6
65+	0.6	2.1
Gender: Male	60.2	45.7
Educational qualification		
Low	35.2	35.4
Middle	40.6	45.1
High	24.2	19.5
Subjective health at first pension receipt: at least fair	93.3	90.2
Subjective health in 2011: at least fair	86.7	83.1
Spouse's health at respondent's first pension receipt: at least fair	89.1	88.7
Registered residence: urban	93.7	94.1
Last workplace before receiving first pension		
Government agencies	9.6	12.5
Public institutions	21.2	22.6
NGO	0.4	0.4
Firms (state, collective and private)	54.7	61.1
Self-employed	12.7	1.7
Others	1.4	1.7
Has grandchildren under 6	39.6	37.8
Economic activity of spouse		
Employed	57.8	63.8
Not employed	29.1	26.4
Farming	13.1	9.8
Total n	*475*	*880*

Note: n = 1,355.
Source: CHARLS data 2011–12.

percentage of rural residents. Among the working pensioners, far more (13 per cent) were self-employed before retirement (and probably still are) than among nonworking pensioners (1 per cent). By contrast, nonworking pensioners were more often employed in firms and somewhat more often in government agencies before receiving their first pension. Regarding the existence of grandchildren younger than six years, there is no marked difference between working and nonworking pensioners.

In sum, working pensioners normally have full-time work. Almost a third of them have jobs in the manufacturing industry, but they also work in many other sectors. Compared with nonworkers, pensioners in paid work are more likely to be male, highly educated, healthy and self-employed before receiving their first pension payment. To explore the determinants of post-retirement work multivariately, we use a binary logistic regression model in the following section.

7.6 Multivariate analysis: The propensity to work despite receiving a pension

In the following, we examine the probability of continuing work or re-entering the labour market among pensioners. Table 7.4 reports the odds ratios of the related binary logistic regression model which includes those older persons who have started to receive a pension. The independent variables include all the variables discussed descriptively above.

Exploratory statistics indicate that pensioners are less likely to continue working or be re-employed as age increases. This is confirmed by the negative odds ratio for age in 2011 in the logistic regression model. Age at receiving a first pension also has an effect on work decisions after retirement. The results support that, in contrast to the first pension receipt age below 50 years, being 65 and older at first pension receipt reduces the probability of working. The odds of this group for re-employment are only 21 per cent compared to the group aged below 50. First pension receipt at age 60 to 64, however, positively affects pensioners' re-employment decisions.

Gender has a significant effect on work past pension age. The odds for men to be employed after having started to receive a pension are 1.7 times those of women. Health status also has a significant impact on working. The odds of working among healthy pensioners are 2.0 times those of unhealthy retirees. The positive effects of being male and being healthy on pensioners' propensity to pursue paid work are consistent with previous studies on older adults aged 60 and over (Qian and Jiang 2006; Lin and Chi 2008).

Table 7.4 Binary logistic regression model for determinants of paid work among pensioners

	Odds ratio	Standard error
Age at first pension receipt (ref. up to 49)		
50–54	0.92	0.17
55–59	1.00	0.22
60–64	1.63*	0.40
65+	0.21*	0.16
Age in 2011	0.68***	0.07
Age^2 in 2011	1.00***	0.00
Gender (ref. female)		
Male	1.66*	0.33
Health at first pension receipt (ref. bad)		
at least fair	1.97*	0.52
Health in 2011 (ref. bad)		
at least fair	1.23	0.22
Educational qualification (ref. low)		
Middle	0.78	0.12
High	1.18	0.22
Registered residence (ref. rural)		
urban	1.20	0.34
Spouse's health at respondent's first pension receipt (ref. bad)		
at least fair	0.87	0.18
Economic activity of spouse at respondent's first pension receipt (ref. employed)		
Not employed	0.86	0.13
Farming	1.10	0.25
Last workplace before receiving first pension (ref. government agency)		
Public institution	1.40	0.31
NGO	1.60	1.51
Firm (state, collective and private)	1.44+	0.30
Self-employed	13.06***	4.62
Other	1.09	0.61
Log of (individual) monthly pension	0.77*	0.10
Grandchildren under 6 (ref. no)		
Yes	0.71	0.16
Gender*grandchildren (ref. woman with grandchild<6)		
Man with grandchild<6	1.70+	0.46
n	*1,355*	
Pseudo R^2	0.090	
Log likelihood	−798.48	

Note: Significance level +$p < 0.1$, *$p < 0.05$, **$p < 0.01$, ***$p < 0.001$.
Source: CHARLS data 2011–12.

Surprisingly, education does not have any significant effect on the propensity to work despite receiving a pension when controlling for all other variables. This is different from previous findings that older adults with higher education are more likely to continue working. We assume that pensioners in our study are younger than the samples in other studies (for example Zhang 1999; Chou 2010), which tend to target pensioners aged 60 and older. It seems that the selective effect of education on late employment mainly operates at higher ages. *Hukou* (residence registration) does not play a significant role for work decisions among pensioners. As in the descriptive results, the health status of the respondent's spouse at the time of the respondent's first pension receipt has no significant impact on the respondent's likelihood to work. Similarly, the economic activity of the respondent's spouse at the time of the respondent's first pension receipt does not play an important role for his/her propensity to work.

The workplace or employer before retirement, by contrast, has a significant effect on the individual inclination to work despite receiving a pension. As expected, those who were self-employed before retirement are the least likely to stop working after retirement. The odds of the self-employed remaining in (or restarting) work are 13 times that of civil servants. Those who were employees in firms before receiving a pension are also somewhat more likely than those from government agencies to go back to work or to continue working. Their odds for re-employment are 1.4 times that of (former) civil servants. The amount of the monthly pension has a significantly negative effect on the probability of pensioners to work. An e-fold (that is, 2.72-fold) increase in monthly pension decreases the odds of re-employment by 23 per cent, when other variables are controlled for. For example, in the case of an increase in monthly pension income from 100 Yuan (16 US$) to 272 Yuan (44 US$), the odds of re-employment decrease by the factor 0.77. This implies that economic necessity leads many pensioners to continue working.

Having grandchildren younger than six years does not affect work decisions of pensioners. When we include an interaction term between gender and having grandchildren, however, we find that women with small grandchildren are less likely to engage in paid work. The odds of working for women with small grandchildren are only 59 per cent (1:1.70) of their male counterparts with grandchildren younger than six years. In other words, the odds of working for men with small grandchildren are almost twice as high as those of women with small grandchildren. This confirms the traditional family-rooted culture in China, leaving more family commitments to women than men.

7.7 Discussion and conclusions

China has gone through a social and economic transformation over the past three decades. Societal change has created both new social risks and opportunities for individuals in the labour market. Urban older workers, on the one hand, often exit from their main career employment prematurely, for example due to socially assigned (pension) schemes aimed at downsizing for efficiency-related or other reasons. On the other hand, many of these (often very young) pensioners have continued to participate in paid work in recent decades. Furthermore, the relaxation of the *Hukou* system has allowed rural residents to work in urban cities, and some rural workers are also covered by a small pension when reaching their pension age of 60. To enhance our understanding of who participates in paid work despite receiving a pension, and why pensioners (both urban and rural) continue to work, this article investigated the determinants of work beyond pension age. We found that being male, in good health, having been self-employed or a firm's employee just before starting to receive a pension all have a positive effect on continuing to work or re-entering the labour market after retirement. By contrast, being female, particularly in combination with having grandchildren under the age of six, and receiving generous pension payments have a negative effect on the probability of working beyond retirement. In contrast to studies on older adults working past their 60s, we did not find that education significantly affects how likely Chinese pensioners are to work. We assumed that given their low average age (approximately 53 years) when they start to receive a pension, most of the investigated pensioners continue working or begin to work again independent of their qualification, as long as they are healthy enough to do so. This might be an indication of the fact that working beyond pension age is not mainly driven by the attractiveness of higher qualified jobs or connected to non-material reasons like self-fulfilment or status. It also shows that work opportunities for those who are better qualified are not significantly superior to those of less qualified people.

The amount of pension payments a retiree receives is also a predictor of labour market participation. The lower a pension is, the more likely it is that pensioners are pushed to continue participating in the labour market – economic necessity plays a major role in working as a pensioner. This raises concerns about the economic well-being of pensioners. In addition, the high probability of working for those who have been self-employed before they started receiving a pension points to the fact that many of the self-employed probably continue working beyond pension age. They can control the continuation or end of their working life to a greater degree, as the official retirement age does not apply to them and they cannot be dismissed. Those who were employed in firms before retirement are also more likely to continue working (or begin to work again), compared to those who have been civil servants. Lastly, it was found that women are less likely to stay in the workforce than men after retirement, also because they are more likely to be involved in the care of young grandchildren. The traditional family-rooted culture of China seems to rely significantly on women's involvement at home after they have retired.

Due to data limitations, we could not examine how pensioners obtain their jobs in the labour market. Zhang and Li (2000) find an effect of social capital on urban retired adults' job seeking. Specifically, strong personal networks (for example family members, relatives and friends) seem to contribute to successfully finding a job. In addition, Xiao (2004) points to the fact that the government fails to provide (the often quite young) pensioners with job training programmes and job information. The state has not been targeting training programmes and employment guidance specifically at (mostly urban) pensioners because their pensions make their later life foreseeable and secure. With an abundant labour supply, labour shortages are not a problem, and the employment of other groups such as jobless graduates and migrant workers is the major concern of the Chinese government, as rising unemployment may stir social unrest among these groups.

In contrast to China, the industrialized nations attempt to keep or integrate older workers in the labour market by discouraging early retirement, encouraging lifelong learning and offering adequate opportunities for training, in order to respond to problems in financing pensions and (projected) labour shortages due to population ageing. In this way, they have replaced or have started to replace 'early retirement' with 'active ageing' (Taylor 2008), although with differing success and timing. In China, however, there have been no workplace-related laws and policies for older workers (Feng and Li 2012), and no measures preventing old-age discrimination.

Population ageing is a global phenomenon that is expected to affect labour supply in China as well and present it with significant financial and social challenges. The Chinese population has been ageing rapidly over the past two decades. Undoubtedly, the transition from an unlimited to a limited labour supply may threaten its industrial structure and competitive advantage in the years ahead. With the pension system facing a looming crisis of sustainability, and declining numbers of younger workers in the future workforce, the development and utilization of vast reserves of older workers will be a viable option. However, the important issue will be how to encourage the labour market participation of older workers *before* the real crisis emerges.

To meet the challenges of population ageing and the potentially declining labour supply, the Chinese government has recently proposed to prolong working lives by increasing pension ages. Although the majority of the population is opposed to it, this option is still under discussion. Even if it is decided, it will not take effect immediately. From the individual perspective, especially of employees who do physical work, health risks connected to hard physical work during prolonged working lives are a serious concern. For them, the individual benefits of working longer that are discussed in terms of combining income with social participation are a myth.

Understanding the motives of working pensioners can serve as an additional perspective on these dissenting voices. As we have shown, there are strong indications that economic motives are very important for working past retirement. Based on this, it is less likely that work after retirement is a choice that is exclusively and positively connected to the idea of social participation.

Although the extension of working lives is currently unpopular in China, it is important to consider and discuss the financial consequences of the low pension age and the growing numbers of people receiving a pension. Reforming the current fragmented pension system, however, is more crucial than extending working lives. Particularly, a unified pension scheme including all rural workers and irrespective of the nature of the work units would be a step towards less inequality and poverty in old age. It would also help to ease the conflicts of interest between state and non-state sectors, and between civil service employees and employees in other sectors.

Regarding the prolongation of working life, policy makers should also consider measures aimed at reducing the health risks of disadvantaged and vulnerable older workers. Creating job opportunities for all disadvantaged older workers is also high on the agenda, together with proper job training for older workers. Age discrimination against older workers should also not be ignored. As long as older workers are dismissed because of their age, their re-employment chances are poor, and if these problems are not addressed, longer working lives will undoubtedly present risks of unemployment and poverty for some older workers.

Notes

1. The Chinese *Hukou* system was implemented in the mid-1950s and has profoundly shaped China's social stratification ever since. It basically divides the Chinese population into two categories belonging to two different sectors: 'agriculture' and 'non-agriculture', based on his/her mother's *Hukou* status when an individual was born. It can be perceived as an ascribed status. Based on this system, many privileges and rights are only granted to urban citizens, such as access to the state formal employment, good urban education for children and housing. Even moving to a city is administratively controlled, although mobility-related restrictions have been somewhat relaxed in the past decades (for more detail see for example Wu and Treiman 2004).
2. The most recent pension reform in China targets civil servant privileges. Public sector employees will contribute to the state pension system (*Financial Times* 2015).
3. For more information, see the CHARLS website: http://charls.ccer.edu.cn/en. The latest wave of the data (2013–14) has just been released.
4. Logarithmic transformations make a positively skewed distribution more normally distributed. Furthermore, the effect of the independent variable on the dependent one can be interpreted in terms of relative changes (for example in per cent).
5. It might be possible that selection effects are at work here: older pensioners who retire from their main career employment early (even earlier than indicated by these averages) may have died early as well so that they are not included in the sample. Despite this, the average age at retirement we calculated based on the CHARLS data still seems to be reliable because it is in line with the results from existing studies (see for example Du and Wang 2010).

References

Agresti, A. (1996), *An introduction to categorical data analysis*, New York: Wiley.

Benjamin, D., Brandt, L. and Fan, J. Z., (2003), *Ceaseless toil? Health and labour supply of the elderly in rural China*, Working Paper 2003–2579, Ann Arbour, MI: William Davidson Institute, Stephen M. Ross Business School, University of Michigan.

Blau, D. M. and Riphahn, R. T. (1999), 'Labor force transitions of older married couples in Germany', *Labour Economics*, 6 (2), 229–51.

Cai, F. (2004), 'The aging trend and pension reform in China: Challenges and options', *China & World Economy*, 12 (1), 36–49.

Cai, F., Du, Y. and Wang, M. (2009), *Employment and inequality outcomes in China* (unpublished manuscript), Beijing: Institute of Population and Labour Economics, Chinese Academy of Social Sciences, http://www.oecd.org/employment/emp/42546043.pdf, date accessed 16 February 2015.

China State Council (1978), *The provisional measures of the State Council on the resettlement of sick, vulnerable, senior and handicapped cadres and the provisional regulations of the State Council on the retirement and resignation of workers*, China State Council document 1978: 104, Beijing: The State Council of the People's Republic of China.

Citylab (2012), *Chinese urbanization, by the Numbers*, by N. Berg, 16 August 2012, http://www.citylab.com/work/2012/08/chinese-urbanization-numbers/2969/, date accessed 26 November 2014.

Chou, R. J. (2010), 'Workforce participation among older adults in China: Current knowledge and future research directions', *China Journal of Social Work*, 3 (2–3), 247–58.

Deller, J. and Maxin, L. (2009a), 'Silver Workers – Eine explorative Studie zu aktiven Rentnern in Deutschland', *Arbeit*, 17 (3), 166–79.

Deller, J. and Maxin, L. M. (2009b), 'Berufliche Aktivitäten von Ruheständlern', *Zeitschrift für Gerontologie und Geriatrie*, 42 (4), 305–10.

Du, Y. and Wang, M. Y. (2010), *Demographic ageing and employment in China, Employment Working Paper No. 57*, Geneva: International Labour Office.

The Epoch Times (2012), *China's pensions decline below warning level*, by Z. Gao, 21 September 2012, http://www.theepochtimes.com/n2/china-news/chinas-pensions-decline-below-warning-level-295039.html, date accessed 24 September 2013.

Feng, Y. and Li, N. (2012), 'Re-employment of pensioners: Labour relation or service relation', *Social Science Front*, 2012 (7), 182–9 (in Chinese).

Financial Times (2015), *China pension reform targets civil servants privileges*, by G. Wildau, 15 January 2015, http://www.ft.com/intl/cms/s/0/3141eb64-9c97-11e4-a730-00144feabdc0.html#axzz3Pu5LENu, date accessed 15 January 2015.

Giles, J., Wang, D. and Cai, W. (2012), 'The labor supply and retirement behavior of China's older workers and elderly in comparative perspective', in: J. P. Smith and M. Majmundar (eds.), *Aging in Asia: Findings from new and emerging data initiatives*, Washington, DC: National Academies Press, 116–47.

Hank, K. and Buber, I. (2007), *Grandparents caring for their grandchildren: Findings from the 2004 survey of health, ageing and retirement in Europe, MEA Discussion Paper 127–2007*, Mannheim: Mannheim Research Institute for the Economics of Ageing.

HelpAge International (2013), *Pension coverage in China and the expansion of the New Rural Social Pension, Pension Watch Briefing No. 11*, London: HelpAge International, http://www.refworld.org/pdfid/5301df5d4.pdf, date accessed 10 November 2013.

Jiang, X. Q. and Du, P. (2009), 'Work among Chinese older adults and policy implications', *Academic Journal of Zhongzhou*, 2009 (4), 109–13 (in Chinese).

Kohli, M. (1986), 'Social organization and the subjective construction of the life course', in: A. B. Sørensen, F. E. Weinert and L. R. Sherrod (eds.), *Human Development and the Life Course: Multidisciplinary Perspectives*, Hillsdale, NJ: Erlbaum.

Lin, D. C. and Chi, I. (2008), 'Determinants of work among older adults in urban China', *Australian Journal on Ageing*, 27 (3), 126–33.

Organisation for Economic Co-operation and Development (OECD) (2013), *Pensions at a Glance 2013. OECD and G20 indicators*, Paris: OECD, http://dx.doi.org/10.1787/pension_glance-2013-en, date accessed 24 October 2014.

Pang, L., de Brauw, A. and Rozelle, S. (2004), 'Working until you drop: The elderly of rural China', *The China Journal*, 2004 (52), 73–94.

Parker, S. C. and Rougier. J. C. (2007), 'The retirement behaviour of the self-employed in Britain', *Applied Economics*, 39 (6), 697–713.

Pozen, R. C. (2013), *Tackling the Chinese pension system*, Chicago: Paulson Institute.
Qian, X. and Jiang, X. C. (2006), 'Factors affecting re-employment intent among older Chinese in urban China', *Population Journal*, 2006 (5), 24–9 (in Chinese).
Salditt, F., Whiteford, P. and Adema, W. (2008), 'Pension reform in China', *International Social Security Review*, 61 (3), 47–71.
Scherger, S., Hagemann. S., Hokema, A. and Lux, T. (2012), *Between privilege and burden. Work past retirement age in Germany and the UK*, ZeS-Working Paper No. 04/2012, Bremen: Centre for Social Policy Research.
Taylor, P. (2008), *Ageing labour forces: Promises and prospects*, Cheltenham: Edward Elgar.
Wang, H. M. (2001), 'An analysis of the determinants about senior citizens' re-employment – The case of the Yan-yuan area of Peking University', *Market and Demographic analysis*, 7 (1), 64–70 (in Chinese).
Watson Wyatt Worldwide (2006), *Work Greater China™2006: A comprehensive study of employee attitudes in greater China*, Asia Pacific Country Report, Hong Kong: Watson Wyatt Worldwide.
Wang, X. J., Wang, Y. and Kang, B. W. (2009), 'Estimates of pension replacement rates of different groups of people in China', *Statistics and Decision*, 2009 (3), 10–12 (in Chinese).
Wu, X. and Treiman, D. J. (2004), 'The household registration system and social stratification in China: 1955–1996', *Demography*, 41 (2), 363–84.
Xiao, S. (2004), 'Investigation and analysis of old folk's human resource exploitation – A case study of Haizhu district in Guangzhou,' *Journal of the Graduates Sun Yat-Sen University (Social Sciences)*, 2004 (1), 72–7 (in Chinese).
Zeng, Y. (2011), 'Effects of demographic and retirement-age policies on future pension deficits, with an application to China', *Population and Development Review*, 37 (3), 553–69.
Zhang, W. J. (2010), 'A study on the effect on labour participation of the Chinese elderly', *Population & Economics*, 2010 (1), 85–9, 92 (in Chinese).
Zhang, Y. (1999), 'Effect of educational level on re-employment on older retirees', *Population Science of China*, 1999 (4), 27–34 (in Chinese).
Zhang, Y. and Li, J. Y. (2000), ' "Strong social network" and the re-employment of older adults', *China Population Science*, 2000 (2), 34–40 (in Chinese).

Part II
Contexts

8
Pension Reform in Europe: Context, Drivers, Impact

Karen M. Anderson

8.1 Introduction

Pension reform has been an important item on national political agendas across Europe for three decades. Governments have recalibrated and sometimes slashed pension benefits not only to reduce public expenditure but also to increase incentives for individuals to retire later. This chapter analyses patterns of pension reform across Europe, focusing on the causes and consequences of reforms.

The chapter begins with a general discussion of the most important dimensions of pension policy in Section 8.2 (including the public-private mix, flat-rate vs. earnings-related benefits and contribution financing vs. tax financing) because these are the parameters that policy makers adjust when they reform pension schemes. In Section 8.3, the chapter then investigates the factors driving pension policy change, such as population ageing, fiscal austerity and changed employment patterns, as well as the pension reforms themselves. There has been large variation in policy responses because of different national institutional starting points and the role of electoral politics. Pension reforms will also have important socio-economic consequences, as Section 8.4 shows. In broad terms, inequality and poverty in retirement are likely to increase somewhat across Europe, although reform effects vary across pension systems. Section 8.5 discusses the impact of the ongoing global financial crisis and euro crisis on the future of pension provision.

The economic and demographic conditions that marked the three decades following the Second World War were unusual: birth rates were high, economic growth was steady and employment levels for full-time breadwinners were high. The introduction and/or expansion of generous, pay-as-you-go pensions was a plausible policy choice in this context. In 1960, the average total fertility rate (average number of children per woman) in Europe was 2.59 (Eurostat 2004: 40) compared to 1.58 in the EU-28 in 2012 (Eurostat 2015a). Similarly, the old-age dependency ratio (that is, the ratio of persons 65 and older to the number of persons aged 15 to 64) has changed substantially, increasing from 15.2 per cent in 1960 to 26.7 per cent in 2012 (Eurostat 2015b).

The expansion of pension systems in the 20th century to cover the working and the middle classes was a breakthrough in social protection. Before the 20th century, with a few exceptions, pensions were only available to a select few: privileged white-collar workers and high-level civil servants, especially military officers. All others worked as long as they could, and when their physical condition prevented them from supporting themselves, they relied on family, charity and punitive poor relief. Even after the first state pensions were introduced in Northern Europe in the late 19th century and the early 20th century, benefits were meagre, and most recipients who could work continued to do so. The first pensions were thus not meant to provide income sufficient to leave the labour market, and most people combined pension benefits with employment income (Thane 2002).

The development of pensions since the Second World War changed this pattern, and retirement began to be a new phase of the life course for many groups. Rather than combining meagre pensions and irregular employment income, the elderly now began to actually retire, largely because they enjoyed sufficient state and occupational pensions, or they received support from their children, who benefitted from employment in the post-war economic boom. Even so, many European governments, for example the UK, encouraged workers to stay employed after the statutory retirement age because of labour shortages and concerns about the fiscal costs of pensions. This orientation was short-lived, however, and by the 1960s, fewer and fewer pensioners combined work with retirement (Thane 2002).

By the mid-1970s, the high-fertility, high-employment, high-growth equilibrium had begun to give way to a period of economic and demographic uncertainty. For most European governments, the initial response was labour shedding; early retirement became the preferred policy response to rising unemployment in many countries (Ebbinghaus 2006). At the same time, however, the first signs of fiscal strain began to be felt as fertility rates fell, and employment failed to return to pre-crisis levels. By the 1980s, European governments began to grapple with the fiscal consequences of post-industrialism and demographic ageing. After four decades of unprecedented expansion, European governments began to slowly take the first steps towards pension reform (Immergut et al. 2007).

8.2 Pension system design

The academic literature typically conceptualizes pension systems along several dimensions: public versus private provision, flat-rate versus earnings-related benefits and capital-funding versus pay-as-you-go financing. The first dimension concerns the actors responsible for pension provision, the second dimension refers to the level of benefits and its relation to employment income, and the third dimension concerns the source of funding. Capital funding means that monetary assets back up pension promises, while in a pay-as-you-go scheme current revenues finance current benefits. The multi-pillar template popularized by the World Bank (1994) relies on these dimensions, identifying three pillars of pension provision. The first pillar comprises state pensions, either flat-rate (all pensioners receive the same benefit level) or earnings-related (the pension is a proportion of previous employment income). The second pillar includes both mandatory and voluntary workplace pensions (usually earnings-related), and the third pillar includes individual, private pension savings. The first pillar may be occupationally fragmented, with some groups (often tenured civil servants) receiving higher benefits than other wage earners. Similarly, the second pillar in most countries typically includes a wide variety of schemes.

In Western Europe, and to some extent, Central and Eastern Europe, the first pillar usually dominates, with second- and third-pillar benefits supplementing state pensions. In Western Europe, many analysts prefer to use the distinction between Beveridge and Bismarck to define pension systems. In Beveridgean systems, the first pillar provides a mainly tax-financed flat-rate pension sufficient to lift pensioners out of poverty. Second- and third-pillar benefits provide earnings-related benefits and individual savings to secure income replacement. Bismarckian systems rely on contribution-financed first-pillar schemes to provide income replacement, so the second and third pillars are relatively small. The new democracies of Central and Eastern Europe do not fit easily within this template. The legacy of state socialism means that the first pillar is typically large, but in many Central and East European (CEE) countries, the second pillar has been expanded substantially for younger age cohorts (Orenstein 2008).

Despite the prevalence of the Beveridge versus Bismarck distinction, the categories it relies on obscure the gendered effects of pension systems. If we follow Esping-Andersen's (1990) regime typology, the Bismarckian category includes both social democratic pension regimes (such as Sweden) based on high female labour market participation and individual, employment-based pension entitlement, and conservative/corporatist regimes (such as Germany) based on moderate female labour market participation and extensive derived pension rights (Sainsbury 1996; Anderson 2015). The Beveridgean category includes liberal regimes, exemplified by the UK.[1]

Pension reform research demonstrates that pension systems are vulnerable to different kinds of economic and demographic pressures, depending on their institutional structure (Hinrichs 2000; Bonoli and Shinkawa 2005; Immergut et al. 2007). Moreover, institutional legacies, or policy feedbacks (Pierson 1994), constrain the menu of options available to reform-minded politicians. Existing public policies confer resources on actors, encourage interest group formation and behaviour, shape voters' opinions about public policies and create incentives for actors to coordinate and adapt their behaviour to existing policies (Pierson 2004). For pension systems dominated by public provision – like the social democratic and conservative/corporatist regimes that comprise the Bismarckian/social insurance cluster – these policy feedbacks usually generate high levels of support and policy stability because of the electoral risks associated with cutting back popular programmes. Electoral risks also constrain retrenchment in liberal regimes, despite the relatively lower level of public provision. However, multi-pillar pension systems in all three types of pension regime (social democratic, conservative/corporatist and liberal) provide opportunities for compensating some of the cuts in public provision by shifting responsibility for retirement provision to the private occupational sector (Trampusch 2006; Anderson 2010; Ebbinghaus 2011).

First-pillar schemes dominated by contribution-financed, employment-related benefits are particularly vulnerable to rising unemployment and unfavourable old-age dependency ratios, because revenues drop as the number of pensioners increases. This problem may be exacerbated by the defined benefit (DB) structure of most public, earnings-related systems before the last decade or so of reform, because the pension level was guaranteed in advance, and contributions are adjusted as necessary. A defined contribution (DC) benefit formula does not guarantee a specific pension level, but instead bases benefits on the rate of return to accumulated contributions. Systems based on flat-rate pensions in the first pillar face a different set of pressures. Basic pensions are typically tax-financed, so rising dependency ratios translate into higher costs for the state.

Second-pillar pensions may be either voluntary or (quasi-)mandatory. Voluntary systems like the pre-2008 British one (the 2008 Pensions Act introduced auto-enrolment) depend on the willingness and capacity of employers to provide pensions, as pension costs are part of the overall wage bill. In (quasi-)mandatory systems (for example the Netherlands and Denmark), the state requires employers to provide workplace pensions, or delegates this task to collective bargaining. Many, if not most, workplace pensions in Europe are capital funded, so they are vulnerable to the performance of financial markets (Ebbinghaus 2011).

8.3 Pension reform

European pension systems emerged over many decades as a result of political and economic decisions taken in specific temporal contexts (Myles 1994). As discussed above, low fertility and uneven employment growth now characterize the post-industrial economies of Europe. In addition, shifts in employment and family structures have produced increasing numbers of people with incomplete employment biographies, and also more people who have insufficient employment income to guarantee a decent pension.

The drop in birth rates and increasing longevity have been especially problematic for defined benefit, pay-as-you-go first-pillar schemes because of their intergenerational design: current workers and taxpayers finance current pensions in the expectation that when they retire, younger generations will finance their pensions. Even capital-funded defined benefit second-pillar pensions are not immune to this vulnerability. In addition to relying on financial market returns, these schemes depend on the willingness of current and future workers to finance the portion of benefits not covered by investment returns. Finally, pensions are political creations: elected politicians adopted pension programmes in order to win votes. Governments have often promised short-term pension gains to voters without fully considering the long-term financial consequences. As Lynch (2006) shows, this type of particularistic electoral strategy produced pension-heavy welfare states in Italy and other Southern European countries.

National responses to these socio-economic challenges have been mediated by political factors and institutional starting points – the pension policy structures discussed in the previous section. How have pension reforms unfolded in specific national contexts in the past two decades? Are there signs of convergence in terms of pension policy design? The pension literature addresses these questions with reference to 'worlds' of pension reform. Bonoli (2003) distinguishes between reform logics in 'social insurance' pension systems (Bismarckian systems or the conservative/corporatist regimes in Esping-Andersen's scheme) and 'multi-pillar' systems (either the liberal or conservative/corporatist regimes in Esping-Andersen's scheme). Social democratic pension regimes are located in both the social insurance (Sweden, Finland, Norway) and multi-pillar (Denmark) clusters. Because of the dominance of pay-as-you-go public provision, reform in social insurance systems is marked by incremental adjustments in benefit generosity and financing to promote financial sustainability. Pension reform that involves the imposition of sizeable losses on current and future pensioners is usually an electorally unpopular strategy, so reformers face strong incentives to undertake piecemeal reforms (Pierson 1994). Reform in multi-pillar systems follows a different logic because of the relatively small size of the first pillar and the importance of the second pillar. These design features mean that fiscal pressures are less severe than in social insurance regimes. However, the importance of workplace pensions in the overall pension package means that multi-pillar regimes perform less well in providing retirement security for atypical workers or workers in sectors not covered by occupational pensions (see also Hinrichs and Jessoula 2012).

More recent scholarship examines pension reform dynamics in CEE countries. In contrast to the mature public pension schemes that characterize most Western European countries, the institutional starting points of the CEE countries' pension regimes are marked by the legacy of state socialism and the transition to democracy and capitalism since 1990. The CEE countries inherited a patchwork of expensive public pension policies in the context of rising unemployment because of the collapse of their centrally planned economies. As in Western Europe, many CEE countries used the pension system to absorb surplus labour, driving up pension costs even more. Orenstein (2008) analyzed pension reform processes in the CEE countries and found that privatization was important in most reforms. The 'mixed' reform strategy involved the expansion of the second pillar and the gradual reduction of the first pillar, whereas the 'parallel' strategy allowed individuals to choose whether to participate in the second and third pillars (Orenstein 2008). The next sections discuss these reforms in more detail.

8.3.1 Pension reform in social insurance countries

As noted above, the social insurance category includes countries in which first-pillar, earnings-related pensions dominate old-age provision. Pensions are typically financed by payroll contributions shared by employers and employees (with some exceptions), and financing is pay-as-you-go. In the conservative/corporatist regimes, the goal of the pension system is status maintenance rather than redistribution, with occupationally fragmented public schemes providing benefits that allow wage earners to maintain their previous standard of living. Because pensions are defined benefit, they are expensive and particularly vulnerable to fluctuations in employment and the old-age dependency ratio.

Hinrichs (2000) identified six trends in pension reform in social insurance countries: tightening the link between contributions and benefits (and thus moving towards defined contributions instead of defined benefits); increasing the pension age; expanding pension rights for care work; increasing the share of tax financing; harmonizing special pension schemes (and integrating them into the general rules); and expanding pension testing in social pension schemes (reducing pensions if a claimant has other sources of income). These reform instruments were designed primarily to increase the financial sustainability of pension schemes, but also to reward (albeit modestly) those with caring responsibilities. Taken together, these reform options have the effect of widening the revenue base for public pensions, 'rationalizing redistribution' (Myles and Pierson 2001) by harmonizing different occupational pension schemes, and increasing work incentives by making pensions more actuarially fair. Since 2000, the menu of pension reform instruments has not changed much in social insurance countries, with one exception: the expansion of the occupational and individual private pensions to compensate for benefit cuts in the first pillar (Ebbinghaus 2011). Moreover, there is increasing variation in the extent to which pension systems have begun to award pension rights to individuals for their caring responsibilities.

The 1990s and 2000s were marked by the adoption of far-reaching pension reforms in several social insurance countries. For example, Sweden (1994), Germany (2001) and Italy (1992, 1994) legislated reforms that reduced public, earnings-related benefits and expanded private- or individual-funded accounts to fill the gap. The comprehensive *Swedish* reform attracted much international and academic attention because it represents a major shift from generous DB benefits to notional defined contribution (NDC)[2] benefits combined with the introduction of individual, funded accounts in the first pillar. A broad parliamentary coalition negotiated the reform in the 1990s and managed to prevent it from becoming an election issue. The reformed pension system has lost some of its redistributive character, and pension outcomes will become more unequal because of the tighter link between contributions and benefits and the effects of individual investment accounts. In addition, the reformed pension system includes automatic stabilizers that adjust pension accrual and payouts to general life expectancy and economic growth (Anderson and Immergut 2007; Wadensjö and Sjögren Lindquist 2011).

Germany is a prominent example of how politicians have adopted incremental reforms mostly in the absence of societal consensus. Beginning in the 1980s, the German pension system has faced recurring financial problems, largely because of persistently high unemployment and the costs of German unification. German reforms have been more incremental than those passed in Sweden and Italy, but they add up to large-scale policy change. The first major changes were adopted under a centre-right (Christian Democratic-Liberal) coalition. A 1989 reform shifted pension indexation from gross to net wages, raised women's pension age, cut early retirement and improved pension credits for child-rearing. Changes adopted in 1997 included a 'demographic' factor which would adjust future pension payments to life expectancy. The unpopularity of the proposal contributed to the downfall of the government; the 1998 election brought a Social Democratic (SPD)-Green Party coalition to power that quickly reversed some of the 1997 cuts. By early 1999, however, the government had changed course and embarked on fundamental reform. In 2001–02, the Red-Green coalition adopted a pension reform designed to slow the growth of pension spending and hold contributions below 22 per cent of income. To achieve this, the reform included modest benefit reductions for future pensioners, the introduction of subsidized voluntary private investment accounts ('Riester' pensions) and incentives for the expansion of occupational pensions (Schulze and Jochem 2007; Anderson and Meyer 2003).

Since 2001, reforms have focused on further stabilizing first-pillar expenditures, increasing incentives for second- and third-pillar pension provision, further reducing early retirement, raising the statutory retirement age and expanding caregiver credits in the pension system. The Red-Green government adopted a 'sustainability' factor in 2004, and legislation adopted in 2007 gradually raises the statutory retirement age to 67. In addition, legislative changes adopted in the 1990s and 2000s have dramatically reduced early exit from employment (Ebbinghaus 2006). The expansion of caregiver credits is an exception to the trend of reducing benefit generosity. Caregiver credits were introduced in 1986; reforms adopted in 1989, 1997 and 2014 increased their value and expanded their applicability. The 2014 legislation also allows retirement at 63 for persons with a 45-year work history, with this age limit rising to 65 in the years to come.

Thus, after two decades of reform, the German public pension system no longer provides future retirees a pension in line with the standard of living achieved during employment (Hinrichs 2012; Meyer 2014). Subsidized voluntary private provision will speed the growth of the second and third pension pillars, but this will only partially compensate for benefit cuts in the first pillar.

Italy is known for the dominance of pension spending in its system of social provision (Lynch 2006; Ferrera and Jessoula 2007), so reforms have focused on reducing the costs of the first pillar and expanding the second pillar. Indeed, Italy spends the highest proportion of its gross domestic product (GDP) on public pension provision in the European Union (EU) (14 per cent in 2007), and pension reform has been an important component of efforts at fiscal consolidation, not least because of fiscal constraints of Economic and Monetary Union in the EU (Stepan and Anderson 2014).

Attempts to reduce costs and expand the second pillar began in earnest in the 1980s, but did not succeed until the 1990s. One of the first targets of reform was to reduce fragmentation across public and private schemes concerning contribution rates, benefit formulae and pension ages. An important facilitator of pension reform has been Italian policy makers' exploitation of opportunities presented by the severance pay system, TFR (*trattamento di fine rapporto*), to expand occupational pensions. All private sector workers are covered by the employer-financed TFR. The 1993 legislation introduced a new regulatory framework for second-pillar pensions, including tax incentives. The reform allowed wage earners to transfer their generous TFR rights to the new occupational schemes. The 1995 Dini reform gradually replaced the old first-pillar DB scheme with an NDC scheme. Reforms in the 2000s strengthen incentives for individuals to allocate their TFR rights to the second-pillar schemes, reduce incentives for early retirement and harmonize men's and women's pension ages, and austerity legislation adopted between 2009 and 2011 continues this trend (Ferrera and Jessoula 2007; Jessoula 2012).

Compared to other social insurance pension systems, *France* has been a model of relative policy stability. To be sure, reforms adopted since the 1990s (the 1993 Balladur Reform, 1999 Jospin Reform, 2003 Raffarin Reform, Fillon Reforms in 2007 and 2010 and the 2012 Ayrault Reform) harmonize benefits across first-pillar schemes, modestly reduce the generosity of first-pillar benefits, raise the minimum pension age and create a public pension reserve fund. However, efforts to expand the second and third pillars have been half-hearted compared to other countries in the social insurance cluster. An important and distinctive feature of the French second pillar is that financing is often pay-as-you-go rather than capital funding. In 2010, 16.5 per cent of workers were covered by an occupational pension (OECD 2013a). The first-pillar schemes, despite modest benefit cuts, continue to dominate the retirement income packages of French pensioners.

The onset of the latest financial crisis in 2008 has not led to a new wave of substantial public pension reforms, except in the countries deeply affected by the euro crisis (Greece, Spain, Ireland, Portugal). *Greece* has been especially hard hit, because of the extreme effects of the sovereign debt crisis there. At the end of 2009, the Greek budget deficit was more than 15 per cent of GDP, and by early 2010 the Greek government began to have trouble borrowing on international financial markets to finance deficits. The Greek government turned to the EU and International Monetary Fund (IMF) for help, receiving the first of several bailout packages in May 2010. The other bailouts came in 2012. Financial assistance from the 'troika' (EU, European Central Bank (ECB), IMF) came with strict conditions, including pension reform. Reform of the fragmented public pension system was an important part of the fiscal consolidation programmes. Initial cuts were modest: holiday bonuses for pensioners were reduced, and those with high pensions had to pay higher taxes. In July 2010, parliament adopted a substantial pension reform that raised the pension age from 60 to 65 for most workers and standardized men's and women's pension ages. The reform included penalties for early retirement and for combining pension receipt with employment. Further benefit cuts would be realized by combining the very high number of existing pension schemes into three funds by 2018. The second bailout in February 2012 also contained modest cuts in pensions and an increase in the retirement age from 65 to 67. Monastiriotis (2013: 7) reports that individual public pension payments have been cut by more than an average of 25 per cent between January 2010 and January 2013.

The sovereign debt crisis also forced major pension reform in *Portugal*. Prior to the onset of the crisis, Portugal had adopted a substantial reform in 2007 that integrated separate, special pension schemes into the statutory pension regime. The reform also included typical parametric changes: a sustainability factor that adjusts pensions to changes in general life expectancy and less generous indexing. Portugal turned to the EU for financial assistance in 2011 and received a bailout in May 2011.

The austerity measures required by the bailout included the cancellation of holiday allowances for pensioners until 2014 and other modest cuts in pension generosity. There has been little interest in Portugal in expanding second- and-third pillar pensions.

8.3.2 Pension reform in multi-pillar regimes

The multi-pillar cluster includes Switzerland, Denmark, the Netherlands, the UK and Ireland. This cluster has faced challenges different from those of the social insurance countries: the need to maintain or even expand minimum pension provision for low-income earners and strengthen the regulation of the occupational pensions system in order to safeguard its income replacement function.

The main advantage of the multi-pillar approach to pension provision is that risks are spread across all three pillars. However, there is considerable heterogeneity in this cluster because minimum pension provision in the UK and Ireland is considerably lower than in Denmark, the Netherlands and Switzerland. This means that public pension spending in Denmark, the Netherlands and Switzerland is much higher than it is in the UK and Ireland, and poverty rates among pensioners are much higher in the two latter countries. These differences reflect central regime characteristics: the UK and Ireland belong to the liberal regime category, Denmark is classified as social democratic, and the Netherlands and Switzerland are attributed to the conservative regime category, reflecting their emphasis on collectively organized but privately provided social policies. Second-pillar pensions in pre-2008 UK and Ireland were voluntary, whereas in the other countries of this cluster, they are either mandatory, as in Switzerland, or quasi-mandatory, as in Denmark and the Netherlands where occupational pensions are central components of collective labour agreements. With the exception of Denmark, all of the multi-pillar systems have changed considerably in the past two decades, either because of conscious reforms or as a result of unintended policy effects. In the liberal cluster, employer voluntarism no longer produces wide occupational pension coverage because of the weakness of both collective bargaining and legislation that encourages or mandates second-pillar provision.

In the UK, the universal state pension has always been fairly low, and occupational provision has been marked by large variation in the design of pension schemes. Since the 1990s, policy change has followed two tracks. First, UK governments have adopted reforms designed to expand supplementary earnings-related pension benefits in both the first and second pillars. Second, employers have retreated somewhat from their role in providing second-pillar pensions even as regulation has become more strict (Bridgen and Meyer 2011).

The Thatcher government's attempt to radically reform the public pension system in the 1980s was only partially successful. The state pension remained largely intact, but the modest state earnings-related pension scheme (SERPS) was radically scaled back. Individuals were permitted to contract out of SERPS and to join an occupational pension scheme instead, in keeping with Thatcher's goal of reducing the role of the state in social provision. A series of scandals in the occupational pension sector in the 1990s, however, revealed the limits of this approach, and regulation was tightened. After the Labour party returned to power in 1997, its approach to pension provision was to improve the coverage of non-state pension provision by introducing a means-tested Minimum Income Guarantee for pensioners in 1999 (expanded and renamed the Pension Credit in 2003) and encouraging the growth of regulated personal pensions (the 'Stakeholder' pension) in 1999. These measures had only a modest impact, and reforms since then have focused on improving the state pension and increasing incentives for employers to offer occupational pensions. The 2008 Pensions Act requires employers to offer workers access to workplace pensions meeting minimum standards, with NEST pensions (National Employment Savings Trust; auto-enrolment defined contribution schemes) being one option. The 2014 Pensions Act introduces an increased single-tier state pension. In addition to these changes, employers have largely switched from DB to DC schemes for new entrants, which will result in lower benefits for many future occupational pensioners (Bridgen and Meyer 2011).

In the conservative/social democratic multi-pillar cluster, both the Netherlands and Switzerland have experienced substantial deteriorations in second-pillar pensions. The state pension in both countries remains fairly stable, but because of its generosity, it is a natural target for retrenchment in periods of fiscal austerity. For example, the *Dutch* state pension provides a generous benefit to all residents that is earned by 50 years of residency between the ages of 15 and 65. State pension spending is correspondingly high, at about five per cent of GDP in 2012, and this has put public pension reform on the political agenda since the mid-1990s. Pension indexation has been the primary vehicle for achieving savings; benefits were partially indexed in the late 1990s and early 2000s. Recent reforms have focused on raising the state pension retirement age (to 66 by 2018 and to 67 by 2021) and adjusting benefits to life expectancy.

The much-praised Dutch second pillar has undergone even more substantial change. Until 2002, most occupational schemes provided a benefit equal to 70 per cent of the final salary, including the state pension, after 35 years of service. Dutch occupational pensions are capital funded, and their defined benefit structure requires that pension fund assets exceed the value of projected liabilities. Many pension funds experienced large losses on financial markets in both financial crises of the 2000s, the dotcom bubble in 2002 and the most recent financial crisis. Pension funds responded by cutting benefits and raising contributions. In the first half of the 2000s, most pension schemes switched from a final salary benefit formula to an average salary benefit formula. The financial crisis that began in 2008 resulted in a further round of benefit cuts: many pension funds suspended the indexation of accrued rights and pension payouts, and some even reduced pensions. A 2014 reform reduces the accrual rate and limits the tax deductability of contributions to incomes below €100,000. The two financial crises have also prompted consolidation in the occupational pension sector; in 2000, there were more than 1,000 pension schemes, and the number is less than 400 today.

The *Swiss* pension system combines a fairly generous state pension with mandatory occupational pensions. The high costs of the state pension mean that it is not immune to pressures caused by rising life expectancy. However, most attempts at introducing modest reductions in benefits have failed, largely because of the veto opportunities offered by the Swiss political system (Bonoli 2007). For example, legislation containing a modest increase in women's pension age failed in 2004 and 2011. Attempts to reform occupational pensions have been slightly more successful. A 2003 reform reduced the annuity conversion rate for occupational pensions from 7.2 per cent to 6.8 per cent. The same legislation expanded the contribution base for occupational pensions and increased accessibility. The minimum interest rate applied to the assets of pension funds is not subject to legislation and is easier to change. Since 2002 the minimum interest rate has decreased from 4 per cent to 1.5 per cent (Bonoli et al. 2013).

8.3.3 Pension reform in the Central and East European states (CEEs)

Pension systems in the CEE cluster face pressures for reform that are very different from those facing Western European pension systems. The transition to democracy and capitalism in the CEEs has had profound consequences for pension provision. The CEE states inherited comprehensive state pensions from the state socialist period, so post-1990 pension reforms have often involved layering private pension arrangements on top of slimmed-down public benefits. In 9 of the 13 new CEE member states of the EU, pension privatization has been a key component of pension reform. As Orenstein (2008) argues, privatization fits well with the economic development strategies of the CEE countries which aimed at liberalization, especially at keeping production costs low. For most CEEs, pension reform has been a balancing act. On the one hand, governments try to ensure pension adequacy for those who participated in the state socialist systems. On the other hand, governments introduce second- and third-pillar reforms that expand pension provision but relieve the state from the primary responsibility of doing so. International actors like the World Bank and the IMF have been important influences for CEE reforms.

The *Polish* experience illustrates these reform tendencies. In 1999, Poland shifted from a pension system dominated by public provision to a multi-pillar system with NDC benefits and funded DC benefits in the first pillar as well as incentives for the development of second- and third-pillar pensions. The Polish case is important because it borrows some of the principles of the 1994 Swedish reform. As such, the reformed pension system in Poland transfers much of the risk of income security in old age from the state to individuals. Even if the state mandates participation in the first-pillar schemes, pension levels depend on returns to investment (Guardiancich 2012).

Pension reform in *Hungary* reveals the limits of the privatization strategy in the CEEs. The Hungarian transition to democratic capitalism included a substantial pension reform based on the resurrection of Bismarckian principles from before the Second World War and adherence to the recommendations of international financial institutions. The post-1990 pension system was based on both pay-as-you-go, earnings-related benefits and funded personal accounts. The first pillar continued to experience financial difficulties, so governments responded with incremental modifications to benefits, including restrictions on early retirement and an increase in the pension age. Reforms adopted between 2009 and 2011 increase the retirement age from 62 to 65 by 2022. In 2010, the government decided to shift contributions from the second pillar to the first in order to cover funding shortfalls. This marked a radical shift from the privatization strategy initiated in 1997. Subsequent measures accelerated the process of nationalizing the second pillar (Gal 2013). Hungary's experiment with pension privatization appears to be over, at least for now.

The *Estonian* experience illustrates still other aspects of pension reform processes in the CEE states. The post-1990 pension system was based on the system introduced under state socialism, but reforms beginning in 1998 introduced second- and third-pillar pension schemes. The first pillar provides both a minimum pension financed by general revenues and an earnings-related benefit based on contributions. Workers are required to participate in the DC schemes in the second pillar, and these are financed by both employees and employers. Recent reforms introduced pension credits for parents, but deficits in the first pillar continued to be a problem, especially during the financial crisis beginning in 2008. Like Hungary, Estonia adopted legislation to temporarily suspend contributions to the second pillar, thereby increasing revenues to the first pillar (this ended in 2011). Unlike Hungary, however, this measure remained temporary and did not result in the nationalization of the second pillar.

8.4 Assessing reform trajectories and outcomes

How have three decades of reform shaped old-age security? Despite different institutional starting points, three policy levers dominate pension reform in Europe: reducing benefit generosity, postponing retirement and expanding private provision (Immergut et al. 2007; Ebbinghaus 2011). These reform trajectories will result in lower public pensions for future retirees compared to the level of benefits that today's pensioners enjoy in many European countries. However, longer working lives, combining employment with pension receipt, and increased reliance on private pension income will compensate for some of the cuts in public and workplace pensions.

Despite similarities in reform trajectories and some degree of convergence, there is still much heterogeneity across European pension systems, at least in terms of spending. Figure 8.1 shows the enormous variation in relative spending on public and private pensions in Europe: public spending continues to dominate in the conservative/corporatist and Eastern European pension regimes, whereas the private share of pension spending is much higher in the different types of multi-pillar systems. Figure 8.2 shows similar results concerning public spending on pensions in 2010: there is no discernible trend towards convergence in public pension spending levels (countries are grouped in the graph according to their pension regime type). However, reductions in benefit generosity are typically unpopular with voters, so they are phased in over a long period, and their impact on public spending is not

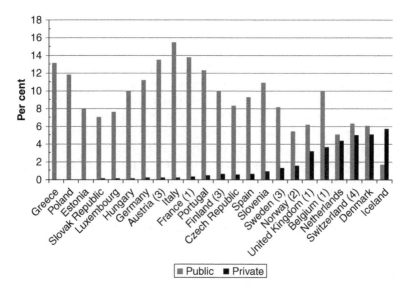

Figure 8.1 Public and private expenditure on pensions in selected European countries, 2012, as per cent of GDP

Notes: All types of private plans are included. Public and private expenditures on pensions refer respectively to 2009 and 2012, except the following: (1) private pension data refer to 2011; (2) private pension data refer to 2008; (3) private pension data refer to 2010; (4) public pension data refer to 2008.

Source: OECD Global Pension Statistics, from OECD (2013b: Data F15), own compilation.

immediately felt. Moreover, even reduced benefit generosity does not result in lower spending if the number of pensioners rises.

One of the great ironies of the financial crisis that began in 2008 is that substantial losses on financial markets and historically low interest rates have begun to undermine the foundations of the multi-pillar pension systems in the Netherlands, Switzerland and Denmark (Häusermann 2010; Anderson 2011; Goul Andersen 2011). Countries whose pension schemes invested heavily in equities in 2007 suffered substantial losses on financial markets in 2008. However, global financial markets have recovered since 2010 and many pension funds have recouped most of their losses.

How has pension reform influenced socio-economic conditions such as poverty and inequality? One of the most important achievements of the expansion of public pension provision since the 1930s is the dramatic reduction in old-age poverty. In 2010, the poverty rate (income less than 50 per cent of median income) for those 65 and over is very

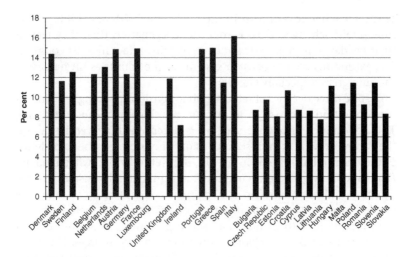

Figure 8.2 Public spending on pensions (as per cent of GDP) in 2011
Note: Includes early retirement pensions and disability pensions.
Source: Eurostat (2015c), own compilation.

close to the rate for the overall population in the OECD (12.8 per cent compared to 11.3 per cent; OECD 2013a). This ratio is likely to change in many European countries as the effects of pension retrenchment kick in over the next two to three decades. For example, Meyer (2014) and Hinrichs (2012) forecast substantial increases in old-age poverty in Germany. Recent research demonstrates similar effects in Italy (Jessoula 2012) and Sweden (Berglund and Esser 2014). Inequality among pensioners is also likely to increase in the next three decades because of the tighter link between employment income (and contributions) and pension benefits, and the larger role played by private and workplace pensions in topping up reduced public pensions, as access to private and workplace pensions is very unequally distributed in many countries, and they are less redistributive.

8.5 Conclusion

Pension reform has been a recurring item on government agendas in Europe since the 1980s. National approaches to reform have ranged from incremental, parametric reforms to less frequent and more far-reaching reforms. Nowhere in Europe has a government attempted radical retrenchment of a public scheme (with the possible exception of the Thatcher government in the UK in the 1970s and 1980s). To be sure, public schemes across Europe are less generous than they were in their heyday, but their basic structure remains intact in most countries. The exceptions are the pension systems in the countries hit hardest by the euro crisis (especially Greece).

Despite three decades of reform, public pension spending remains stable or continues to rise in most countries, largely because of ageing. This appearance of relative stability is deceptive, however, because of the future growth of non-state pension provision. As the experiences of the multi-pillar systems in the UK, the Netherlands, Switzerland and Denmark show, private occupational and individual pension schemes can grow quickly relative to public provision, depending on how they are designed. To the extent that pension reforms have included incentives for the growth of non-state social provision, future retirees will receive more of their pension income from private sources. If pension savings are collectively organized and administered, as they are in the Netherlands, Switzerland, Denmark and to some extent the UK, risks will be collectively shared. However, the shift from DB plans to DC plans means that individuals bear more of the risks associated with future pension payouts. This is the logic behind the NDC reforms of public pensions in Sweden, Italy and Poland, as well as the shift from DB to DC in private occupational pensions. Even if the shift towards non-state pension provision and financially sustainable public pensions addresses some of the challenges facing Europe's pension systems, it will almost certainly mean increasing inequality between pensioners and workers, and among pensioners. This means that the main challenge of future pension policies will be to maintain adequacy in terms of minimum state pension provision, ensure equal access to private pension provision (and prevent market failure in this area) and deal with the old-age inequalities arising from unequal (especially gendered) employment opportunities and precarious employment.

Reductions in public pension generosity will also shape the relationship between employment and retirement. As discussed, important goals of pension reform have been to reduce early retirement, raise the statutory retirement age and link pension benefits to changes in longevity. These measures fundamentally alter the relationship between employment and retirement by reconfiguring the relative weight of both phases of the life course: periods of employment will become longer relative to periods of retirement. Moreover, pension reforms in many European countries blur the boundary between employment and retirement by permitting senior citizens to combine pension receipt with paid employment (see the other contributions in this book). As longevity continues to increase, many seniors in good health will prefer to continue working beyond the statutory retirement age (if there is one). However, many seniors with insufficient pension income will find themselves forced to combine paid work with pension receipt.

Notes

1. The social democratic regime is based on universal, individual entitlement to state-provided social benefits financed mainly by taxes and social contributions. The conservative/corporatist (or Christian democratic) welfare state relies on employment-based, contribution-financed social benefits that protect wage earners and their families. Social provision is also occupationally fragmented and status preserving. In the liberal regime, a tax-financed, means-tested tier of social provision provides minimum protection, whereas most wage earners receive social provision through their employer or modest state schemes (Esping-Andersen 1990).
2. In an NDC scheme, contributions are credited to an individual's pension account, but the assets in the account are not capitalized.

References

Anderson, K. M. (2010) 'Promoting the multi-pillar model? The EU and the shift toward multi-pillar pension systems', in: Y. Borgmann-Prebil and M. Ross (eds.), *Developing solidarity in the EU: Citizenship, governance and new constitutional paradigms*, Oxford: Oxford University Press, 216–34.
Anderson, K. M. (2011), 'Occupational pensions in the Netherlands: Adapting to demographic and economic change', in: B. Ebbinghaus (ed.), *The varieties of pension governance: Pension privatization in Europe*, Oxford: Oxford University Press, 292–317.
Anderson, K. M. (2015), *Social policy in the European Union*, Basingstoke: Palgrave.
Anderson, K. M. and Immergut, E. M. (2007), 'Sweden: After social democratic hegemony', in: E. M. Immergut, K. M. Anderson and I. Schulze (eds.), *The handbook of West European pension politics*, Oxford: Oxford University Press, 349–95.
Anderson, K. M. and Meyer, T. (2003), 'Social democracy, unions, and pension politics in Germany and Sweden', *Journal of Public Policy*, 23 (1), 23–54.
Berglund, T. and Esser, I. (2014), *Modell i förändring. Landrapport om Sverige*, Oslo: Fafo.
Bonoli, G. (2003), 'Two worlds of pension reform in Western Europe', *Comparative Politics*, 35 (4) 399–416.
Bonoli, G. (2007), 'Switzerland: The impact of direct democracy', in: E. M. Immergut, K. M. Anderson and I. Schulze (eds.), *The handbook of West European pension politics*, Oxford: Oxford University Press, 203–47.
Bonoli, G. and T. Shinkawa (eds.) (2005), *Ageing and pension reform around the world*, Aldershot: Edward Elgar.

Bonoli, G., Braun, D. and Trein, P. (2013), *Pensions, health and long-term care – Switzerland, Asisp (Analytical support on social protection reforms and their socio-economic impact) country document*, Brussels: European Commission/Gesellschaft für Versicherungswissenschaft und -gestaltung e. V., http://socialprotection.eu/files_db/1433/CH_asisp_CD13.pdf, date accessed 29 January 2015.

Bridgen, P. and Meyer, T. (2011), 'Britain: Exhausted voluntarism – The evolution of a hybrid pension regime, in: B. Ebbinghaus (ed.), *The varieties of pension governance: Pension privatization in Europe*, Oxford: Oxford University Press, 265–92.

Ebbinghaus, B. (2006), *Reforming early retirement in Europe, Japan and the USA*, Oxford: Oxford University Press.

Ebbinghaus, B. (ed.) (2011), *The varieties of pension capitalism*, Oxford: Oxford University Press.

Esping-Andersen, G. (1990), *The three worlds of welfare capitalism*, Princeton: Princeton University Press.

Eurostat (2004), *Population statistics*, Luxembourg: Office for Official Publications of the European Communities.

Eurostat (2015a), *Table: Total fertility rate (code: tsdde220)*, Luxembourg: Eurostat, http://ec.europa.eu/eurostat/tgm/table.do?tab=table&init=1&language=en&pcode=tsdde220&plugin=1, date accessed 27 January 2015.

Eurostat (2015b), *Dependency ratios, 1950 and 2012 (% of the population aged 15–64)*, Luxembourg: Eurostat, http://ec.europa.eu/eurostat/statistics-explained/index.php/File:Dependency_ratios,_1960_and_2012_(%25_of_the_population_aged_15–64).png, date accessed 3 March 2015.

Eurostat (2015c), *Table: Expenditure on pensions (code: tps00103)*, Luxembourg: Eurostat, http://ec.europa.eu/eurostat/tgm/table.do?tab=table&plugin=1&language=en&pcode=tps00103, date accessed 3 March 2015.

Ferrera, M. and Jessoula, M. (2007), 'Italy: A narrow gate for path shift', in: E. M. Immergut, K. M. Anderson and I. Schulze (eds.), *The handbook of West European pension politics*, Oxford: Oxford University Press, 396–453.

Gal, R. I. (2013), *Pensions, health and long-term care – Hungary, Asisp (Analytical support on social protection reforms and their socio-economic impact) country document*, Brussels: European Commission/Gesellschaft für Versicherungswissenschaft und -gestaltung e. V., http://socialprotection.eu/files_db/896/asisp_ANR10_Hungary.pdf, date accessed 3 March 2015.

Goul Andersen, J. (2011), 'Denmark: The silent revolution towards a multipillar pension system', in: B. Ebbinghaus (ed.), *The varieties of pension governance: Pension privatization in Europe*, Oxford: Oxford University Press, 183–209.

Guardiancich, I. (2012), 'Poland: Between flexible labour markets and defined contributions. In K. Hinrichs and M. Jessoula (eds.), *Labour market flexibility and pension reforms: Flexible today, secure tomorrow?*, London: Palgrave, 93–122.

Häusermann, S. (2010), *The politics of welfare state reform in continental Europe. Modernization in hard times*, Cambridge: Cambridge University Press.

Hinrichs, K. (2000), 'Elephants on the move: Patterns of public pension reform in OECD countries', *European Review*, 8 (3), 353–78.

Hinrichs, K. and Jessoula, M. (eds.) (2012), *Labour market flexibility and pension reforms. Flexible today, secure tomorrow?*, Basingstoke: Palgrave.

Hinrichs, K. (2012), 'Germany: A flexible labour market plus pension reforms means poverty in old age', in: K. Hinrichs and M. Jessoula (eds.), *Labour market flexibility and pension reforms*, Basingstoke: Palgrave, 29–61.

Immergut, E. M., Anderson, K. M. and Schulze, I. (eds.) (2007), *The handbook of West European pension politics*, Oxford: Oxford University Press.

Jessoula, M. (2012), 'A risky combination in Italy: "Selective flexibility" and defined contribution pensions', in: K. Hinrichs and M. Jessoula (eds.), *Labour market flexibility and pension reforms*, Basingstoke: Palgrave, 62–92.

Lynch, J. F. (2006), *Age and the welfare state. The origins of social spending on pensioners, workers, and children*, Cambridge: Cambridge University Press.

Meyer, T. (2014), *Beveridge statt Bismarck! Europäische Lehren für die Alterssicherung von Frauen und Männern in Deutschland*, Berlin: Friedrich Ebert Stiftung.

Monastiriotis, V. (2013), 'A very Greek crisis', *Intereconomics*, 48 (1), 4–9.

Myles, J. (1994), *Old age in the welfare state. The political economy of public pensions*, Boston: Little, Brown and Company.

Myles, J. and Pierson, P. (2001), 'The comparative political economy of pension reform', in: P. Pierson (ed.), *The new politics of the welfare state*, New York: Oxford University Press, 305–33.

OECD (Organisation of Economic Co-operation and Development) (2013a), *Pensions at a glance*, Paris: OECD.

OECD (2013b), *Pension markets in focus (No.10)*, Paris: OECD, http://www.oecd.org/finance/private-pensions/pensionmarketsinfocus.htm, date accessed 3 March 2015.

Orenstein, M. (2008) 'Out-liberalizing the EU: Pension privatization in Central and Eastern Europe', *Journal of European Public Policy*, 15 (6), 899–917.

Pierson, P. (1994), *Dismantling the welfare state*, Cambridge: Cambridge University Press.

Pierson, P. (2004), *Politics in time*, Princeton: Princeton University Press.

Sainsbury, D. (1996), *Gender, equality and welfare states*, Cambridge: Cambridge University Press.

Schulze, I. and Jochem, S. (2007), 'Germany: beyond policy gridlock', in: E. M. Immergut, K. M. Anderson and I. Schulze (eds.), *The handbook of West European pension politics*, Oxford: Oxford University Press, 660–710.

Stepan, M. and Anderson, K. M. (2014), 'Pension reform in the European periphery: The role of EU reform advocacy', *Public Administration and Development*, 34 (4), 320–31.

Thane, P. (2002), *Old age in English history*, Oxford: Oxford University Press.

Trampusch, C. (2006), 'Industrial relations and welfare states: The difference dynamics of retrenchment in Germany and the Netherlands', *Journal of European Social Policy*, 16 (2), 121–33.

Wadensjö, E. and Sjögren Lindquist, G. (2011), 'Sweden. A viable public-private pension system', in: B. Ebbinghaus (ed.), *The varieties of pension governance: Pension privatization in Europe*, Oxford: Oxford University Press, 240–61.

World Bank (1994), *Averting the old age crisis*, New York: Oxford University Press.

9
The Transition to Retirement: The Influence of Globalization, Public Policy and Company Policies

Victor W. Marshall

The concept of the 'tripartite life course' has a lengthy history in both North American and European theory and research, postulating that work is the central organizing domain that structures the life course, which can be described as having three stages: preparation for work, working years and retirement. However, research has generated sufficient evidence to question the utility of the tripartite life-course construct, except as a foil against which to ascertain departures from the simple structure of the life course, their causes and consequences. I first outline the emergence and critique of this theoretical concept. Then, focusing on the changing transition from work to retirement in North America since the middle of the 20th century, I examine factors that have caused these changes in the work to retirement transition, and their consequences. European life-course researchers have a stronger history than North American researchers of paying attention to macro-level factors and public policy as shaping transitions in the working life course. However, a few have studied corporate factors that shape work and retirement transitions. In the following, I therefore stress the importance of some much neglected structural issues, in particular globalization, public policy and corporate policy, in affecting retirement transitions. I conclude with a short discussion about the challenges for future research in this area.

9.1 The life course theorized over time: From tripartition to the risk society

Leonard D. Cain, Jr., first formalized the life-course perspective, defining life course 'to refer primarily to those successive statuses individuals are called upon to occupy in various cultures and walks of life as a result of aging' (Cain 1964: 277–8). Cain stressed the life course as organized around work. His contribution predated Martin Kohli's first usage of the term in English which divided the life course into 'periods of preparation, "activity", and retirement' (1986: 72). As late as 1991, Kohli and Rein (1991: 21) had claimed that this tripartition of the life course had been 'firmly established by the 1960s' and that 'old age had become synonymous with the period of retirement: A life phase structurally set apart from "active" work life'. In retrospect, that judgement was not entirely accurate. Prior to Cain, Wilensky (1960: 555) argued that 'it is likely that with continuing industrialization careers are becoming on average more discrete and are characterized by more numerous stages, longer training periods, less fluctuation in the curve of rewards (amount, timing, duration), a more bureaucratic setting, and more institutionalization'.

This speaks to institutionalization coupled with individualization of occupational careers. Kohli (2007: 257–8) now concludes: 'With hindsight it is easy to see that the institutionalized life course that I discerned as the provisional endpoint of modernization corresponded to a rather specific historical period – the 1960s when many of its features culminated in what is now often termed the "Fordist" model of social structure and the life course.' In macro-sociological terms, this model was based on rapid and seemingly stable economic growth, low unemployment and expansion of the welfare state. In much current life-course research, we consider *departures* from the standardized life course as setting the themes for our investigations. We have come to understand that life-course transitions are much more complex *in the transition to work* (Hogan 1978, 1980; Rindfuss et al. 1987; Settersten et al. 2005; Heinz 2009), *in the transition to retirement* (Kohli et al. 1991; Marshall and Marshall 1999) and *in transitions across the working years themselves* (DiPrete 2002; Brückner and Mayer 2005; Haviland et al. 2010).

The 'risk society' concept (Beck 1992; Beck et al. 1994) also makes us aware of the historical evolution over the past few decades in which the state does less than it used to, to protect people from risk. This marks a return to an earlier era in which states assumed little responsibility to assume risk from individuals, and a yet earlier era in which states did not exist. Nevertheless, it is important to emphasize the importance of the state in establishing conditions which create or deny to individuals the possibilities for a good retirement. Some retirement-related risks, such as income loss with retirement, can be anticipated but not with certainty. For example, unexpected 'early retirement' can occur because of job loss or the need to provide family care. A highly institutionalized life-course regime at the state level can protect against chance events (DiPrete 2002). O'Rand (2011), while acknowledging disagreement on the extent of de-institutionalization of the life course, describes global ageing as related to the individualization of risk and changes in the institution of the life course that shift the locus of responsibility for protecting the life course from the public sector to the individual and family. She notes, 'Others argue that structural forces such as globalization and the service economy are acting in concert with demographic trends (especially declines in fertility) to erode traditional institutions of industrial societies and, in turn, to de-institutionalize the life course' (O'Rand 2011: 693).

In what follows, as I focus on the work to retirement transition, out of necessity I neglect outstanding research in the *family* life course (with attention to gender influences) and the *health* life course, as well as research efforts studying phenomena such as crime through this perspective. We need to look at the *intersection* of various life-course domain trajectories and their mutual causation. Also neglected here are a growing literature on *agency* (Marshall and Clarke 2010; Settersten and Gannon 2005) and a small but important literature on *chance events* and their impact on the life course (Shanahan and Porfelli 2006; Marshall and Bengtson 2011).[1] As a final introductory remark, I am convinced that to understand the complexity of work and retirement transitions, we need not only complex quantitative, longitudinal and comparative analyses, but also more systematic qualitative investigations and multiple-methods studies, including case studies of workers in specific firms and comparative case study analyses. I briefly touch on these research design issues below.

9.2 The transition to retirement in North America

Two landmark volumes (Schmähl 1989; Kohli et al. 1991), published a quarter-century ago and relying mostly on European data, focused on how *public policies* encouraged people to leave work *earlier* – whereas in the current situation it is important to examine not only public policy but other factors that have shaped labour force participation up to the current period. As in the European countries, the key benchmark age when considering retirement in the USA is 65, probably because that age was adopted in 1935, when the government pension plan, Social

Security was begun. The normal age for receipt of full Social Security payments was recently raised to 67. It can be taken early with reduced benefits; delay beyond the standard eligibility age leads to increased benefits. If either option is taken it is considered 'early' or 'late'.[2] Age 65 is still the year when Americans become eligible for Medicare, the government's health care insurance plan. Thus, there is a historically grounded cultural understanding of age 65 as 'time for retirement' in North America that is closely related to government pension and health insurance regulations.

Public policies in the USA have thus set the timing of the retirement transition normatively at age 65, and this age is still largely consensual, despite the trends that can be seen in Figure 9.1, which shows labour

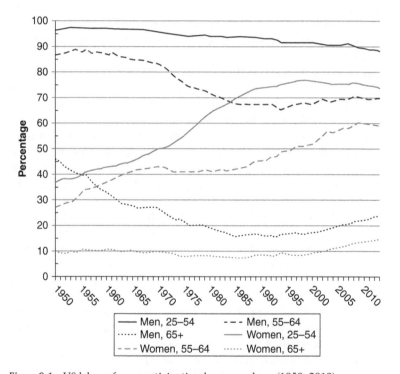

Figure 9.1 US labour force participation by age and sex (1950–2013)
Note: Civilian labour force, unadjusted yearly averages.
Source: Labor Force Statistics from the Current Population Survey (http://www.bls.gov/cps/), U.S. Bureau of Labor Statistics (2013b).

force participation rates in the USA for men and women in three broad age categories, over the period 1950 to 2013.

As the graph shows, with the exception of men aged 25 to 54, there have been dramatic changes in participation. Men in the two older age brackets experienced a large decrease in labour force participation from the middle of the last century until the mid-1980s, followed by rising participation in the last decade of the century and up to the present. So we see for the USA what Kohli et al. (1991) referred to as early exit from the labour force in their landmark book. In the USA, the vast majority of American men aged 55 to 64 were in the labour force in the 1950s and 1960s. Male labour force participation rates fell dramatically in the 1970s, and then fell more gradually, until reaching a low of 65.5 per cent in 1994. As the last century drew to a close, participation rates turned very gradually upwards, reaching a high of 70.4 per cent in 2008. While this was happening, the gender revolution in workforce participation saw dramatic increases for women in all three age categories.

Importantly, the data in the chart and accompanying discussion do not distinguish between full-time and part-time workers. In the USA, the percentage of older workers (aged 65 and older) working full-time doubled between 1995 and 2007, while part-time workers increased by only 19 per cent. By 2007, more than half of older workers, by this definition, worked full-time, and the U.S. Bureau of Labor Statistics predicted that this trend would continue (Sloan Center on Aging & Work 2013). This is a dramatic change over the past 20 years, and one that appears to be continuing.

In recent history, the European and North American nations have experienced greater *diversity in work to retirement patterns*, such that, with regard to the second and third stages of the life course, the initial concept of the *tripartite* life course is now seen to be historically limited in its applicability. Increasing proportions of older people are again[3] in retirement, but so many are not that 65 can no longer be considered institutionalized as the age at which people retire. This diversity is more than simply variability in the timing of retirement, hours worked while retired or gender differences. There is also diversity in the *pathways to retirement*, in the *causes of variability* in labour force participation by age, and in the *motivations* and *social constraints* that lead some people to wish to retire earlier and many to work longer (Han and Moen 1999; Mermin et al. 2007; Raymo et al. 2010). Among social constraints are the obligations that workers may have through their 'linked lives' to spouses, parents and children (Zissimopoulos and Karoly 2007).

Variation in pathways has been well documented for the USA in a series of papers by Cahill, Giandrea and Quinn (2012). Using the longitudinal Health and Retirement Study, these authors found that of respondents aged 51 to 61 in 1992, who had any work after age 50, a large percentage of men (12 per cent) and women (32 per cent) had not had a full-time career job since age 50. In addition, it is important to recognize that many people transition to self-employment at older ages, some people are simultaneously earning wages and self-employed, and self-employment rates are higher among older than younger workers. For American men, the self-employment rate at age 60 is 20 per cent but rises to 39 per cent by age 66 (Zissimopoulos and Karoly 2007). Thus, '[...] many middle-aged and older workers move from a full-time career job through other transitions before finally leaving the labour force. Such transitions include reduction in hours worked and movements from full-time to part-time work, as well as transitions from wage and salary employment to self-employment' (Zissimopoulos and Karoly 2007: 273).

A recent survey by the Conference Board showed that more than 60 per cent of American workers aged 45 to 60 plan to delay retirement (Washington Post 2013). That figure is up from 40 per cent two years earlier. About half of the rise comes from workers who, after the 2008 'Great Recession', are responding to job loss (Johnson and Park 2013), shrunken savings and declines in home equity value (Butrica et al. 2010). Changing economic conditions result from theoretically definable conditions, but they are experienced by individuals as 'chance' factors.

One result of these developments is that more people are working beyond the traditional retirement age because they feel they cannot afford to retire (Kingson and Morrissey 2012). As well, those who have lost their jobs have faced difficult labour market conditions. Displaced older workers are less likely to get a new job, and it takes longer to do so (Rix 2011). Many become discouraged workers, that is, they give up searching and decide to retire. As in the case of earlier unemployment, this can go together with (sometimes considerable) decreases in well-being, for example the onset of depression or declining cognitive health (Marshall et al. 2001; He et al. 2003, 2006; Clarke et al. 2012). 'Bridge jobs' are generally not designed by employers, but rather constructed or improvised by employees. Such 'contingent work' provides little security. It is often contract or temporary work without long-range prospects. Another factor cited to explain the increase in retirement age is that fewer employers offer retiree health insurance, so many stay on the job to retain existing health insurance (Cahill et al. 2012). This reflects a combination, in the USA, of the effects of public and corporate policy – a combination that I will discuss in the following sections.

9.3 Institutional factors associated with changes in work to retirement patterns

I now consider three factors that cause diversity in retirement transitions. I begin with a discussion of globalization, following which I turn to public policy and finally to the corporate level, especially the changing nature of work and employment.

9.3.1 Globalization and the changing nature of work

Globalization of jobs to developing countries, often associated with downsizing in the developed countries, is a major factor driving the breakdown of the stable working life course that had been considered 'normal' in highly developed countries. Globalization has shifted the work of (often unionized) employees in long-term jobs – that is, having 'careers' – to contract or non-contracted casual workers in developing countries. Thus, globalization acts on and through companies that employ people, but in different ways in countries that have different levels of economic development, governmental safety nets and public policies.

According to the International Monetary Fund website (www.imf.org), economic globalization 'refers to the increasing integration of economies around the world, particularly through trade and financial flows. The term sometimes also refers to the movement of people (labour) and knowledge (technology) across international borders' (IMF 2002). Standing (2008) refers to a new 'globalisation class structure', with growing numbers of people detaching themselves, or being detached, from national regulatory and protective systems because they engage in 'casual' or 'informal' labour. As he points out, many people like casual jobs. However, a life course based solely or largely on casual labour has many disadvantages for one's later years. It may mean no, or delayed retirement, and a working life course accompanied by financial insecurity and economic deprivation. Standing offers a typology of eight social classes, of which three can be considered likely to enjoy economic security in the later years. These are the Elite, the Salariat (in civil service and corporate bureaucracies) and Core Workers (formerly known as 'the working class') who hold full-time, typically unionized jobs and are likely to have a smooth and predictable pathway to retirement. Four additional types are likely to experience no, or disrupted and unpredictable retirement pathways. These are the Proficiens (consultants and those on short-term contracts), Flexiworkers, the Unemployed and the Detached who are cut off from mainstream social benefits. Standing's typology can be useful in examining not only different pathways to retirement but also different possibilities of ever having one.

A sophisticated theoretical model showing how globalization works through institutional filters such as pension and welfare systems, occupational systems and employment-based policies affecting continuation of employment or early exit is provided by Hofäcker (2010: 118). He argues that processes of globalization 'have triggered a rising importance but at the same time also an increasing volatility of markets', with global markets highly dependent on a global field of 'random shocks', such as natural disasters, terrorist attacks or regional economic crises. These make it difficult for global firms to make long-term plans, necessitating that they maintain flexibility in the size of their labour forces, working time, labour contracts and wages.

9.3.2 Public policy, the working life course and their impact on retirement transitions

Mayer has articulated the role the nation state plays in imposing both order and constraint on individual lives. At the individual level, the state 'legalizes, defines and standardizes most points of entry and exit; into and out of employment, marital status, into and out of sickness and disability, into and out of education. In doing so, the state turns these transitions into strongly demarcated public events and acts as gatekeeper and sorter' (Mayer 1986: 167). Public policies directly and indirectly shape and reshape career patterns over the entire working life course (Cooke 2006).

Retirement is one of these transitions whose development was strongly influenced by government (Myles 1989; Myles and Street 1995). Consequently, the difficulties experienced by workers anticipating, hoping for or despairing of the possibility of retirement cannot be attributed solely to individual characteristics, the nature of the economy or the organization of work. Governments and the state apparatus set policies and practices that can open or close opportunities for people to have a decent retirement – as shown by exemplary pioneering European studies (Guillemard 1986, 2000; Schmähl 1989; Kohli et al. 1991).

For the USA, Blau and Goodstein (2010) found that improved Social Security did not contribute strongly to the decline in men's labour force participation from the 1960s to the mid-1980s. Increases in the full pension age and the Delayed Retirement Credit explain about a quarter of the increases since the 1990s, but increasing educational attainment was more important. Higher education is associated with higher pay, less arduous and more fulfilling jobs. Maestas and Zissimopoulos (2010) note several other factors that can affect age at retirement in the USA. Increased life expectancy, along with more years in good health and federal requirements for workplace accommodation of persons with disabilities, probably led some people to decide to delay exit from the workforce. Changes to the 1968 Age Discrimination in Employment Act may have increased the retention of employees but in some cases are thought to have increased age discrimination in hiring by making it more difficult to dismiss older workers.[4]

The dramatic switch from defined benefit to defined contribution pension plans may also play a small role. This is a change in the private sector that ultimately results in greater reliance on state pension benefits. According to the U.S. Bureau of Labor Statistics (2013a), coverage in any corporate pension plan was 67 per cent in 1992–93, falling to 63 per cent in 2007. However, the type of plan changed dramatically: the percentage of employees covered by defined benefit pension plans fell from 35 per cent in 1990–91 to 18 per cent in 2011, with a partly compensating rise in defined contribution plan coverage from 32 per cent to 43 per cent (see also Wiatrowski 2012). However, at peak in the 1980s only about half of American workers had defined benefit plans, which encourage staying employed until the traditional retirement age (65 in the USA), but provide little incentive to remain working beyond, while defined contribution plans contain no penalties for extending employment.

9.3.3 The corporate level

Globalization has an impact not only on government programmes or public policy affecting the working life course, but also at the corporate level (Blossfeld et al. 2011). As Heinz (2001: 7) states the issue, 'We are accustomed to explaining people's varied individual voyages in work careers by social origin, gender, educational level, and occupational skills. However, employers' recruiting and promotion policies, on the one hand, and the implications of "linked lives", on the other, are at least as important as individual variables.' We are moving away from the Fordist life course that had been created after the Second World War. This development has been intertwined with globalization (IMF 2002). In the Fordist work regime, workers were promised a secure employment over their life course until retirement. Agreements between unions and management promised lifetime employment, progress through the ranks in rigid seniority systems, some protection from unemployment (the first hired, last fired principle), regular raises in pay to meet or exceed the rising cost of living and generous retirement benefits, which justified retirement on a predictable schedule (Quadagno et al. 2001, 2003). Needless to say, not all workers experienced that regularity of the life course, but it was a normative model, and those who did experience the Fordist life course conformed quite closely to the tripartite life course as described by Cain and Kohli.

Quadagno and colleagues (2003) saw Fordism in the USA as declining from the 1980s, with changes in the manufacturing sector, globalization and the offshoring of jobs, and the move from defined benefit to defined contribution pension plans. Declines in the heavy manufacturing industries of the 'rust belt' (such as Detroit, Michigan and the steel and manufacturing regions of Ohio, Pennsylvania and Indiana) contributed to the move away from the Fordist life course. Cappelli (1999) describes the breaking of the contract between management and labour as heavy manufacturing declined in the 1980s, with corporate downsizing and restructuring becoming the new fashions in corporate management. The new employment relationship is one in which

> [...] managers seek simply to lower the expectations of employees by limiting the employer's obligations on job security (for example, shifting from defined benefit to defined contribution pension plans) and career development – the dreaded 'employability' doctrine that pushes responsibility for careers onto employees – while assuming that most other aspects of the relationship, including high levels of employee performance, will continue.
>
> (Cappelli 1999: 1)

Quadagno and her colleagues (2001, 2003; Hardy et al. 1996) provide a detailed account of this shift in the American automobile industry, as well as of the consequences of corporate takeovers and downsizing in the banking industry. The contract was broken for both blue-collar and white-collar jobs. In efforts to be more nimble, companies downsized and sought cheaper labour offshore, often leading to early retirement or shifts to bridge jobs as a pathway to full retirement, and leaving workers with no or lower retirement income.

Decreasing commitment between individuals and employers in the 'new economy' is accompanied by greater vulnerabilities for workers in an era of increasing health care costs, with greater longevity leading to greater retirement expense, and the weakening position of labour. This shift in risk from employer to employee occurs at a time when states have become less important in assisting individuals to address problems in life-course management. The guarantee of a financially secure retirement is no longer made by the employer. The shift to defined contribution plans exemplifies the move to the risk society, with individuals responsible for managing their retirement portfolios. The roughly one-third of adult Americans who do not have any private pensions must rely on Social Security, savings and personal assets.

Canadian research with small, medium and large corporations in Canada and the USA in the Issues of an Aging Workforce (IAW) programme (Marshall and Marshall 1999, 2003) has shown how this happens at the company level, leading to an individualization of career and a move away from the expectation that employees will continue working with a company until age 65, at which point they retire with a defined benefit pension.[5] A sociologist might think of *career as a property of social structure*: Go to work for a company and a set of graded steps lies in front of you, ending in retirement at a fixed, predictable time, as in the Fordist paradigm. We found this to be the case in a medium-sized Canadian steel company (Slater Steels) that was unionized, and in Sun Life, the largest life and health insurance company in Canada. Slater Steels had modernized its production processes and needed fewer workers. Because it was unionized, it fulfilled the contract of secure employment by avoiding layoffs. However, it downsized its complement by greatly reducing hiring. As a result, its workforce was older than would otherwise be the case, average duration of employment with the company was 19 years and employees could anticipate retirement with pension at age 65. Sun Life was a stable company when we studied it, providing structured careers leading to retirement at an institutionalized age.

We found a vivid contrast in the case studies of small garment industry firms in Montreal and New York City.[6] In both cities, the garment industry was in decline due to globalization of the industry. Inexpensive labour in developing countries and offshoring of production was made possible by easier and faster shipping of goods between developing countries with lower wage costs than Canada and the USA.[7] Employees were over 70 per cent female and similarly about 70 per cent immigrant background, and their average age older than that in most industries, due to the downturn in new entrants in this declining industry. They had little human capital. One strategy in Montreal was to close the company and subsequently reopen it as a non-unionized shop. As the

Canadian director of the International Ladies' Garment Workers' Union (ILGWU) told us, 'It is too easy to open and close a company in the apparel industry [...]. The workers [...] lose conditions of work, they lose their paycheck, they lose vacation pay, they lose their health benefits, they lose their pension benefits, they lose and lose and they are getting older' (McMullin and Marshall 2001: 119; see also McMullin and Marshall 1999).

This strategy adversely affects older workers and can precipitate unemployment or retirement, but we also found that large corporations used a different 'downsizing' strategy which changed the nature of job and career. A vice-president of Human Resources at Bell Canada, a large national corporation in telecommunications that had reduced its employee complement between 1992 and 1996 and was undergoing another major downsizing during the time of our study, told us, 'We don't want to be thought of as "Mother Bell" anymore.' We surveyed former employees of Bell, which would not agree to a study of current employees due to its current 'Termination Incentive Program' (TIP). Some retirees from earlier incentive programmes referred to having been 'TIPd' into retirement. The company had experienced several waves of downsizing, and our survey found that 51 per cent of retirees subsequently found part-time work, 17 per cent found full-time work and 32 per cent commenced self-employment, suggesting the importance of bridge jobs as a pathway from work to retirement (Singh and Verma 2001).

IAW case studies of three other major corporations showed a contrasting approach to career management. As mentioned earlier, Sun Life retained a Fordist human resources policy with retirement at age 65 or an optional, scheduled 'early retirement'. It had not undergone corporate 'downsizing' that was quite prevalent in Canada at that time. Its employees could foresee a traditional, stable employment career leading to retirement at a fixed age. In contrast, Prudential Life Insurance Company of America, the largest health and life insurance company in the USA, was undertaking a wave of downsizing when we studied it. It did not allow its human resources department to use the word 'career' in any of its documents and discouraged the notion of long-term employment. Nova Corporation, a Calgary-based multinational corporation in petrochemical manufacturing and gas pipelines, retained the use of the term 'career', but through its Employee Transition and Continuity Program and in its newsletter it actively sought to change the meaning of 'career'. Employees were encouraged to 'take charge of your career' and to 'invest in yourself', while receiving the message that continuing change at Nova made it unlikely that employees could depend on lifetime employment. One of its workshops for employees was called, 'Life is a highway (and you're in the driver's seat)' (Marshall and Marshall 1999: 60–61). Career at Nova Corporation was cast not as a property of the firm – not a set of positions through which one can progress up a stable corporate ladder – but as a property *within the individual*. Corporate policies did encourage early retirement using conventional techniques such as incentives, but another exit route through early retirement provided financial assistance and corporate management to help employees develop a business plan, enabling them to establish their own companies and contract their work to Nova (Marshall and Marshall 1999, 2003).

The IAW research program demonstrated that not only public but also corporate policy affects the stability of workers' careers. The more recent international research of the Workforce Aging in the New Economy (WANE) project addressed global and corporate-level factors affecting career possibilities and patterns in small and medium-sized firms in the ICT sector. Platman (2009: 65) says that such firms

> [...] face formidable pressures in order to deliver profitable services and products. There are limits to how much flexibility and security a firm can offer in a climate of fluctuating demand, fierce cost competition and uncertain technological progression [...]. Developing systems or procedures which were designed to retain staff for longer, and beyond retirement age, was well down the list of priorities if the business was to survive [...].[8]

9.4 Conclusions

We will continue to make research progress if we address retirement transition phenomena from both institutional and social psychological approaches, develop theory to link macro- and micro-level phenomena and use a diverse toolkit to gather and analyse both qualitative and quantitative data. I have focused on macro-level social structural issues. While life-course researchers, particularly in Europe, have paid considerable attention to these issues, they have, with some notable exceptions (for example chapters in Heinz and Marshall 2003 and Marshall et al. 2001), emphasized public policy rather than corporate policy. North American life-course research has for the most part taken the individual as the unit of analysis.

I illustrated the effects of corporate policies on working careers and retirement trajectories and transitions, drawing primarily on research employing case studies, including comparative case study analysis. National-level longitudinal surveys in Europe and North America now commonly generate excellent data on employment and life-course transitions, including the transition to retirement (for example Clarke et al. 2011, 2012), but they only rudimentarily place the respondent in a textured work environment. By design, they cannot address the complex dynamics of public and corporate policies and the work environment, as these shape careers.

There are three general types of inquiry in the general area of work and retirement to address the issues I have just listed. The first type of inquiry would continue to focus on individuals and their life courses as affected by public and corporate policies, as well as by structural changes in the economy and the sphere of production. The second type of inquiry would continue to focus on the dynamics of social structure and public policy in relation to what might be called the opportunity structure to work and to experience an economically viable transition to retirement. The third type of inquiry would be to simultaneously address the micro and macro levels.

I addressed two promising areas for future research. The first is globalization and its impacts on the structuring of work, which act perversely to render more precarious the very possibility of an individual to achieve a structured career. Globalization can be seen as a step beyond the partial breakdown of the Fordist work regime. The second promising area for future research is continuing examination of the impact of the public policy shift towards the risk society, evident in government changes throughout the OECD countries to the age of eligibility for the receipt of the state pension. State policies can provide safety nets against poverty in retirement and also enable people to recover from chance events, such as natural disasters and corporate bankruptcies. In the move to the risk society, these safety nets have become much weaker. While people generally engage in little long-range planning, chance events can seriously disrupt any plans that have been made, often leading to job loss and other changes to working careers and plans for retirement. This is evidenced by the effect of the recent recession on labour force participation in Europe and in North America, as well as in the less developed nations in the new global economy.

Theorizing the life course has come a long way from the early emphasis on the tripartite life course, which postulated three life-course stages with crisp transitions from one to the other. The tripartite life-course concept is still generally valid (Brückner and Mayer 2005), but the complexity of life-course transitions in work and from work to retirement are now well recognized. Understanding this complexity, its causes and consequences, constitutes the challenge for the new generation of life-course researchers. Empirically, continuing work will benefit from studies employing both quantitative and qualitative data about individuals, workplaces, and both corporate and public policies and their linkages.

Notes

1. I cannot address these two topics here, but see Marshall and Clarke (2010); Marshall (2011); Marshall and Bengtson (2011).
2. For a brief account of changes in the American system, see Kingson and Morrissey (2012).
3. It is easy to forget that in the recent past, while not many people lived beyond age 60, many of those who did so worked, as there was no standardized age for retirement.
4. Ageism is more prevalent in hiring than in human relations in the workplace itself. It is easier to discriminate against people you do not know (see for example Lahey 2008).
5. The value of case studies at the company level is in being able to describe the immediate work context of ageing and retiring. Large-scale longitudinal surveys of individuals, which I have also used (Clarke et al. 2011, 2012), decontextualize social action as people navigate their careers, but allow complex comparative and multivariate analyses and, increasingly, cohort analyses. IAW case study reports and a summarizing comparative report are available from the author. The methodology for comparative analysis is described in Marshall (1999).
6. All employees were unionized because we recruited companies through the International Ladies' Garment Workers Union (ILGWU, now called UNITE).
7. For example, employment in the Canadian garment industry declined by 13 per cent over the period 1986 to 1990. Union membership declined by 25 per cent over this period (Marshall and Marshall 1999: 57).
8. For other examples from WANE, see Platman (2009); Haviland et al. (2010); Marshall (2010); McMullin and Marshall (2010); Charness (2013); Taylor et al. (2013). For British research on work and the life course at the somewhat neglected corporate level in other sectors, see Vickerstaff et al. (2003).

References

Beck, U. (1992), *Risk society: Towards a new modernity*, London: Sage.
Beck, U., Giddens, A. and Lash, S. (1994), *Reflexive modernization. Politics, tradition and aesthetics in the modern social order*, Stanford: Stanford University Press.
Blau, D. M. and Goodstein, R. M. (2010), 'Can social security explain trends in labor force participation of older men in the United States?', *Journal of Human Resources*, 45 (2), 328–63.
Blossfeld, H.-P., Buchholz, S. and Kurz, K. (eds) (2011), *Aging populations, globalization and the labor market. Comparing late working life and retirement in modern societies*, Cheltenham: Edward Elgar.

Brückner, H., and Mayer, K. U. (2005), 'De-standardization of the life course: What does it mean? And if it means anything, whether it actually took place?', *Advances in Life Course Research*, 9, 27–53.

Butrica, B. A., Smith, K. E. and Toder, E. J. (2010), 'What the 2008 stock market crash means for retirement security', *Journal of Aging & Social Policy*, 22 (4), 339–59.

Cain, L. D. (1964), 'Life course and social structure', in: R. E. L. Faris (ed.), *Handbook of modern sociology*, Chicago: Rand McNalley, 272–309.

Cahill, K. E., Giandrea, M. D. and Quinn, J. F. (2012), 'Older workers and short-term jobs: Patterns and determinants', *Monthly Labor Review*, May 2012, 9–32.

Cappelli, Peter (1999), *The new deal at work: Managing the market-driven workforce*, Boston: Harvard Business School Press.

Charness, N. (2013), 'Job security in an insecure world: Adaptations of older workers in the IT industry', in: P. Taylor (ed.), *Older workers in an ageing society. Critical topics in research and policy*, Cheltenham/Northampton: Edward Elgar, 109–14.

Clarke, P., Marshall, V. W. and Weir, D. (2012), 'Unexpected retirement from full time work after age 62: Consequences for life satisfaction in older Americans', *European Journal of Ageing*, 9 (3), 207–19.

Clarke, P., Marshall, V. W., House, J. and Lantz, P. (2011), 'The social structuring of mental health over the adult life course: Advancing theory in the sociology of aging', *Social Forces*, 89 (4), 1287–314.

Cooke, M. (2006), 'Policy changes and the labour force participation of older workers: Evidence from six countries', *Canadian Journal on Aging*, 25 (4), 387–400.

DiPrete, T. A. (2002), 'Life course risks, mobility regimes, and mobility consequences: A comparison of Sweden, Germany, and the United States', *American Journal of Sociology*, 108 (2), 267–309.

Guillemard, A.-M. (1986), *Le déclin du social*, Paris: Presses Universitaires de France.

Guillemard, A.-M. (2000), *Aging and the welfare-state Crisis*, Newark/London: University of Delaware Press/Associated University Presses.

Han, S.-K. and Moen, P. (1999), 'Clocking out: Temporal patterning of retirement', *American Journal of Sociology*, 105 (1), 191–236.

Hardy, M. A., Hazelrigg, L. and Quadagno, J. (1996), *Ending a career in the auto industry: 'Thirty and Out'*, New York/London: Plenum Press.

Haviland, S. B., Morgan, J. C. and Marshall, V. W. (2010), 'New careers in the new economy: Redefining career development in a post-internal labor market industry', in: J. A. McMullin and V. W. Marshall (eds.), *Aging and working in the new economy: Changing career structures in small IT firms*, Cheltenham/Northampton: Edward Elgar, 39–62.

He, Y., Colantonio, A. and Marshall, V. W. (2003), 'Later life career disruption and self-rated health: An analysis of the general social survey', *Canadian Journal on Aging*, 22 (1), 45–58.

He, Y., Colantonio, A. and Marshall, V. W. (2006), 'The relationships among career instability and health condition in older workers: A longitudinal data analysis of the Survey of Labour and Income Dynamics', in: L. Stone (ed.), *New Frontiers of Research on Retirement*, Ottawa: Statistics Canada, 321–42.

Heinz, W. R. (2001), 'Work and the life course: A cosmopolitan-local perspective', in: V. W. Marshall, W. R. Heinz, H. Krüger and A. Verma (eds.), *Restructuring work and the life course*, Toronto: University of Toronto Press, 3–22.

Heinz, W. R. (2009), 'Structure and agency in transition research', *Journal of Education and Work*, 22 (5), 391–404.

Heinz, W. R. and Marshall, V. W. (eds.) (2003), *Social dynamics of the life course: Transitions, institutions, and interrelations*, New York: Aldine de Gruyter.

Hofäcker, D. (2010), *Older workers in a globalizing world. An international comparison of retirement and late-career patterns in Western industrialized countries*, Cheltenham/Northampton: Edward Elgar.

Hogan, D. P. (1978), 'The variable order of events in the life course', *American Sociological Review*, 43 (4), 573–86.

Hogan, D. P. (1980), 'The transition to adulthood as a career contingency', *American Sociological Review*, 45 (2), 261–76.

IMF (International Monetary Fund) (2002), *Globalization: Threat or opportunity?, Issue Brief, 12 April 2000, corrected January 2002*, Washington, D C: IMF, https://www.imf.org/external/np/exr/ib/2000/041200to.htm, date accessed 19 July 2014.

Johnson, R. W. and Park, J. S. (2013), *Labor force statistics on older Americans 2012*, Retirement Security Data Brief No. 8 (January), Washington, DC: Urban Institute Program on Retirement Policy.

Kingson, E. and Morrissey, M. (2012), *Can workers offset social security cuts by working longer?, EPI Working Paper, Briefing Paper No. 343*, Washington, DC: Economic Policy Institute.

Kohli, M. (1986), 'The world we forgot: A historical review of the life course', in: V. W. Marshall (ed.), *Later life: The social psychology of aging*, Beverly Hills: Sage, 271–303.

Kohli, M. (2007), 'The institutionalization of the life course: Looking back to look ahead', *Research in Human Development*, 4 (3–4), 253–71.

Kohli, M. and Rein, M. (1991), 'The changing balance of work and retirement', in: M. Kohli, M. Rein, A.-M. Guillemard and H. van Gunsteren (eds.), *Time for retirement: Comparative studies of early exit from the labor force*, Cambridge: Cambridge University Press, 1–35.

Kohli, M., Rein, M., Guillemard A.-M. and van Gunsteren, H. (eds.) (1991), *Time for retirement: Comparative studies of early exit from the labor force*, Cambridge: Cambridge University Press.

Lahey, J. N. (2008), 'Age, women and hiring: An experimental study', *Journal of Human Resources*, 43 (1), 30–56.

Maestas, N. and Zissimopoulos, J. (2010), 'How longer work lives ease the crunch of population aging', *Journal of Economic Perspectives*, 24 (1), 139–60.

Marshall, V. W. (1999), 'Reasoning with case studies: Issues of an aging workforce', *Journal of Aging Studies*, 13 (4), 1–13.

Marshall, V. W. (2010), 'A life course perspective on information technology work', *Journal of Applied Gerontology*, 30 (2), 185–97.

Marshall, V. W. (2011), *Risk, vulnerability and the life course, Lives Working Paper Series 2011 (1)*, Lausanne: Swiss National Centre of Competence in Research and Swiss National Science Foundation.

Marshall, V. W. and Bengtson, V. L. (2011), 'Theoretical perspectives on the sociology of aging', in: R. A. Settersten, Jr. and J. L. Angel (eds.), *Handbook of sociology of aging*, New York: Springer, 17–33.

Marshall, V. W. and Clarke, P. J. (2010), 'Agency and social structure in aging and life course research', in: D. Dannefer and C. Phillipson (eds.), *International handbook of social gerontology*, London: Sage, 294–305.

Marshall, V. W., Clarke, P. J. and Ballantyne, P. J. (2001), 'Instability in the retirement transition: Effects on health and well-being in a Canadian study', *Research on Aging*, 23 (4), 379–409.

Marshall, V. W. and Marshall, J. G. (1999), 'Age and changes at work: Causes and contrasts', *Ageing International*, 25 (2), 46–68.

Marshall, V. W. and Marshall, J. G. (2003), 'Ageing and work in Canada: Firm policies', *The Geneva Papers on Risk and Insurance*, 28 (4), 625–39.

Mayer, K. U. (1986), 'Structural constraints on the life course', *Human Development*, 29 (3), 163–70.

McMullin, J. A. and Marshall, V. W. (1999), 'Structure and agency in the retirement process: A case study of Montreal garment workers', in: C. D. Ryff and V. W. Marshall (eds.), *The self and society in aging processes*, New York: Springer, 305–38.

McMullin, J. A. and Marshall, V. W. (2001), 'Ageism, age relations, and garment industry work in Montreal', *The Gerontologist*, 41 (1), 111–22.

McMullin, J. A. and Marshall, V. W. (eds.) (2010), *Aging and working in the new economy: Changing career structures in small IT firms*, Cheltenham/Northampton: Edward Elgar.

Mermin, G. B. T., Johnson, R. W. and Murphy, D. P. (2007), 'Why do boomers plan to work longer?', *Journal of Gerontology: Social Sciences*, 62B (5), 286–294.

Myles, J. (1989), *Old age and the welfare state*, Lawrence: University of Kansas Press.

Myles, J. and Street, D. (1995), 'Should the economic life course be redesigned? Old age security in a time of transition', *Canadian Journal on Aging*, 14 (2), 335–69.

O'Rand, A. M. (2011), '2010 presidential address: The devolution of risk and the changing life course in the United States', *Social Forces*, 90 (1), 1–16.

Platman, K. (2009), 'Extensions to working lives in the information economy', in: A. Chiva and J. Manthorpe (eds.), *Older workers in Europe*, Maidenhead: McGraw Hill Open University Press, 53–68.

Quadagno, J., MacPherson, D., Keene, J. D. and Parham, L. (2001), 'Downsizing and the life-course consequences of age and gender in employment and income security', in: V. W. Marshall, W. R. Heinz, H. Krüger and A. Verma (eds.), *Restructuring work and the life course*, Toronto: University of Toronto Press, 303–18.

Quadagno, J., Hardy, M. and Hazelrigg, L. (2003), 'Labour market transitions and the erosion of the Fordist lifecycle; discarding older workers in the automobile manufacturing and banking industries in the United States', *The Geneva Papers on Risk and Insurance*, 28 (4), 640–51.

Raymo, J. M., Warren, J. R., Sweeney, M. M., Hauser, R. M. and Ho, J.-H. (2010), 'Late-life employment preferences and outcomes: The role of midlife work experiences', *Research on Aging*, 32 (4), 419–66.

Rindfuss, R. R., Swicegood, C. G. and Rosenfeld, R. A. (1987), 'Disorder in the life course: How common and does it matter?', *American Sociological Review*, 52 (6), 785–801.

Rix, S. E. (2011), *The unemployment situation, March 2011: Little change in unemployment rate for older persons, but rise in unemployment duration*, Fact Sheet 220 (April), AARP Policy Institute.

Schmähl, W. (1989), *Redefining the process of retirement: An international perspective*, New York: Springer.

Settersten, R. A., Jr., Furstenberg, F. F. and Rumbaut, R. G. (eds.) (2005), *On the frontier of adulthood: Theory, research and public policy*, Chicago: The University of Chicago Press.

Settersten, R. A., Jr. and Gannon, L. (2005), 'Structure, agency, and the space between: On the challenges and contradictions of a blended view of the life course', *Advances in Life Course Research*, 10, 35–55.

Shanahan, M. J. and Porfelli, E. J. (2006), 'Chance events in the life course', *Advances in Life Course Research*, 11, 97–119.

Singh, G. and Verma, A. (2001), 'Is there life after career employment? Labour-market experience of early "retirees" ', in: V. W. Marshall, W. R. Heinz, H. Krüger and A. Verma (eds.), *Restructuring work and the life course*, Toronto: University of Toronto Press, 288–302.

Sloan Center on Aging & Work (2013), *Fact of the week: Half of workers aged 65+ work full time*, http://www.bc.edu/research/agingandwork/archive_facts/2013/13-04-01.html, date accessed 6 November 2014.

Standing, G. (2008), 'Economic insecurity and global casualization: Threat or promise?' *Social Indicators Research*, 88 (1), 15–30.

Taylor, P., McLoughlin, C., Brooke, E., Biase, T. D. and Steinberg, M. (2013), 'Managing older workers during a period of tight labour supply', *Ageing and Society*, 33 (1), 16–43.

United States Bureau of Labor Statistics (2013a), 'The last private industry pension plans', *The Economics Daily*, Washington, DC: US Department of Labor, http://www.bls.gov/opub/ted/2013/ted_20130103.htm, date accessed 30 December 2013.

United States Bureau of Labor Statistics (2013b), *Labor Force Statistics from the Current Population Survey (Table Tool)*, Washington, DC: US Department of Labor, http://data.bls.gov/pdq/querytool.jsp?survey=ln, date accessed 15 August 2014.

Vickerstaff, S., Cox, J. and Keen, L. (2003), 'Employers and the management of retirement', *Social Policy & Administration*, 37 (3), 271–87.

Washington Post (2013), *Fiscal troubles ahead for most future retirees*, by M. A. Fletcher, 16 February 2013, http://www.washingtonpost.com/business/economy/fiscal-trouble-ahead-for-most-future-retirees/2013/02/16/ae8c7350-5905-11e2-88d0-c4cf65c3ad15_story.html, date accessed 18 July 2014.

Wiatrowski, W. J. (2012), 'The last private industry pension plans: A visual essay', *Monthly Labor Review*, December 2012, 3–18.

Wilensky, H. L. (1960), 'Work, careers, and social integration', *International Social Science Journal*, 12 (4), 543–60.

Zissimopoulos, J. M. and Karoly, L. A. (2007), 'Transitions to self-employment at older ages: The role of wealth, health, health insurance and other factors', *Labour Economics*, 14 (2), 269–95.

10
Companies and Older Workers: Obstacles and Drivers of Labour Market Participation in Recruitment and at the Workplace

Jutta Schmitz

10.1 Introduction

For a long time, the labour market participation of older workers in many European countries was low. However, during the past decade it has increased substantially in most countries (Dietz and Walwei 2011: 363). This trend reversal was, amongst others, brought about by the discussion of demographic scenarios. A number of changes in national labour market and pension policies have resulted from this, stimulating changes in individual retirement behaviour and leading to later retirement. Much empirical research has been done on all these factors to show that (early) retirement regulations can discourage older workers from remaining in the labour market (Immergut et al. 2007; Saint-Paul 2009; Hamblin 2010), or that personal choices are motivated by individual factors and social contexts (Schils 2008; Engelhardt 2012: 550). Recent studies also broach the issue of macroeconomic conditions (Szinovacz et al. 2014; van Rijn et al. 2014) or carve out the influence of cultural settings (Jansen 2013).

Within this discourse, companies have for a long time remained a 'black box' (Brussig 2007: 200). Newer labour market research refers to a major gap between the public discussion of demographic problems, on the one hand, and the lack of or only rudimentary operational action on the level of organizations, on the other (Tullius et al. 2012: 113). Even if European guidelines strongly emphasize the need to raise the labour market participation of older workers, and nearly every government has set up employment or activation programmes for that target group, empirical surveys show that the majority of European employers still 'does not expect the ageing of their staff to affect the productivity level within their organization' (Conen et al. 2012: 642). In this context, it can be assumed that the access to employing organizations is one of the main barriers to a better labour market integration of older people.

To substantiate this thesis, the influence and activities of companies with respect to older workers will be discussed in the following sections.[1] After taking stock of the labour market situation of older workers in Europe, the following section shows the influence of companies as gatekeepers, and specifically how companies control the access to as well as retention in employment. Then it is shown what companies are (not) doing to integrate older workers and what they could do. The focus lies on the age group from around 55 years up until to retirement. Nonetheless, working up until pension age also increases the probability of working beyond pension age. The article finishes with a conclusion. Here, the question of how the labour market participation of older workers can lead to a higher number of working retirees is also touched upon.

10.2 Labour market participation of older workers in Europe

To meet the challenge of an ageing population, the European Council decided at the beginning of this millennium to set the target value for raising the average European employment rate of older people aged between 55 and 64 years to 50 per cent by 2010. Given an employment rate lower than 38 per cent at that time, this seemed to be an ambitious aim. In actual fact, with 46.3 per cent the intended goal was nearly achieved in 2010.

Between 2003 and 2013, labour market participation[2] of older workers between 55 and 64 years increased in nearly every European country, even though the starting position and the degree of increase differ significantly (see Figure 10.1; Eurostat 2015a). Germany achieved the largest growth in employment rates of older people in this period. While the labour market participation of older workers in Germany was at the European average in 2003, it was surpassed only by Sweden ten years later, with Sweden's initial value being already very high in 2003. This development can be attributed to Germany's comparatively good performance during the world economic crisis, which had a dampening effect especially on the male employment rate in most other countries. However, Germany's positive development is also due

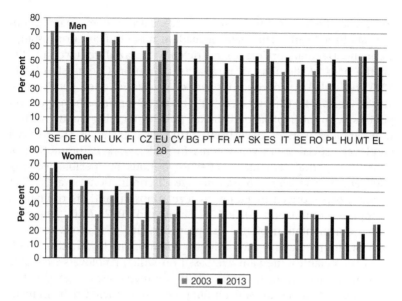

Figure 10.1 Employment rates of people aged 55 to 64 in selected EU countries, 2003 and 2013 (by sex)
Source: Eurostat (2015a) based on EU Labour Force Survey, own compilation.

to cohort effects and specific shifts in the age structure within this age group.

On average, the employment rate of older women in the 28 EU countries has risen from 30.7 per cent in 2003 to 43.3 per cent in 2013, and – at a considerably higher level – the rate of men increased from 49.7 to 57.5 per cent. Initial gender differences in employment were particularly high in Eastern European countries like Czech Republic, Poland and Slovakia, but also in Southern countries like Malta, Cyprus and Greece. Nevertheless, the gender gap has decreased during these ten years, mainly as a result of the general increase in female labour market participation across cohorts and the increase of the general pension age for women, which was below the male pension age in many countries. In 2003, the female employment rate among older Europeans was only 62 per cent of the male one, while ten years later this figure had grown to 75 per cent.

The drivers behind the change in old-age employment rates are complex and multidimensional. Institutional as well as individual factors influence the length of employment and employment exit. The raising of the general state pension age in many European countries is probably the most obvious institutional influence. Furthermore, many countries abolished early retirement or made it more difficult (for instance Germany, Denmark or Poland), restricted withdrawal from working life on the basis of disability or incapacity (European Commission 2010: 121) and scaled up contributory periods needed for a full pension (see also Anderson in Chapter 8 of this book). Over and above pension policies, a number of labour market policies focusing on the activation of older workers have been introduced. Examples are reduced employer's social security payments for older workers, promotion of phased or partial retirement schemes, options for flexible working and general campaigns to prevent age discrimination towards older workers, partly driven by European regulations on anti-discrimination (European Commission 2010: 121; Eurofound 2012: 7–10). The concrete implementation of the European equality regulations to avoid age discrimination varies according to country. The UK, in 2011, abolished the default retirement age based on which employers were able to dismiss older workers. Although in Germany such a general law never existed, there are still binding age limits included in individual contracts or collective agreements (Mahlmann 2011; O'Dempsey and Beale 2011). Even though these contractual agreements are very controversial, they do not necessarily violate the regulations of the anti-discrimination law if there is a valid reason for having an age limit.

On the individual level, the ability to work is a necessary precondition for a prolonged working life. A recent meta-analysis demonstrates the significance of health, identifying poor health as a major risk factor for early exit from paid employment (van Rijn et al. 2014). Furthermore, personal qualifications, preferences and motivational aspects (Peter and Hasselhorn 2013) or the financial situation of the household and family circumstances (Schils 2008) also have a significant impact on the decision when to retire. Persons with low educational qualifications are often confronted with low incomes and bad working conditions, for example work-related health risks (Wahrendorf et al. 2013), so for specific groups, the cumulative effect of several barriers often leads to an early transition into retirement. On the European average, only 36.4 per cent of the people aged 55 to 64 with a lower educational background had a job in 2013. In contrast, approximately one half of those with a medium educational qualification were employed, and the share of those working with a high educational qualification was even markedly higher (68.3 per cent) (Eurostat 2015b).

10.3 The role of organizations, gatekeepers and age-related perceptions

Besides institutional barriers and individual factors as well as historical events or cultural settings, individual employment is also affected by organizations. Within companies, persons with the function of gatekeepers navigate individual transitions, memberships and roles by implementing the organization's staffing policy, including older employees or job applicants.

Gatekeepers usually come into play if scarce goods (in this case jobs and positions) need to be distributed. In this perspective, the possibilities of employment are linked to the specific labour demand of private companies, administrations, non-commercial services or other employers (Goedicke 2006: 503). Gatekeepers operate as intermediates between individual and organization with an exclusive power and the duty to select among the generally entitled or available candidates (Struck 2001). Their selection activities are not necessarily limited to the recruitment of employees. In companies, there are at least two important groups of gatekeepers. The first group consists of personnel managers who mainly deal with the recruitment or dismissal of employees. The second group is composed of line managers (of production or other departments), who work with (older) employees on a daily basis and oversee working arrangements of or personnel measures for them (Buss and Kuhlmann 2013: 352–4). Even if the two groups overlap, this differentiation is useful to look at company policies for older workers. The first group plans (external) strategies concerning the access to the company, while the latter deals with (internal) strategies concerning the handling of older workers. On both levels, the gatekeepers' decisions are mainly affected by three factors: the organizational context, the attributes of employees and the disposition of managers (Karpinska et al. 2012: 3).

The *context of the organization* comprises a range of external conditions as well as internal parameters. On the one hand, general economic factors such as economic booms, recessions or labour force shortages play an important role. In addition, the regulatory framework, institutionalized practices or demographic patterns have an external impact on managerial decisions (Struck 2001: 16). On the other hand, internal factors such as staffing goals as well as economic, social and demographic company structures resulting from prior decisions influence on the gatekeepers' attitudes towards older workers.

Gatekeepers also rate the individual *attributes of employees* to decide over retention, retirement or recruitment, as these attributes are connected to their (perceived) contribution to business objectives. Characteristics such as qualifications, health status, occupational flexibility, motivation or work-related behaviour are taken into consideration (Karpinska et al. 2012: 4–6). Overall, the perceived productivity of older workers is crucial. In this context, the assumption of a wage-productivity gap applying to older workers is widespread, implying that their wages often surpass their productivity. The employment of older workers consequently appears to entail probable financial losses for companies (Van Dalen et al. 2010: 1018). In fact, the thesis of a general age-related loss of productivity has been disproved repeatedly (Börsch-Supan et al. 2006; Börsch-Supan and Weiss 2013). Moreover, empirical evidence shows that there is a trade-off between age-related constraints and experience: even if physical fitness as well as some cognitive, motor and sensory functions decline with increasing age, this process usually also includes a gain in accuracy and work-related know-how. Depending on profession and workload within the life course, this trade-off may differ of course (Mümken 2014: 3).

Nonetheless, Conen et al. (2012) show that internal mindsets of employers with regard to age and ageing are not free from stereotyping. Their analysis of data from comparative surveys carried out among employers in Denmark, France, Germany, Italy, the Netherlands, Poland and Sweden comes to the result that a 'majority of European employers [...] does not expect the ageing of their staff to affect the productivity level within their organization', in general. However, many European employers still expect growing labour costs as a consequence of an ageing staff. By studying the ideas about labour costs and productivity together, the study reveals that about half of the employers associate ageing personnel with a net cost increase (Conen et al. 2012: 642). These results show that not only facts but also beliefs of gatekeepers matter. That is why the *attitudes of gatekeepers* are of great importance when it comes to the retention or recruitment of older workers. Actually, like gender and race, age is a major basis for stereotypes and discrimination (Desmette and Gaillard 2008: 170). Stereotypical perceptions of older workers often include negative ascriptions such as poor health, an inflexible attitude, resistance to change and low trainability. Other evidence shows that age-based ascriptions can also be positive, such as that older workers are more reliable, loyal and better communicators than the younger. Especially in the case of working retirees, positive age-based ascriptions of managers might facilitate the employment of people in pension age, at least for certain jobs and in certain situations.

10.4 Recruitment of older workers

As shown above, the labour market participation of older workers is influenced by companies' gatekeepers in two ways. Companies may keep ageing employees, on the one hand, or hire older applicants, on the other. Only the combination of both strategies is able to meet the challenges of the ageing workforce successfully.

At the moment, re-employment chances of unemployed older workers are very poor in most European countries. Brussig, for example, shows for the German case that the rise of the employment rate of older workers during the past decade was mainly due to a longer retention of workers by employers, but not due to improved re-employment chances. The share of re-employed older people in all newly hired people in 2009 stood at only 13 per cent, whilst the overall proportion of older workers was twice as high (26 per cent) (Brussig 2011:1). Kidd et al. (2012) observe the same situation for the UK, where the improvement in the employment rate of older people also mainly results from a longer duration of employment among this group, and not from their increased re-employment after job loss (Kidd et al. 2012: 530). Overall, the number of people who start a new job decreases constantly with increasing age.

The reasons for the poor re-employment chances of older workers are often seen in institutional parameters such as seniority wages or high reservation wages. Seniority wages are usually based on the work experience within a certain company and tend to serve as an incentive for (younger) workers through their progressing career. A related factor is that 'productivity of younger employees is higher than their wages, while for older workers the opposite is true' (Bellmann and Brussig 2007: 4). This leads to economic inducements to choose younger applicants over older: if employers want to treat their employees equally, they have to pay similar wages to workers of the same age, even if the wages exceed the (assumed) productivity of newly hired older employees. If there are strong unions, works councils or collective bargaining, companies may not even have a choice and need to pay equal wages for employees with similar characteristics or similar tasks (Zwick 2008: 5).

Nonetheless, seniority wages are part of the country-specific settings of labour markets and their importance differs widely between European countries. The same applies to reservation wages. They indicate the lowest wage a worker is willing to accept within a certain institutional setting. Reservation wages are determined by individual factors such as individual job preferences and previous earnings, as well as institutional parameters such as the generosity of unemployment benefits or job search obligations due to benefit receipt. In the case of older workers, individual reservation wages may be generally higher than those of younger jobseekers, because commonly they have received higher wages before becoming unemployed. Additionally, the exact institutional framing also varies with age, as the duration of unemployment benefits, the legal job search obligations or the availability of pensions depend on age (Dietz and Walwei 2011: 371). This way reservation wages of older workers might be barriers to entering the labour market, even though most European governments have reformed their labour market policies during the past two decades to promote a stricter activation of older workers (Nivorozhkin et al. 2013: 518).

The micro-level analysis mentioned above emphasizes both the importance of company strategies and jobseekers' behaviour within the recruitment process of older people. On the basis of a representative survey carried out among companies in Germany (IAB establishment panel), Bellmann and Brussig (2007) show the correlation between firm characteristics and their recruitment behaviour towards persons aged 50 and over. The results indicate that the majority of companies (75 per cent) that hired new personnel in 2004 did not receive any applications from older jobseekers. Of the remaining firms that did receive applications from older jobseekers, only half recruited them. Surprisingly, employers who had rejected older applicants stated that the popular assumptions 'too high wages', 'limited ability to make redundant' and 'no long-term perspective' were not their reasons for rejecting them. Instead, the most frequently mentioned age-related ascription was 'low flexibility and versatility' (30 per cent) (Bellmann and Brussig 2007: 8). The results also show that the companies' attitudes towards older people are not the only cause of their poor re-employment chances. The evidence points to a major problem of information and coordination by showing a remarkable gap between firms which received applications and those who were willing to recruit older job candidates. Looking at company size reveals, for example, that small and medium-sized firms rarely received applications from older people. At the same time, the majority of older persons are recruited by companies with up to 250 employees. In order to avoid such a mismatch, the authors recommend that older applicants should 'broaden their search spectrum [...] to smaller and lesser known companies' (Bellmann and Brussig 2007: 16).

10.5 Measures for older workers within companies

Retention strategies are another aspect of how companies affect the labour market participation of older workers. A meaningful indicator of companies' willingness to keep their ageing staff is the number of personnel measures for older employees, such as training, health protection, age-specific workplace design or career planning which also covers the transition to retirement. Empirical analyses for Germany illustrate that such measures are rarely used. In 2011, only about 18 per cent of all companies employing persons aged 50 or older realized specific measures for this group. However, there are huge differences between companies of different sizes and branches. While bigger firms and those who belong to the sectors of public administration/social insurance or mining/energy/water use measures for older workers far more often than other sectors, the opposite is true for companies of small or medium size or those from agriculture, construction and trade. In the last ten years, both the general prevalence and the distribution of measures among companies of different sizes and sectors remained nearly the same (Leber et al. 2013: 3–4).

These results indicate that the public discussion of demographic change has not yet led to extensive changes on the company level, at least not in Germany. In addition, the goals of measures for older workers often differ from those for younger workers and do not necessarily include the *integration* of older workers. In the case of the German example, the most widespread measure used is partial retirement, an instrument that does not serve to retain older workers, but to retire them earlier (Tullius et al. 2012: 119), and that only became less important recently due to the end of public subsidies. A number of reasons exist for the limited interest of German, but also European companies in measures for helping to integrate older workers into the labour market. On the one hand, because of their size and the missing knowledge about adequate tools, companies might not *be able* to implement measures for older workers. On the other hand, firms that face the pressure of cost and rationalization might not *want* to invest in the (usually long-term oriented) measures for older workers. In addition, remaining stereotypes of older employees still may lead to the higher popularity of severance packages compared to retention measures *unconsciously* (Brussig 2007; Latniak et al. 2010; Tullius et al. 2012).

At the same time, there are a number of compelling reasons to overcome these organizational barriers. The experience of individual companies shows a wide range of (economic) benefits from retaining and training older workers, such as keeping or evolving extra experience and skills, preventing increasing costs (of hiring and training new employees or of sick pay) or gaining competitive advantage in times of 'regional labour bottlenecks for qualified personnel' (Naegele and Walker 2006: 5). Within the academic discourse, company case studies have shown examples of good practice for at least a decade now. In general, these good practices consist of 'combating age barriers, either directly or indirectly, and providing an environment in which each individual is able to achieve his or her potential without being disadvantaged by their age' (Walker and Taylor 1998: 3). The research crucially shows how the employment of older workers can be facilitated and how this benefits the employers.

The exchange of knowledge and experience between countries in this and other policy areas is supported by the European Union's open method of coordination as a political strategy for mutual learning. In this context, the European Foundation for the Improvement of Living and Working Conditions (Eurofound), for example, released a database of more than 200 good practice cases in age management (Eurofound 2009). Similarly, the Organisation for Economic Co-operation and Development (OECD) regularly publishes country-based reviews (OECD 2014). These sometimes very extensive examinations of cases shed light on the specific organizational context of the successful integration of older workers, on age-related staffing activities and on how arising problems can be overcome. However, the findings cannot easily be generalized to other organizations, and the concentration on a small number of selective good-practice examples makes it difficult to find cases that cover different economic, political, social or sectorial settings. Therefore, the implementation of good practice still is 'rather limited, particularly as far as smaller firms and the public sector are concerned' (Frerichs et al. 2012: 667). Hence newer research uses case studies as a basis for meta-level analysis with the aim to produce a set of more generic types of (human resources) strategies for older workers (Frerichs et al. 2012; Spoket 2012) or examines such policies in a comprehensive way using quantitative methods (van Dalen et al. 2014).

All in all, previous research finds a variety of aspects which are crucial for successful age management. To systemize them the current state of research is summarized in the following sections by outlining four universal dimensions of age management.

10.5.1 Employability and training

The fact that inactivity and unemployment are heavily concentrated amongst the less educated and the less skilled underlines the necessity for (older) workers to continuously adapt to change. Even if there is no common definition of 'employability', this widespread catchword implies continuous skill development and training of workers in order to adapt to technological change, new job requirements and to offset the employee's changing personal resources (like health, functional capacity or attitudes), with the aim of keeping up their work ability. For this aim to be achieved, a preventive and permanent investment in training starting at earlier career stages is of great importance and should be continued in old age.

However, current practices of adult education show that training is still much more common among young adults and those with a good basic education. Averaged across the EU-27 countries, hardly one in four workers aged 55 or older participated in internal training in 2005, compared to one in three among younger employees (Schönfeld and Behringer 2013: 171). Looking at the distribution of further training across age groups, the participation of workers in in-house as well as in off-the-job training decreases significantly with age in every European country except Denmark. This correlation between participation in training and higher age can be explained by basic assumptions of human capital theory: for both employers and employees, investments in labour-market relevant qualifications decrease continuously because the time period shrinks in which the resulting increases in productivity can pay off. To invert this correlation, companies have to 'prolong the remaining time horizon of older workers' (Funk 2004: 21) and generally increase the incentives to invest in human capital. Furthermore, it can be more economical to develop the skills of older workers instead of bearing the costs of premature retirement and hiring new employees.

The motivation to participate in further training is shaped by individual choices, specific institutional conditions or financial support. On the company level, it is also an employee's line manager who has to ensure that older workers are given full and equal access to training programmes and that these programmes are tailored to their specific needs. In practice, managers tend to exclude older workers from training not only due to economic reasons, but also because they often assume that older workers are unlikely to learn (Woolever 2013: 126). This shows that age discrimination plays a role in nearly every decision of companies' gatekeepers. However, other, more general obstacles to further training exist and not only concern older workers. For example, advanced internal training is not provided by all companies: on European average, only 56 per cent of all companies offer further training – of course widely varying according to country, industrial sector and the size of enterprise (Eurostat 2015c).

10.5.2 Health promotion and work environment

Even if it is well known that gainful employment under good conditions can preserve or even benefit workers' health (through keeping their fitness, structuring daily routines or ensuring social contacts), it can also undermine health, and often does (see also Matthews and Nazroo in Chapter 12 of this book). If work is physically or mentally strenuous, unbalanced or poorly organized, it can damage cognitive or functional capacity or health and decrease work ability. In order to keep older employees in the labour market, work should not jeopardize health, and that applies to the *whole* work life (Ilmarinen 2005: 112). With regard to this aim, two main aspects of the work setting matter: the physical demands and the psychosocial environment. Physical job requirements potentially damage health if they, for example, include heavy lifting, repetitive movements or unilateral postures. Health problems also may arise from noise, fumes or defective working appliance. In the psychosocial work environment, they may come from time pressure, stress and lack of opportunities for autonomy and personal development (Fraser et al. 2009: 262). In addition, loss of reputation due to higher age, demoralization in disagreements amongst employees or age-related discrimination can lead to psychological distress. Discriminatory practices may include exclusion from training programmes or promotion as well as age-based stereotypes of line managers, supervisors or colleagues. Thus, age-based decreases in motivation may result from poor leadership or an 'age-unfriendly corporate culture' (Naegele and Spoket 2010: 456).

In order to minimize these risks, good practices in health protection and the arrangement of the work environment consist of both preventive measures and procedures that may compensate potential physical decline. Key practices in this field are for example organizational health reports and working groups on health, regular health checks, training supervisors in health management techniques, campaigns against age discrimination and creating opportunities for the (re-)integration of incapacitated workers through less demanding jobs or ergonomic workplace design (Naegele and Walker 2006: 17; Frerichs et al. 2012: 672).

10.5.3 Flexibility and internal mobility

According to popular assumptions, 'flexibility' is a key to a company's economic success. Concerning the retention of older workers, internal numerical flexibility is crucial. It relates to the length, position or distribution of working time and covers concrete measures such as overtime hours, flexitime wage records or short-time work. These internal practices may correspond to both the companies' (economic) interest as well as the older workers' interest in retaining their work ability, provided they follow the workers' needs and preferences. Flexibility in working time possibly enables older workers to stay in employment as it can reduce work-related physical strain and might imply a stepwise transition to retirement. A Danish hospital, for example, offers their older workers the possibility to opt out of shift work or reduce their working time to part-time, while the same amount of pension contributions is paid as before (Eurofound 2007). As an example of an age-friendly policy in the interest of the employer, the Greek branch of an international energy company can be cited, whose human resources policy is characterized by flexible working conditions and continued training for all employees (Eurofound 2005). Due to the economic crisis, the company was obliged to downsize and employees were made redundant regardless of their age but instead accounting for their knowledge, role and value to the company. This practice actually supported older workers as their dismissal would have meant a greater loss of know-how and company-specific skills. Additionally, the company allowed flexible working practices. Older workers were, for example, allowed to adjust their working times according to their wishes in order to adapt them to any caring responsibilities or to particular health problems. In specific types of jobs (with contact with petrochemicals for example), some workers could also leave the workforce at an earlier age. Another approach in this firm was to allow otherwise retired workers to work periodically on temporary contracts. This constituted a form of flexibility for the company and was of help to financially needy or lonely retirees (Eurofound 2005). It also shows how internal strategies may fuel growth of work beyond pension age.

Beyond variations of working time, further forms of flexibility emphasize that there is not just one job that suits an employee for the whole of his or her working life. Two paths of job-related mobility are discussed in the literature. On the one hand, vertical promotion may not necessarily follow pre-determined pathways but could be linked to the specific expertise of older workers. In this case, professional advancement follows its own seniority structure with corresponding growth of salary, status, competences and so on. Frerichs names an Austrian pilot project as a good example in which special expert jobs for older workers were created in the area of nursing (Frerichs 2013: 190). However, only larger companies may have the possibilities to arrange such age-specific vertical career paths as they usually offer a greater variety of positions than smaller firms. On the other hand, horizontal tracks, which mean job rotation or redeployment on the same hierarchic level, can be a worthwhile form of flexibility for older workers (Naegele and Walker 2006). Here, older workers are mobile in so far as they change into positions that reduce pressure and focus on their specific strengths. Classic examples are changes between manufacturing jobs and the service division (Frerichs 2013: 190). Again, such measures can be in both the company's and the individual worker's interest, as they can help employees to stay more easily in work until pension age. Göbel and Zwick found that such measures make a difference to a company's success, as the relative productivity of older workers is significantly higher in such companies that offer either age-specific positions for older workers or specifically equipped jobs (Göbel and Zwick 2013).

10.5.4 Career management

In order to establish an overall strategy for career management, the mobility concepts for older workers just discussed can be combined with the measures for training and health management. Line managers should actively accompany staff members during their whole working career and evaluate strengths and weaknesses in order to tackle potential barriers to constant employment together with them (Karpinska et al. 2012: 3). In this manner, the cooperative career planning of both employers and employees (of all ages) may help to develop a career perspective and develop work-related long-term goals (Naegele and Walker 2006: 13).

Over and above career development, career management also covers the end of the working life. On that account, supportive institutional settings (for example allowing flexible transitions to retirement) can be accompanied by internal measures. The practice of Stanley Consultants, the winner of the American Association of Retired Persons'[3] 'Best Employers for Workers over 50' award in 2013, is a good example for an organizational programme that facilitates a smooth transition to retirement but leads to work beyond pension age as well (AARP 2014). Within the company, employees approaching pension age can participate in the company's 'open-ended' phased retirement programme to meet their specific needs with regard to work-life balance. After the transition to retirement, the company stays in touch with its retirees, offers a 24/7 nurse hotline, provides a disease management programme, gives monetary incentives for having an annual physical exam and regularly invites retirees to company events. To complete the possibility of a gradual transition to retirement, temporary work assignments, consulting or contract work, telecommuting, as well as part-time work are all employment opportunities offered to retirees (AARP 2014). Internal strategies such as these may also act as pull factors for work beyond pension age. However, access to such measures is very unequally distributed between employees of different companies and only open to 'insiders' who are already employees of a company or similar organization.

10.6 Conclusion: Prospects and limits of organizational age management

As shown above, the labour market participation of older workers is shaped by institutional settings, organizational contexts and individual attributes and behaviours – and their interplay. Whilst the European member states try to increase the employment of older workers through corresponding labour market and pension policies, and older persons themselves seem to make a move towards a longer working life, many obstacles remain in particular on the company level. This conclusion is underpinned by the evidence on companies' decisions that concern the recruitment of older workers, on the one hand, and the limited access to internal measures that aim to retain them, on the other.

To overcome these barriers, measures of employability and training, health promotion and workplace design, flexibility, mobility and career management need to be combined to an overall strategy of organizational age management. However, as national legislation and welfare regulations, the size, sector, workload and corporate cultures of companies differ, the discussed measures cannot be easily drawn together to a 'one-size-fits-all' (Frerichs 2007: 78) approach. Hence, many of the case studies are just 'shining examples' or 'beacons of light' (Walker and Taylor 1999: 68) that cannot simply be copied. In the future, each organization will face the challenge to come up with its own specific combination of measures that corresponds to its situation.

As a side effect, age(ing)-friendly companies and organizations may increase the employment of retirees; additionally, if more people work up to pension age, more people will probably also work beyond this age. Whether these retirees are a well-paid reserve of skilled staff or whether they are exploited as cheap labour (either because they want to work or because they are in financial need) depends not only on the circumstances on the company level but also on institutional factors – and it will differ depending on country and historical situation. Against the background of the unutilized capacities of older workers before regular pension age there is no structural need to promote retirees as an objective of internal staffing policies; such a strategy could also be critically questioned from a perspective defending a strong concept of a work-free retirement for all (see Hagemann and Scherger in Chapter 11 of this book).

Notes

1. The article has benefitted from many helpful comments from Simone Scherger, whom I would like to thank for her invaluable assistance.
2. Persons in employment in the Eurostat data are those who worked at least one hour for pay during the reference week or were temporarily absent from such work. The following figures are based on own calculations with the Eurostat data (Eurostat 2015a).
3. AARP is a nonpartisan, nonprofit association based in the USA that represents older persons. With its annual Best Employers Award it honors organizations that value the skills and talents of experienced workers above 50 years.

References

AARP (2014), *2012 AARP best employer: Stanley Consultants*, Washington, DC: AARP, http://www.aarp.org/work/2013-aarp-best-employers/stanley-consultants-aarp-best-employers, date accessed 7 August 2014.

Bellmann, L. and Brussig, M. (2007), *Recruitment and job applications of older job-seekers from the establishments' perspective*, IZA Discussion Paper No. 2721, Bonn: Institute for the Study of Labor, http://repec.iza.org/dp2721.pdf, date accessed 19 November 2014.

Börsch-Supan, A. and Weiss, M. (2013), *Productivity and age: Evidence from work teams at the assembly line*, MEA Discussion Paper 148–2007, Munich: Munich Center for the Economics of Aging, http://mea.mpisoc.mpg.de/uploads/user_mea_discussionpapers/1057_148-07.pdf, date accessed 19 November 2014.

Börsch-Supan, A., Düzgün, I. and Weiss, M. (2006), 'Sinkende Produktivität alternder Belegschaften? Zum Stand der Forschung', in: J. Prager and A. Schleiter (eds.), *Länger leben, arbeiten und sich engagieren. Chancen werteschaffender Beschäftigung bis ins Alter*, Gütersloh: Verlag Bertelsmann Stiftung, 85–102.

Brussig, M. (2007), 'Betriebliche Personalwirtschaft in einer alternden Erwerbsbevölkerung. Formen, Verbreitung und Ursachen', *Zeitschrift für Management*, 2 (2), 198–223.

Brussig, M. (2011), *Neueinstellungen im Alter: Tragen sie zu verlängerten Erwerbsbiografien bei?*, Altersübergangsreport 03/2011, Düsseldorf/Berlin/Duisburg: Hans-Böckler-Stiftung/Forschungsnetzwerk Alterssicherung/Institut Arbeit und Qualifikation, http://www.iaq.uni-due.de/auem-report/2011/2011-03/auem2011-03.pdf, date accessed 7 August 2014.

Buss, K.-P. and Kuhlmann, M. (2013), 'Akteure und Akteurskonstellationen alter(n)sgerechter Arbeitspolitik', *WSI-Mitteilungen*, 2013 (5), 350–9.

Conen, W. S., Henkens, K. and Schippers, J. (2012), 'Employers' attitudes and actions towards the extension of working lives in Europe', *International Journal of Manpower*, 33 (6), 648–65.

Desmette, D. and Gaillard, M. (2008), 'When a "worker" becomes an "older worker". The effects of age-related social identity on attitudes towards retirement and work', *Career Development International*, 13 (2), 168–85.

Dietz, M. and Walwei, U. (2011), 'Germany – No country for old workers?', *Journal for Labor Market Research*, 44 (4), 363–76.

Engelhardt, E. (2012), 'Late careers in Europe: Effects of individual and institutional factors', *European Sociological Review*, 28 (4), 550–63.

European Commission (2010), *Joint report on social protection and social inclusion 2010*, Luxembourg: Publications Office of the European Union.

Eurofound (European Foundation for the Improvement of Living and Working Conditions) (2005), *International energy company, Greece: Flexible working practices and training*, Dublin: Eurofound, http://eurofound.europa.eu/observatories/eurwork/case-studies/ageing-workforce/international-energy-company-greece-flexible-working-practices-and-training, date accessed 7 August 2014.

Eurofound (2007), *Aalborg Hospital, Denmark: Flexible working practices*, Dublin: Eurofound, http://eurofound.europa.eu/observatories/eurwork/case-studies/ageing-workforce/aalborg-hospital-denmark-flexible-working-practices, date accessed 7 August 2014.

Eurofound (2009), *Ageing workforce – Database*, Dublin: Eurofound, http://eurofound.europa.eu/areas/populationandsociety/ageingworkforce, date accessed 7 August 2014.

Eurofound (2012), *Employment trends and policies for older workers in the recession*, Luxembourg: Publication Office of the European Union, http://www.eurofound.europa.eu/publications/htmlfiles/ef1235.htm, date accessed 7 August 2014.

Eurostat (2015a), *Employment (main characteristics and rates) – annual averages [lfsi_emp_a]*, Luxembourg: Eurostat, http://appsso.eurostat.ec.europa.eu/nui/submitViewTableAction.do, date accessed 4 February 2015.

Eurostat (2015b), *Employment rates by sex, age and highest level of education attained (%) [lfsa_ergaed]*, Luxembourg: Eurostat, http://appsso.eurostat.ec.europa.eu/nui/show.do?dataset=lfsa_ergaed&lang=e, date accessed 4 February 2015.

Eurostat (2015c), *Training enterprises as % of all enterprises [trng_cvts01]*, Luxembourg: Eurostat, http://appsso.eurostat.ec.europa.eu/nui/submitViewTableAction.do, date accessed 4 February 2015.

Fraser, L., McKenna, K., Turpin, M., Allen, S. and Liddle, J. (2009), 'Older workers: An exploration of the benefits, barriers and adaptions for older people in the workforce', *Work*, 33 (3), 261–71.

Funk, L. (2004), *Employment opportunities for older workers: A comparison of selected OECD countries, CESifo DICE Report 2/2004*, Munich: Institute for Economic Research.

Frerichs, F. (2007), 'Arbeitsmarktpolitik für ältere ArbeitnehmerInnen im Wohlfahrtsstaatenvergleich', *WSI-Mitteilungen*, 2007 (2), 78–85.

Frerichs, F. (2013), 'Alternsmanagement im Betrieb – Herausforderungen und Handlungsansätze', in: G. Bäcker and R. Heinze (eds.), *Soziale Gerontologie in gesellschaftlicher Verantwortung*, Wiesbaden: Springer Fachmedien, 185–95.

Frerichs, F., Lindley, R., Aleksandrowicz, P., Baldauf, B., Galloway, S. (2012), 'Active ageing in organisations: a case study approach', *International Journal of Manpower*, 33 (6), 666–85.

Göbel, C. and Zwick, T. (2013), 'Are personnel measures effective in increasing productivity of old workers?', *Labour Economics*, 2013 (22), 80–93.

Goedicke, A. (2006), 'Organisationsmodelle in der Sozialstrukturanalyse: Der Einfluss von Betrieben auf Erwerbsverläufe', *Berliner Journal für Soziologie*, 16 (4), 503–23.

Hamblin, K. (2010), 'Changes to policies for work and retirement in EU15 nations (1995–2005): An exploration of policy packages for the 50-plus cohort', *International Journal of Ageing and Later Life*, 5 (19), 13–43.

Ilmarinen, J. (2005), *Towards a longer worklife. Ageing and the quality of worklife in the European Union*, Helsinki: Finnish Institute of Occupational Health/Ministry of Social Affairs and Health.

Immergut, E. M., Anderson, K. M. and Schulze, I. (eds.) (2007), *The handbook of West European pension politics*, Oxford: University Press.

Jansen, A. (2013), 'Kulturelle Muster des Altersübergangs: Der Einfluss kultureller Werte und Normen auf die Erwerbsbeteiligung älterer Menschen in Europa', *Kölner Zeitschrift für Soziologie und Sozialpsychologie*, 65 (2), 223–51.

Karpinska, K., Henkens, K. and Schipper, J. (2012), *Retention of older workers. Compact of managers' age norms and stereotypes*, Netspar Discussion Paper No. 04/2012-018, Tilburg: Network for Studies on Pensions, Aging and Retirement.

Kidd, M., Metcalfe, R. and Sloane, P. (2012), 'The determinants of hiring older workers in Britain revisited: An analysis using WERS 2004', *Applied Economics*, 44 (4), 527–36.

Latniak, E., Voss-Dahm, D., Elsholz, U., Gottwald, M. and Gerisch, S. (2010), *Umsetzung demografiefester Personalpolitik in der Chemischen Industrie. Inhaltliche und prozessuale Analyse betriebliche Vorgehensweisen*, Düsseldorf: Hans-Böckler-Stiftung.

Leber, U., Stegmaier, J. and Tisch, A. (2013), *Altersspezifische Personalpolitik. Wie Betriebe auf die Alterung ihrer Belegschaften reagieren*, IAB-Kurzbericht 13/2013, Nürnberg: Institut für Arbeitsmarkt- und Berufsforschung der Bundesagentur für Arbeit, http://doku.iab.de/kurzber/2013/kb1313.pdf, date accessed 20 November 2014.

Mahlmann, M. (2011), *Report on measures to combat discrimination. Directives 2000/43/EC and 2000/78/EC. Country report 2010: Germany*, Utrecht/Brussels: European network of legal experts in the non-discrimination field (Human European Consultancy/Migration Policy Group), http://www.non-discrimination.net/content/media/2010-DE-Country%20Report%20LN_FINAL_0.pdf, date accessed 18 March 2014.

Mümken, S. (2014), *Arbeitsbedingungen und Gesundheit älterer Erwerbstätiger, Altersübergangs-Report 2014–03*, Düsseldorf/Berlin/Duisburg-Essen: Hans-Böckler-Stiftung/Forschungsnetzwerk Alterssicherung/Institut Arbeit und Qualifikation, http://www.iaq.uni-due.de/auem-report/2014/2014-03/auem 2014-03.pdf, date accessed 30 January 2015.

Naegele, G. and Spoket, M. (2010), 'Perspektiven einer lebenslauforientierten Ältere-Arbeitnehmer-Politik', in: G. Naegele (ed.), *Soziale Lebenslaufpolitik*, Wiesbaden: VS Verlag für Sozialwissenschaften, 449–73.

Naegele G. and Walker, A. (2006), *A guide to good practice in age management*, Dublin: Eurofound.

Nivorozhkin, A., Romeu Gordo, L. and Schneider, J. (2013), 'Job search requirements for older workers: The effect on reservation wages', *International Journal of Manpower*, 34 (5), 517-35.
O'Dempsey, D. and Beale, A. (2011), *Age and employment*, Brussels: European Commission (Directorate-General for Justice), http://ec.europa.eu/justice/discrimination/files/age_and_employment_en.pdf, date accessed 18 March 2014.
OECD (Organisation for Economic Co-operation and Development) (2014), *Working better with age. OECD review of policies to improve labour market prospects for older workers*, http://www.oecd.org/els/emp/ageingandemploymentpolicies.htm, date accessed 20 November 2014.
Peter, R. and Hasselhorn, H. M. (2013), 'Arbeit, Alter, Gesundheit und Erwerbsteilhabe', *Bundesgesundheitsblatt*, 56 (3), 415-21.
Saint-Paul, G. (2009), *Does the welfare state make older workers unemployable?*, IZA Discussion paper No. 4440, Bonn: Institute for Labor Studies.
Schönfeld, G. and Behringer, F. (2013), *Betriebliche Weiterbildung in Deutschland und im europäischen Vergleich. Ergebnisse der dritten europäischen Erhebung zur betrieblichen Weiterbildung (ECTS3), Wissenschaftliches Diskussionspapier Heft 141*, Bonn: Bundesinstitut für Berufsbildung.
Schils, T. (2008), 'Early retirement in Germany, the Netherlands and the United Kingdom: A longitudinal analysis of individual factors and institutional regimes', *European Sociological Review*, 24 (3), 315-29.
Spoket, M. (2012), 'Positive organisationale Altersbilder – Acht Beispiele einer guten Praxis im Altersmanagement', in: F. Berner, J. Rossow and K.-P. Schmitzer (eds.), *Altersbilder in der Wirtschaft, im Gesundheitswesen und in der pflegerischen Versorgung*, Wiesbaden: VS Verlag für Sozialwissenschaften, 44-82.
Struck, O. (2001), 'Gatekeeping zwischen Individuum, Organisation und Institution. Zur Bedeutung und Analyse von Gatekeeping am Beispiel von Übergängen im Lebenslauf', in: L. Leisering, R. Müller and K. F. Schumann (eds.), *Institutionen und Lebensläufe im Wandel*, Weinheim/München: Juventa Verlag, 29-54.
Szinovacz, M. E., Martin, L. and Davey, A. (2014), 'Recession and expected retirement age: Another look at the evidence', *The Gerontologist*, 54 (2), 245-57.
Tullius, K., Freidank, J., Grabbe, J., Kädtler, J. and Schroeder, W. (2012), 'Perspektiven alter(n)sgerechter Betriebs- und Tarifpolitik', *WSI-Mitteilungen*, 2012 (2), 113-23.
van Dalen, H., Henkens, K. and Schippers, J. (2010), 'How do employers cope with an ageing workforce? Views from employers and employees', *Demographic Research*, 22 (32), 1015-36.
van Dalen, H., Henkens, K. and Wand, M. (2014), *Recharging or retiring the older worker: Human resource strategies of European employers, NEUJOBS Working Paper No. D17.4*, Brussels: Center for European Policy Studies.
van Rijn, R. M., Robroek, S. J. W., Brouwer, S. and Burdorf, A. (2014), 'Influence of poor health on exit from paid employment: a systematic review', *Occupational and environmental medicine*, 71 (4), 295-301.
Wahrendorf, M., Dragano, N. and Siegrist, J. (2013), 'Social position, work stress, and retirement intentions: A study with older employees from 11 European countries', *European Sociological Review*, 29 (4), 792-802.

Walker, A. and Taylor, P. (1998), *Combating age barriers in employment: A European portfolio of good practice*, Dublin: Eurofound.

Walker, A. and Taylor, P. (1999), 'Good practice in the employment of older workers in Europe', *Ageing International*, 25 (3), 62–79.

Woolever, J. (2013), 'Human resource departments and older adults in the workplace', in: P. Brownell and J. Kelly (eds.), *Ageism and mistreatment of older workers. Current reality, future solutions*, Heidelberg/New York/London: Springer Dordrecht, 111–35.

Zwick, T. (2008), *The employment consequences of seniority wages, ZEW Discussion paper No. 08–039*, Mannheim: Center for European Economic Research.

11
Concepts of Retirement: Comparing Unions, Employers and Age-Related Non-Profit Organizations in Germany and the UK

Steffen Hagemann and Simone Scherger

11.1 Introduction

Retirement and its related welfare arrangements are a historically unique and defining feature of modern and contemporary welfare states. In this chapter,[1] we aim to answer the question of which role *moral* reasoning related to the life phase of retirement plays in the legitimation (or de-legitimation) of current pension reforms, and what form it takes. The interests of the actors involved are on their own not sufficient to explain these reforms. We posit that moral reasoning plays a particularly important role when established and popular welfare institutions like the pension system are reformed and sometimes questioned in their very foundations. More specifically, we explore the meaning that important German and British collective actors attribute to the concept of retirement by examining expert interviews and a selection of documents dealing with policies in the fields of pensions and work in old age, including post-retirement work. We pay particular attention to how the actors embed their arguments into the 'moral economy' of retirement (Kohli 1987), thereby balancing their vested interests and the welfare culture they belong to. How they describe and evaluate post-retirement work, as an exception from 'normal' retirement, can help us to further unpack their often implicit views on what retirement *should* be. Based on this approach we expect to gain valuable insights into the moral ideas incorporated in welfare institutions and negotiated in the political process. These moral ideas are inextricably linked to individual life courses and their experience (see Kohli 1986a and 1986b; Leisering 2004). Furthermore, these ideational dimensions of welfare policies directed at old age are at least one important driver of the related political processes (for example see van Oorschot et al. 2008).

In the following sections, we elaborate on the moral dimension of retirement and its connection to welfare cultures and collective actors (11.2), summarize the relevant institutional background for our country cases (11.3) and our methods and sample (11.4). In the second, empirical part of our chapter, we analyse the concept of retirement held by selected actors: union confederations (11.5), employer organizations (11.6) and very briefly non-profit interest organizations (11.7). Finally, we recapitulate our results and identify perspectives for future research (11.8).

11.2 The moral economy of retirement, welfare cultures and collective actors

Retirement, defined by withdrawing from the labour market and receiving an old-age pension, is part of the modern institutionalized life course (Kohli 1986a). It is a model example of the 'functions', that is, (contingent) answers that the institutionalized life course provides to the structural problems of organizing labour in modern economies (Kohli 1987: 129): succession on the labour market and in companies, stable productivity of the workforce, integration of different life domains and social control in contemporary, rather individualized societies.

The ways in which retirement and pensions are related to questions of integration and social control can be understood more clearly if we consider contemporary market economies as 'work societies' (Kohli 1987: 128): work is not only a means to produce goods and services, but also a central mechanism of social integration and thus a central norm people align their lives with. Hence, the fundamental question arises how a prolonged (and growing) period without work can be justified in a society where work is a central part of life which defines people's identities. Kohli's (1987) answer refers to the notion of a 'moral economy' (Thompson 1971). In this view, retirement and pension systems are not purely rational or 'utilitarian' forms of organizing the exchange of work and income (Kohli 1987: 128), but they

> [...] contain a clear element of reciprocity based on morally bounded claims and expectations [...]. They mix instrumental elements in the

sense of calculable returns for investments with reciprocal elements in the sense of a normative system of mutual obligations. The decisive point, however, is that the former elements are 'embedded' in the latter; therefore, it is feasible to interpret retirement in terms of the moral economy.

The exact form, content and mix of different 'instrumental' and 'reciprocal' elements vary in different welfare regimes, and even the strictly 'instrumental elements' are in fact part of a specific normative order and thus morally loaded (Fourcade and Healey 2007). Historically, the varieties of welfare regimes can be traced back to differences in the evolution of the (welfare) state, with different political actors shaping these processes, and also to wider cultural traditions such as religious ones (see, for example, Esping-Andersen 1990; van Oorschot et al. 2008; for pensions: Kohli and Arza 2011). Welfare regulations are to some extent the expression of welfare cultures, which comprise 'stocks of knowledge, values and ideals' (Pfau-Effinger 2005: 4) and thus also the moral economy. Welfare cultures are not necessarily consistent and homogeneous, and they are continuously changing. While there is a dominating or hegemonic welfare culture in every welfare system – that is, a culture that is closely interlinked with the existing regulations and built on long-standing historical traditions – the welfare culture at large encompasses a broader stock of ideas and values. The existing welfare regulations are the always temporary result of compromises between many different actors who each have their own vested interests, values and beliefs. In the struggles about which beliefs should be 'fixed' in the form of concrete regulations (that is, institutions), some usually more marginal actors strive for reconfigurations of welfare institutions and the dominating moral economy.

Collective actors, when morally justifying their (reform) goals, have to connect and balance three different tasks when deploying their arguments: First, they have to represent the interests of their members which are at the core of their identity. Representing these interests is often closely linked to specific moral arguments. Second, even actors in favour of radical reforms have to embed their arguments into the dominating moral economy of retirement, at least to a degree, so that they are more likely to be understood and accepted. Third, specific arguments in concrete contexts also have to follow situational dynamics and interactions at specific occasions.

The ideas constituting a welfare culture are similar to a stock of building blocks of different shapes and sizes which can be used flexibly to construct buildings – ideas can be applied flexibly to justify or to de-legitimize welfare arrangements and reforms. However, the stock of ideas that are available in a specific welfare culture are neither indefinite nor arbitrary with regard to the policies that can be justified. Thus the policies pursued (or the buildings constructed) differ between countries, while some components might be similar or even the same.

Concrete welfare regulations vary with regard to the central characteristic of which role the state assumes regarding important life-course risks. Where the state plays a strong role in protecting against individual (life-course) risks, individuals have to rely less on (labour) markets to secure their living (Esping-Andersen 1990). As to retirement and pensions, public elements of the pension system ('first pillar') can be more or less important in different welfare regimes. Here, relying on markets refers not only to the necessity to sell one's labour on the labour market, but also to having to rely on private (pension) insurance or financial markets to secure one's living in old age.

Corresponding to the concrete regulations, the specific concept of retirement also varies, that is, the ideas and perceptions of what retirement and pensioners are or rather what they should be. In the concept of retirement, ideas about 'how social security and employment should be connected' and about the relationship between the state, markets and the family come together (Pfau-Effinger 2005: 8). The higher the degree of the decommodification of old age is (and thus the aim of maintaining pre-retirement living standards), the more retirement becomes a distinct phase of life of its own right. This very often coincides with a strong public pension system. As such a form of retirement is more decoupled from the core of a work society, the need to justify retirement as a period which is free of work is more imminent.

We assume that there are a number of different and variable arguments which can serve to morally justify the existence of retirement as a distinct phase of life. They all adopt the notion of a work-free retirement, although not necessarily to the same degree. A first kind of argument directly relates to the function of generational succession, conceived either in a very general way or, more specifically, by arguing that retiring from the labour market is good for society as jobs are freed up for younger workers. This line of reasoning was, for example, used in order to justify generous arrangements for early retirement in the 1980s and early 1990s (for Germany: Jacobs et al. 1991; for the UK: Laczko and Phillipson 1991).

Second, the justification can be related to the 'work society' itself in positing retirement as 'a just reward for a life's toil' (Leisering 2004: 209). This makes more sense in (typically Bismarckian) systems where public pension payments are based on collectively organized contributions and higher contributions are connected to higher pension payments. In Beveridgean systems, by contrast, the state only provides a very basic provision in old age and means-tested benefits are (relatively) more important. Here, the reference to a vulnerable subgroup of (current or future) pensioners should be stronger, for whom the state provides the basic security net only based on their needs.

The normative and regulative link between 'well-deserved' retirement and a hard and long working life can be more or less strong. In an extreme case and as a third kind of argument, the right to a work-free retirement can be claimed relatively detached from its links to paid work: in this case, everyone has a right to retirement in which they can do whatever suits them. There might be wider or narrower definitions of who has deserved retirement through which kind of work, including unpaid social contributions like caring for children to differing degrees. Such a concept of retirement comes closest to the idea of retirement 'as a matter of social and political rights'. Leisering contrasts this idea to a view on retirement 'as a matter of civil rights' (Leisering 2004: 221). The latter implies that there is no (or less of a) positively defined concept of retirement, and retirement tends to be seen in the context of the right to work and to not be excluded from the work society – a right that is also defended in old-age discrimination debates. According to Leisering, this view is more important in the USA than in continental Europe, whereas the idea of retirement as a matter of social rights is more typical in Bismarckian regimes. The UK, as a (relatively) liberal welfare state, can be regarded as being between these two extremes.

These are only the most important examples of moral reasoning about why retirement is deserved or justified. The arguments can complement each other or be complemented by additional arguments, and they can be applied in flexible ways and combinations. All of them should be relatively more important in pension systems which are organized publicly to a higher degree and provide a higher degree of decommodification in old age.

11.3 Germany and the UK: Institutional background and reforms

With regard to the structure and the roots of the pension system, our two country cases can be seen as at least very dissimilar cases; they should thus provide a strong contrast when it comes to their moral justification. Germany has traditionally been the exemplary case of a

Bismarckian welfare regime with a higher degree of old-age decommodification (see Kohli 1987; Schulze and Jochem 2007; also Anderson in Chapter 8 of this book). This first pillar of the pension system, public social insurance, provides the majority of current pensioners' incomes. It aims at maintaining their pre-retirement living standards and is characterized by a strong relation to earnings before retirement, so that differences in pension income should reflect differences in lifetime income. The second pillar of occupational pensions only plays a minor role for some current pensioners, whilst the third pillar of private pensions is negligible except for the self-employed.

The pension reforms since the beginning of the 1990s have weakened the role of the first pillar in Germany (see Schulze and Jochem 2007; Ginn et al. 2009). As a first step, most early retirement routes (Jacobs et al. 1991) were closed or have been coupled with considerable reductions in pension payments. Furthermore, the gradual increase of the statutory pension age to 67 has started, and the level of future pension payments will be lowered. This was complemented by the introduction of subsidized private pensions which are supposed to compensate future pensioners for their lower public pensions. These and connected labour market-related changes (see Hinrichs 2012) have moved the German welfare system partly into a more 'liberal' direction.

The UK pension system, by contrast, is characterized by the greater importance of private pension provision and a weak first pillar (Schulze and Moran 2007). The basic State Pension covers all employed and self-employed people and provides a flat-rate benefit based on the number of years of contributions. It can be topped up by an earnings-related second layer in the first pension pillar, the State Second Pension (S2P); until recently, opting out of this pension scheme and replacing it with an occupational pension was possible. Under the Pensions Act 2014, S2P will (amongst other changes) be phased out from 2016 onwards, while at the same time the rate of the new single-tier State Pension will increase considerably. Occupational pensions (the second pillar) in particular, but also private pension schemes (third pillar), are much more important in the UK than in Germany, although they still do not cover the majority of the population. This is also the reason why many more pensioners have to claim means-tested old-age benefits, pensioner poverty is more widespread and old-age incomes are more unequally distributed in the UK than in Germany (Zaidi 2010). These problems in the UK have been aggravated by the shift from defined benefits to less generous defined contribution schemes in the second pillar, whose outcomes are less secure (and, for example, affected by the financial crisis).

Recent reforms (for overviews see Taylor-Gooby 2005; Schulze and Moran 2007; Ginn et al. 2009) have introduced obligatory pension provision in the second pillar. The increase in the state pension age to 65 for women will be completed in 2018, and it will then be raised stepwise to an age of 68 for the whole population.[2] Finally, the default retirement age (DRA) was abolished in 2011. It had only been introduced in 2006 and allowed employers to dismiss older workers because they had reached state pension age, with the latter having the right to request working longer. In Germany, retirement ages are often (and only) fixed at the level of specific occupations, in collective labour agreements and on company level, but constitute a strong norm (for example Mahlmann 2011: 82–6).

11.4 Methods and sample

The empirical basis for our analysis are transcripts of semi-structured expert interviews conducted in 2011 and 2012 in Germany and the UK, and selected documents published by the sociopolitical actors who were interviewed (for more details regarding methods and data, see Scherger and Hagemann 2014). For this chapter, we selected 8 from our overall sample of 24 sociopolitical actors: The central employers' organization and the unions' umbrella organization in each country represent strongly competing interests in the reform fields of pensions and labour markets (Ebbinghaus 2006). Furthermore, two non-profit organizations were included in each country, whose positions will only be presented very briefly here. They represent the interests of pensioners, older people or (as in one German case) a more broadly defined group of people. The exact role and importance of all these (non-legislative) actors in the two countries are not the same, because the political systems differ considerably, for example regarding the number of veto points or corporative traditions (Schulze and Moran 2007; Schulze and Jochem 2007). The power of the German unions and their confederation is, despite their relative decline, much wider than that of their more fragmented and less centralized counterparts in the UK (see Ebbinghaus 2006: 773–4; Flynn et al. 2013).

Both documents and interviews cover the organizations' evaluation of recent pension reforms and labour market policies for older workers. The interviews also included a question on post-retirement work. The expertise of the interviewees comprises not only specific substantive knowledge of policy and reform contents but also interpretative knowledge on how to deploy ideas in such a way that the three tasks mentioned above are balanced, that is, representing their interests, embedding their arguments into the dominating moral economy and following situational dynamics. The documents are mostly position papers and press releases (for a complete list, see Scherger and Hagemann 2014). As some actors publish documents infrequently and only for specific occasions, the documents cover the time period of 2005 to 2013, with the majority being from 2010 and 2011.

11.5 The unions' view: Retirement as a social right under threat

Both the German Deutscher Gewerkschaftsbund (DGB) and the British Trades Union Congress (TUC) represent member unions from different sectors. Both confederations pursue a relatively inclusive policy approach guided by the value of solidarity and claim to represent all employees. When questioned about the significance of retirement, the expert from the DGB elaborates on state pension age as a social norm; he underlines the role of retirement as a phase which people plan for and look forward to from their 50s onwards. This phase of life is supposed to be free of duties of any kind; it is 'the phase when you are free to decide what to do and perhaps even free to decide to do nothing' (19–20).[3] The expert considers retirement as being under threat, amongst others by the emerging and highly morally charged discourse of obliging older people to be active (in particular volunteering and working) longer. He criticizes that the (often conservative) proponents of this discourse attempt to 'redefine' 'this free phase of life' (42). He regards this as inappropriate because, implicitly, any engagement in retirement should be based on the freedom to do whatever one wants (50).

Although neither the expert nor the documents explicitly mention generational succession in the labour market, the related argument that more older people working makes the labour market situation more difficult, especially for younger people, lingers in some side notes in the DGB's documents (DGB 2006: 8, 2011: 6–7). The DGB emphasizes the constraints that older people are faced with when approaching pension age, and that many older people are not able to work longer due to health-related problems, especially the low-skilled (128–129; DGB 2011: 14). The expert criticizes that there is a growing gap between the actual end of work and the beginning of pension payments due to the increase in the pension age and related labour market reforms (64), which increases the threat of old-age poverty (see also DGB 2011: 3). Correspondingly, the DGB position on pensions in its written statements focuses on the level of pensions (DGB 2006: 2; DGB 2011).

Asked about why some people work in old age, the expert summarizes that 'some want to, the others have to' (409). He elaborates on possible reasons, including social reasons (such as social contacts) and the potential mix of financial and non-material reasons. Despite this conscientious and balanced view, he stresses that the cases of working pensioners he personally knows, among them family members, all work because of 'damn low' pensions (430–8). Evaluating the trend of increased work beyond pension age, the expert states that in his opinion it is very risky and missing the point to think that employment in old age could help to tackle old-age poverty (423–30, 438–9).

Regarding the UK, the expert from the British Trades Union Congress (TUC) underlines that there is 'a long standing trade union commitment to a poverty-free retirement as a right for everyone' (33). His main concern is that this right is not available for everyone – so similarly to the DGB, the TUC sees retirement under threat and opposes raising the pension age. In this context, the expert emphasizes the unequally distributed healthy life expectancy (36–8). As most workers leave the workforce earlier than state pension age and 'involuntarily' (203), because of deteriorating health or age discrimination, the gap between the actual end of paid employment and the start of pension payments is growing. Thus, the TUC is in favour of measures against old-age discrimination and age management policies which enable people to work up to state pension age (200–19). This is complemented by an 'output orientation' with regard to pensions, which favours improvements in the state pension (TUC 2008: 2) and occupational pensions.

The TUC generally welcomes the abolition of the default retirement age and the subsequent right to continue working in several press releases (TUC 2011a, 2011d). However, the expert underlines that this decision has to be a 'genuine voluntary choice' (119) and implicitly doubts that working pensioners work longer out of their free choice (see also TUC 2011d). The TUC believes that people who desire to work beyond pension age should be allowed to do so. However, in the eyes of the expert, 'it's far more important to ensure that people have got a right to retire at 65, there's far more people in need who are affected by not having that right than are affected by being forced to retire early' (115–17). This perspective also shapes the TUC's evaluation of post-retirement work, which the expert identifies as 'one of the ways of dealing with old age poverty' on the individual level (418).

Accordingly, he cites inadequate pensions as the most common reason to continue working, especially for women. Besides financial reasons, the expert refers to a group of people who 'love their jobs and don't want to leave them' (398), and continuing in a tongue-in-cheek tone he states that 'mostly they're journalists writing articles for the newspapers and magazines about pensions' (401–2).

All in all, the British and the German union confederations have a similar perspective. They stress the social right to retirement, which they see as threatened by recent policies regarding pensions and the pension age. Their positions do not include an explicit reference to 'deserving' this right after a long working career – which is somewhat surprising in the German case.[4] If at all, only some undertones of this argument can be found. Both actors stress that many older people cannot work longer than they currently do, thus underlining the protective function of retirement. Whereas the DGB still (at least implicitly) draws a connection between longer employment careers of older people and the labour market chances of younger people, the TUC explicitly rejects this argument (TUC 2011c). Minor differences in their arguments relate to their take on what exactly jeopardizes retirement as a social right, apart from their perception that most people cannot and should not be obliged or pushed to work longer: Whereas the British TUC's position focuses more strongly on old-age poverty, for the DGB expert poverty is only one among several points, and he additionally mentions the discourse trying to oblige retired people to volunteer. Whereas the British expert calls for a poverty-free retirement, the German expert goes further in his demands: retirement should also be free of any duties, if one wishes so. This suggests that he has a somewhat stronger view on what retirement should be, correspondingly to what is incorporated in the German welfare culture. The tendency towards a more modest definition of retirement in the UK of course also mirrors the more pressing problems of old-age poverty.

11.6 The employers' view: Fixed retirement as outdated and costly

The German Bundesvereinigung der Deutschen Arbeitgeberverbände (BDA – Confederation of German Employers' Associations) and the Confederation of British Industry (CBI) represent the interests of employers. Both experts who were interviewed underline that what is good for the employers is good for economic growth, which ultimately benefits society as a whole, including older and retired people (BDA interview 475–81; BDA 2009, 2010; CBI interview 738–47, 784–803). Asked about the social significance of retirement, the expert from the German Bundesvereinigung der Deutschen Arbeitgeberverbände (BDA), after a longer moment of reflection, rather talks about the transition into retirement. Underlining that this is not a thought through statement, he defines retirement as

> a very *classical* concept [...] which basically draws a dividing line between employment and retirement, a concept which is basically very outdated, I would say, so the idea that one eventually, from a certain age onwards, which is even defined by the legislator and not by oneself, virtually withdraws from one's occupational activity, or *must* withdraw and then suddenly ceases doing anything at all, this is actually not the concept which we as BDA have at the back of our minds [continues and talks about improved health of people in pension age today], and in this respect of course the question arises, also given the demographic changes with qualified employees becoming scarce, whether such a *stubborn* age limit or [...] such an abrupt exit from working, whether this is still appropriate today. (41–52)

As can be seen from the quotation, the BDA's 'concept' of retirement does not contain a substantive idea on what retirement as a distinct life phase is, but is shaped by labour market-related reflections. Although the expert has been asked about the social significance of the phase of retirement (which is clearer with the German word 'Ruhestand' used here, which literally means 'status of rest'), his statement that retirement is outdated seems to be mainly related to the process of retirement. In addition to this critique of age limits as inflexible, he later mentions the BDA's general disapproval of age limits, as these are opposed to the organization's 'liberal view on society' (578).

Correspondingly, the BDA is clearly in favour of increasing pension age and of the recent policies to cut the (future) level of pension provision (671, 779–81; BDA 2005: 3, 11). Sustaining these policies makes the pension system more viable, the expert claims (438–81; BDA 2010), and will prevent old-age poverty – which is not a problem at the moment – in the future. The BDA prioritizes a stable or decreased level of pension contribution rates over the level of pension provision. The expert thus welcomes the beginning shift to a system in which the funding principle becomes relatively more important for pensions (compared to the dominating pay-as-you-go principle) and in which individuals themselves are primarily responsible for securing their livelihood in pension age (776–9). At the same time, both the expert and the documents examined disapprove of most redistributive pension components and see the need to strengthen the 'equivalence principle' in the German social insurance pension, that is, the link between (continuous and high) contributions and (higher) pension payments (see, for example, 669–79; BDA 2005: 3). This line of reasoning is furthermore embedded in a very positive view of older people's capabilities. As older people today are in better health in comparison to the past (47–50; BDA 2011: 2), increasing the state pension age is not seen as a problem. In the view of the BDA, most older people are able to work longer, and very many also want to. Complementing measures, such as the flexibilization of age boundaries, continued education and learning, and workplace adaptation (BDA 2009, 2010), are seen as the shared responsibility of employers, unions (215–22) and employees, while too strict and inflexible regulations by the state in these areas are rejected (BDA 2009).

Asked about why people might still work beyond retirement age, the expert first refers to his father, who still works despite being of pension age, because he enjoys his work and the feeling of being needed (531–4). He then mentions a number of non-financial reasons why people generally stay in or go back to employment, such as fun and not wanting to feel useless and old (539–43). Only after all these reasons does he touch on potential financial reasons and at the same time distances himself from the 'advocates of the thesis of old age poverty' who 'probably say that people will go working in droves because they simply cannot afford this anymore' (543–5); only at the end does he briefly mention that 'earning extra' might play a role for a few older workers.

Even more clearly than the German expert, the one from the Confederation of British Industry (CBI) relates the question about the social significance of retirement mainly to the transition into retirement (with the word 'retirement' meaning both the process of retiring and the resulting state). He states that the CBI does not 'necessarily like to talk about retirement as a sort of black and white situation. We think that there is some moment in time where obviously people are living longer and therefore it is natural that people will work longer' (28–30). Whilst the CBI is in favour of increasing the state pension age, extending working lives and flexibilizing the transition into retirement, the organization regards the abolition of the default retirement age (DRA) as very problematic (CBI 2010). According to this view, the DRA, which was being phased out at the time of the interview in 2011, facilitated succession planning for the employers because it enabled them to dismiss people reaching state pension age. Under the new retirement regulations, the employers have to give a reason if they want to dismiss an employee reaching pension age, a process which the employers perceive as difficult and (too) complicated (143–51), the expert reports. Furthermore, the old regulations also enabled employees to 'retire with dignity' and without a 'long and arduous' performance management process (CBI 2010: 8). The CBI defends the employers' power to dismiss employees at pension age and demands acknowledgement of the fact that some older workers do not perform well anymore (CBI 2010: 3).

Similar to the BDA, the CBI is more concerned about the contributions for and the costs of pensions than about the level of provision. The CBI generally underlines that people 'should take responsibility for their retirement and take ownership of the tools that would allow them to save adequately for a pension' (CBI 2011: 1), amongst others occupational pensions, which should not be over-regulated (CBI 2009). This highly individualized notion of retirement implies that the individual has to prepare financially for retirement (701). The expert recognizes that the UK has a problem with old-age poverty (586–8) and therefore welcomes the (planned) reforms of the state pension.[5]

Asked about why some people work beyond pension age, the expert first refers to the social interaction at work and the difficulty to stop working from one day to the next. He also elaborates on economic reasons in connection to pension losses due to the financial crisis (501–2). Nonetheless, he sees work beyond retirement mostly as a matter of individual choice, welcomes people's wish and ability to work longer and considers this as part of welcome flexible retirement arrangements (507–13; CBI 2009: 36).

Most of the differences between the two actors are due to structural and institutional differences between the countries: Negating the existence of old-age poverty would potentially discredit the CBI's position, as old-age poverty is widely perceived as a serious problem in British society. Similarly, being in favour of a slightly improved state pension is only possible for the CBI because provision is low. By contrast, the BDA clearly supports the cuts in the (still) more generous German pension system. Both actors tend to see work beyond pension age as a matter of choice and taking individual responsibility for retirement, although this is less explicit in the German case. They also see older people as generally capable of working longer, and those who cannot are considered an exception. However, the British employers are caught between a rock and a hard place in this respect because at the same time they would have liked to maintain their power to dismiss older people. This is justified by 'soft' and meso-level-related arguments of generational succession, relating to the practical side of human resource management within companies (CBI 2010: 7). On the whole, however, similarities between the two positions prevail. Both actors do not really feel responsible for questions of retirement, and neither of them has a strong concept of retirement as a work-free phase of life which should be a social right. On the contrary, costly pension systems are potential barriers to economic growth and retirement is mostly seen in its function for the labour market: people beyond pension age are a potentially useful employment reserve.

11.7 Non-profit interest organizations: Between the right to work-free retirement and the right to work

The non-profit organizations we chose for our analysis are unavoidably different, with Germany only having very few large organizations representing mainly or explicitly older people. The German Bundesarbeitsgemeinschaft der Senioren-Organisationen (BAGSO, roughly 'Federal Consortium of Seniors' Organizations') is the (federal) umbrella organization of organized seniors in political parties, unions and non-profit organizations; it is a 'weak' and cross-party umbrella organization. The Sozialverband Deutschland (SoVD, roughly 'Social Association Germany') does not explicitly or solely represent older people, but the interests of pensioners, patients, members of the statutory health insurance, those in need of care and disabled people. The National Pensioners Convention (NPC) has a position close to that of the unions. Age UK is the largest charity for older people in the UK and addresses all kinds of issues connected to old age (such as health and care, pensions, old-age poverty and volunteering).

In general, all four non-profit organizations represent the interests of pensioners; however, they do it in very different ways and with very different foci: the German SoVD and the British NPC oppose most of the recent reforms, defend a strong and meaningful concept of retirement and also underline that very many people approaching state pension age are not able to continue working, be it up to the (new) state pension age or beyond. SoVD and NPC see themselves as advocates of disadvantaged and poor pensioners. Whereas the German BAGSO stresses retirement as a reward for a long working life, Age UK sees the 'empowering' right to work as key to the well-being of many pensioners. This is combined with demands to better protect those who cannot work beyond pension age and who are likely to be poor in old age. The NPC spells out a comprehensive concept of retirement as a distinct phase of life which should be a social right for everyone. However, the expert at the same time embeds this into a (more) liberal welfare culture in which individual responsibility is important and retirement is above all a question of costs. He does this by referring to the economic value of the contributions that retirees make, an argument which potentially undermines the idea of retirement as a social right. The view of Age UK is, in comparison, closer to seeing retirement in the context of work as a (civil) right, while at the same time it contains elements of a stronger concept of retirement.

For the German non-profit actors, it is easier to substantiate a strong concept of a social right to work-free retirement because such a concept is part of the traditional setup of the German pension system. At the same time, both the BAGSO and, in a more general and loose way, the SoVD connect their comparably inclusive approach to the emphasis on contributions and the equivalence principle that is ingrained in the German system.

11.8 Conclusions

Our (necessarily selective) analysis has demonstrated how exactly justifications of retirement are applied, exemplified, weighed and connected to each other by different actors in the two welfare cultures of Germany and the UK. Not all actors have a strong and substantive concept of retirement. In particular, retirement as a distinct phase of life does not mean a lot to the employers. Although they are far from opposing the idea of a work-free life phase in old age, they subordinate their reasoning about it to what is beneficial in terms of macroeconomic functioning. The other actors all use a mix of the arguments mentioned in the first part of the chapter in order to justify retirement. Retirement as a social right is strongly advocated by the unions, who see this particular right under threat by recent reforms (in the case of Germany) or by a deplorable state of the pension system which calls for reforms (in the case of the UK). This goes together with a focus on the 'output' side of pensions, their level and distribution. Similar views are shared by most non-profit actors, although the German BAGSO and Age UK have a more moderate take on retirement. References to retirement as 'deserved' by a long career are more common in the German system of social insurance, but tend to be short and formulaic, and are never substantiated with regard to who should be excluded on these grounds. Succession arguments never take centre stage in the arguments of the actors, except for Age UK who takes a decisive stance against their labour market-related version.

The evaluation of post-retirement work by the experts corresponds to their take on retirement. While all experts acknowledge the variety of reasons that exist for working in retirement, most of them accentuate those reasons that go well with their (weak or strong) take on retirement. For example, the unions particularly refer to financial reasons, and the employers concentrate on non-financial reasons. None of the experts is in favour of legal age boundaries for working or is opposed to letting people work who would like to do so. However, for those stressing the social right to a work-free retirement, this question of being 'allowed' to continue working is rather a side theme than a pressing problem.

Figure 11.1 summarizes how the concept of retirement is associated with the most important (varieties of) ideas we have reconstructed. The latter include both normative ideas of what should be (done) and descriptive ideas of what is assumed to be true or real. This 'alignment' or specific clustering of the concept of retirement and related ideas can change, but is not arbitrary; furthermore, the table conveys that contrasting ideas tend to be the opposite ends of a range of arguments, not mutually exclusive categories.

Our analysis was also aimed at exploring the relation between the actors' positions and interests, on the one hand, and the welfare cultures they belong to, on the other. By and large, the moral arguments of actors of the same kind from different countries (especially employers

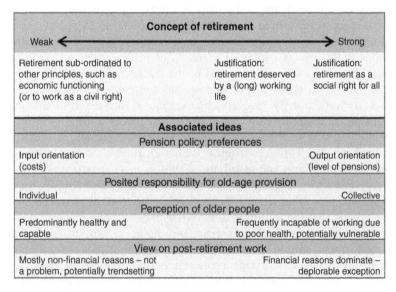

Figure 11.1 Concepts of retirement and associated ideas
Source: Own compilation.

and unions) are more similar to each other than the positions of different actors within one country. However, the actors have to refer to the institutions and ideas which characterize the German or British welfare culture. These differences appear in the form of partly diverging blocks in the 'building' of the specific argument – such as the stronger reference to retirement being deserved by a long working career in Germany. These differences rather constitute diverging tendencies in the moral justification of retirement than different lines of reasoning altogether, also because the actors' positions cut right across the country differences. The commonalities may also indicate that the German and the British welfare regimes overlap in their stock of ideas, as both are members of an increasingly interconnected EU and have to adapt to ageing populations.

Connected to the concept of retirement and its justification, images of older people emerge as either capable and healthy or vulnerable and in need of protection, with these only being the two extremes of a broad spectrum of images. In most cases, the actor's position with regard to prolonged working lives is closely related to the question of what side of this spectrum the actor emphasizes; nonetheless, all experts are well aware of the fact that older people live in a huge variety of situations. The analysed views on retirement can be seen as ideational trade-offs between the protection of (vulnerable) older people and the attempt to install policies that do not discriminate against older people through stereotyping. It is certainly a sign of progress in more recent policies on old age that they underline the 'potential' of older people. Pluralized and less stereotyping images of old age open new possibilities of how to live (and perhaps work) in old age. At the same time, in light of persisting or even increasing inequalities in old age, the protective function of retirement as a genuine social achievement has not lost its relevance. Old age should not be reduced to its potentials for mostly economic-instrumental reasons, because this involves new forms of exclusion. To (re-)negotiate the balance between the aim of protecting older people and the aim of realizing their potential will remain a crucial task in the field of old-age policies.

Notes

1. This chapter is a short version of Scherger and Hagemann (2014) and was written in the context of the Emmy Noether research group 'Work beyond retirement age in Germany and the UK', funded by the German Science Foundation (Deutsche Forschungsgemeinschaft).

2. In late 2013, the British conservative-liberal government announced plans to further increase state pension age to 70 by the 2060s, closely linking the pension age to rises in life expectancy. Under the Pensions Act 2014, a framework for regular reviews of the state pension age has been defined.
3. All German quotations have been translated by the authors. The numbers after the quotes from the expert interviews indicate the line number of the quote; for the documents, the organization and the year of publication is given before the page number of the quote or reference (for more details, including the references of the analysed documents, see Scherger and Hagemann 2014).
4. Although this should not be overstated, this is in some contrast to specific member unions.
5. In the meantime, these reforms have been enacted as the Pensions Act 2014.

References

Ebbinghaus, B. (2006), 'The politics of pension reform: Managing interest group conflicts', in: G. L. Clark, A. H. Munnell and M. Orszag (eds.), *The Oxford handbook of pensions and retirement income*, Oxford: Oxford University Press, 759–77.

Esping-Andersen, G. (1990), *The three worlds of welfare capitalism*, Cambridge: Polity Press.

Fourcade, M. and Healey, K. (2007), 'Moral views of market society', *Annual Review of Sociology*, 33, 285–311.

Flynn, M., Upchurch, M., Muller-Camen, M. and Schroder, H. (2013), 'Trade union responses to ageing workforces in the UK and Germany', *Human Relations*, 66 (1), 45–64.

Ginn, J., Fachinger, U., and Schmähl, W. (2009), 'Pension reform and the socio-economic status of older people', in: A. Walker and G. Naegele (eds.), *Social policies in ageing societies. Britain and Germany compared*, Basingstoke: Palgrave Macmillan, 22–45.

Hinrichs, K. (2012), 'Germany: A flexible labour market plus pension reforms means poverty in old age', in: K. Hinrichs and M. Jessoula (eds.), *Labour market flexibility and pension reforms. Flexible today, secure tomorrow?*, New York: Palgrave Macmillan, 29–62.

Jacobs, K., Kohli, M. and Rein, M. (1991), 'Germany: The diversity of pathways', in: M. Kohli, M. Rein, A.-M. Guillemard and H. v. Gunsteren (eds.), *Time for retirement. Comparative studies of early exit from the labor force*, Cambridge: Cambridge University Press, 181–221.

Kohli, M. (1986a), 'The world we forgot: A historical review of the life course', in: V. W. Marshall (ed.), *Later life. The social psychology of aging*, Beverly Hills: Sage, 271–303.

Kohli, M. (1986b), 'Social organization and subjective construction of the life course', in: A. B. Sørensen, F. E. Weinert and L. R. Sherrod (eds.), *Human development and the life course*, Hillsdale: Lawrence Erlbaum, 271–92.

Kohli, M. (1987), 'Retirement and the moral economy: An historical interpretation of the German case', *Journal of Aging Studies*, 1 (2), 125–44.

Kohli, M. and Arza, C. (2011), 'The political economy of pension reform in Europe', in: R. H. Binstock and L. K. George (eds.), *Handbook of aging and the social sciences*, New York: Elsevier, 251–64.

Laczko, F. and Phillipson, C. (1991), 'Great Britain: The contradictions of early exit', in: M. Kohli, M. Rein, A.-M. Guillemard and H. v. Gunsteren (eds.), *Time for retirement. Comparative studies of early exit from the labor force*, Cambridge: Cambridge University Press, 222–51.

Leisering, L. (2004), 'Government and the life course', in: J. T. Mortimer and M. J. Shanahan (eds.), *Handbook of the life course*, New York: Kluwer, 205–25.

Mahlmann, M. (2011), *Report on measures to combat discrimination. Directives 2000/43/EC and 2000/78/EC. Country report 2010: Germany*, Utrecht and Brussels: European network of legal experts in the non-discrimination field, Human European Consultancy, Migration Policy Group, http://www.non-discrimination.net/content/media/2010-DE-Country%20Report%20LN_FINAL_0.pdf, date accessed 18 March 2014.

Pfau-Effinger, B. (2005), 'Culture and welfare state policies: Reflections on a complex interrelation', *Journal of Social Policy*, 34 (1), 3–20.

Scherger, S. and Hagemann, S. (2014), *Concepts of retirement and the evaluation of post-retirement work. Positions of political actors in Germany and the UK, ZeS-Working Paper No. 4/2014*, Bremen: Centre for Social Policy Research, http://www.zes.uni-bremen.de/veroeffentlichungen/arbeitspapiere/?publ=5023, date accessed 1 November 2014.

Schulze, I. and Jochem, S. (2007), 'Germany: Beyond policy gridlock', in: E. M. Immergut, K. M. Anderson and I. Schulze (eds.), *The handbook of West European pension politics*, Oxford: Oxford University Press, 660–710.

Schulze, I. and Moran, M. (2007), 'United Kingdom: Pension politics in an adversarial system', in: E. M. Immergut, K. M. Anderson and I. Schulze (eds.), *The handbook of West European pension politics*, Oxford: Oxford University Press, 49–96.

Taylor-Gooby, P. (2005), 'UK pension reform: A test case for a liberal welfare state', in: G. Bonoli and T. Shinkawa (eds.), *Ageing and pension reform around the world*, Cheltenham: Edward Elgar, 116–36.

Thompson, E. P. (1971), 'The moral economy of the English crowd in the eighteenth century', *Past & Present*, 50 (1), 76–136.

van Oorschot, W., Opielka, M. and Pfau-Effinger, B. (eds.) (2008), *Culture and welfare state. Values and social policy in comparative perspective*, Cheltenham: Edward Elgar.

Zaidi, A. (2010), *Poverty risks for older people in EU countries – An update (European Centre Policy Brief, January 2010)*, Vienna: European Centre for Social Welfare Policy and Research.

Part III
Consequences

12
Later-Life Work, Health and Well-Being: Enduring Inequalities

Katey Matthews and James Nazroo

12.1 Introduction

Increasing life expectancy means the UK, like other countries, is facing a rapidly growing population of those of pension age relative to those of working age, placing strain on public expenditure as periods of retirement and economic inactivity lengthen. A major strategy in policy responses to ageing populations is to extend working lives. Longer periods of employment, leading to greater pension contributions and shorter periods of retirement, are proposed to provide potential solutions to such problems. One means of encouraging longer working lives is to increase state pension age (SPA). In 2012, a report by the Office for National Statistics showed the average age of withdrawal from the labour market in the UK had risen to 64.6 years for men and 62.3 years for women, from 63.8 years and 61.2 years, respectively, in 2004. Although the average in 2012 was close to or even above the SPA (65 for men and between 60 years 10 months and 61 years 4 months across the 2012 period for women), a substantial proportion of the population are still leaving the workforce before reaching SPA (Phillipson and Smith 2005). If increases to the SPA are to be successful, there needs to be an understanding of barriers to remaining in work, of who the older workforce comprises, and of the factors which might encourage older people to remain in employment. Extended working has been shown to be linked to better health and working conditions, while mental and physical strain from poor conditions lead to early work exit, often due to poor mental health or accumulated physical disability (Phillipson and Smith 2005).

To understand the heterogeneous work and retirement patterns of older people, attention must be paid to social inequalities both prior to and following the transition into either extended work or retirement. Previous research suggests social and health inequalities in later life are the product of inequalities that have existed across the life course (Blane et al. 1993; Marmot and Nazroo 2001). If policy to encourage continued working is to succeed, inequalities and their effects in older age should be considered in terms of a life-course perspective. The human capital approach states a 'cumulative advantage' of extended working may be observed among those who have worked long careers in high quality employment following high educational attainment (Dannefer 2003). However, the same approach must consider the 'cumulative disadvantage' to older workers brought about by lifelong employment in poor conditions, often preceded by lower educational attainment. If implemented correctly, extending working lives in favourable conditions could lead to a cost-effective 'compression of morbidity' among people of pension age and over, whereby good health is maintained through working for longer in good quality employment, and periods of morbidity thereafter are delayed and subsequently shorter (Fries 2002). However, the potentially negative implications for those who are forced to work for longer in poor conditions, or who retire into poorer circumstances, must also be considered.

12.2 Questions emerging from the literature

Recent studies concerning the effects on health of later-life work provide a mixed set of findings. Even studies using methods specifically aimed to eliminate selection biases to produce stronger causal conclusions provide contradictory results. Some show continued working is beneficial (Adam et al. 2007; Rohwedder and Willis 2010; Behncke 2012; Bonsang et al. 2012; Calvo et al. 2013), and others find it to be detrimental to health and well-being (Coe and Zamarro 2011), or show no significant relationship at all (Coe and Lindeboom 2008; Coe and Zamarro 2011; Behncke 2012; Calvo et al. 2013). One important reason for these differences is the heterogeneous nature of the studies and their samples. The definition of 'later-life work', as well as retirement, varies, with many studies including all people working beyond age 50 and fewer focusing exclusively on those of SPA and over. Additionally, employment is often regarded as a homogeneous state, without consideration of differences in working environments. If early retirement is linked with poorer working conditions, which may be linked with lower social status and wealth, we are observing effects of social inequalities which may have persisted across the life course, rather than of only work or retirement.

A more detailed consideration of these studies shows that various factors influence effects of later-life working and retirement. Effects appear to differ by gender. Men are often considered to suffer detrimental effects of retirement, because they endure longer periods of employment than women and therefore place greater importance on their working role (Moen 1996). Those who participate in an activity alongside paid work in older age, such as volunteering, often see an association with better health outcomes than those who are employed only (Hao 2008). The timing of retirement appears important, with earlier retirees experiencing higher depression levels than those who retire at standard ages (Buxton et al. 2005; Butterworth et al. 2006). Quality of work and occupational type play a role in determining the health and well-being of older employees, with better quality and sedentary work associated with better health (Siegrist et al. 2004b; McMunn et al. 2009) and lower quality and manual work associated with poorer health (Chandola et al. 2007). Finally, financial status plays an important role in determining health outcomes among older workers and retirees. Many people continue working in 'job-lock', whereby they cannot afford to leave employment despite unfavourable working conditions (Ekerdt et al. 1983), and many people retire without sufficient wealth to participate in enjoyable activities, both of which are likely contributors to poorer health and well-being.

In order to demonstrate the importance of accounting for the heterogeneous nature of older working populations, Matthews et al. (forthcoming a; also Matthews 2014) used meta-analysis to consider the overall effect of studies examining the impact of later-life work on depression and self-rated health. When all studies were combined, regardless of sample characteristics, findings suggested a beneficial effect on both outcomes. However, when studies were broken down on the basis of some individual characteristics, such as timing of retirement and gender, results were shown to vary by group. Additionally, the importance of considering the heterogeneous nature of older workers was demonstrated by how the results depended on whether fixed or random effects models were used. Fixed effects models, which assume all studies draw from a single homogenous population, showed continued working to be significantly better for depression and self-rated health than retirement. However, random effects models, which acknowledge studies may draw from various populations with differing characteristics, showed no significant differences in the health of workers and retirees.

The following section of the chapter uses evidence from three pieces of analysis, conducted using the English Longitudinal Study of Ageing

(ELSA), to further examine the effects of working and retirement on health in later life, with a particular focus placed on the influence of class inequalities. ELSA (Steptoe et al. 2013) is a panel study of people living in England who were aged 50 or over in 2002 and who had previously responded to the Health Survey for England between 1998 and 2001. The first wave of the survey was conducted in 2002–03, with a response rate of 67 per cent, and the fifth wave was conducted in 2010–11. The analyses discussed within this chapter use waves 1 to 5.

12.3 The effect of working circumstances on health

As suggested by the previous section, the background characteristics of individuals, such as their health, social class, financial status and working conditions, are likely to influence both the likelihood of continuing work in later life and health outcomes regardless of employment status. This difference in individual characteristics leading to increased likelihood of group membership (in this instance continued working as opposed to retiring specifically at SPA) is referred to as selection bias, and the results of studies which do not specifically account for these characteristics are likely to be biased and therefore unreliable. Matthews et al. (forthcoming b; also Matthews 2014) use propensity score matching as a statistical method to specifically deal with these biases and get a more accurate assessment of the effect of working beyond retirement age. Propensity score matching works by calculating an individual 'score' predicting the likelihood of belonging to the 'exposure' group (those working beyond SPA), regardless of whether or not they actually belong to that group, for each case in the data. Subsequently, through matching and comparing those with similar scores but different membership of the 'exposure' group, an unbiased assessment of the differences in outcome for the 'exposed' and 'not exposed' groups can be established. Further details on the method can be found in Rosenbaum and Rubin (1983).

Table 12.1 shows the key findings of the study by Matthews et al. (forthcoming b), which contrasts those people who worked beyond the traditional UK State Pension Age (SPA) (65 for men and 60 for women) with those who worked directly until reaching traditional retirement age and then took retirement. Those who retire prior to reaching SPA are excluded from the analysis in this instance as early retirement is often taken by those with poorer health to begin with. The sample includes a total of 1,167 cases, 482 of whom retire at SPA and 685 of whom continue to work. Further information on the selection and characteristics of the sample can be found in Matthews et al. (forthcoming b; also Matthews 2014). The initial analysis considers workers as a homogenous group, and two subsequent analyses stratify work on the basis of its quality. 'Poor' work quality is work which the individual feels is poorly reciprocated in terms of adequate salary, security, support and prospects, and 'good' work quality represents work which the individual feels is well reciprocated. An effort-reward imbalance scale is used to measure these concepts (Siegrist 1996).

The three outcomes discussed within this piece of analysis are depression, self-rated health and cognitive function. Depression is measured using an eight-item version of the Center for Epidemiologic Studies Depression scale (CES-D, Radloff 1977), which asks respondents to state whether they have experienced eight symptoms of depression within the past week and has a range of 0 to 8 (0 represents no symptoms of depression). Self-rated health is measured using responses to the question of how the individual rates their overall general health, with the five options for response of excellent (represented by 1), very good, good, fair or poor (5). Cognitive function is measured using a combined executive and memory function scale, with a potential range of 0 to 44 (0 represents the poorest possible memory function and 44 the best). Through the use of propensity scores, all models account for gender, age, marital status, wealth, National Statistics Socio-Economic Classification (NS-SEC), education, housing tenure, partner's working status, whether the respondent is self-employed, whether or not the respondent receives or will receive upon retirement a private pension, difficulties with activities of daily living and baseline (that is, wave 1) outcomes of the three outcome measures.

The first panel of Table 12.1 shows the effects of any work beyond SPA on the health outcomes of interest, with the standard errors in parentheses. The two subsequent panels show the effect of work in comparison to retirement when it is stratified on the basis of whether or not it is deemed to be of poor or good quality in terms of the balance between effort and reward. The table presents both unmatched and matched results. The unmatched results are the basic means differences between the workers and retirees, and the matched results are the means differences between the workers and retirees after they have been matched on the basis of the similarity of their propensity scores, that is, when selection biases have been minimized.

When all workers are compared to retirees, continued working is associated with significantly better self-rated health and cognitive function (but there is not a significant difference in depression scores). However,

Table 12.1 Results of propensity score matching: The effect of later-life working on health outcomes (means differences and their standard errors)

		Depression	Poorer self-rated health	Cognitive function
All work vs. retirement	Unmatched	0.012 (0.09)	−0.129* (0.07)	1.249*** (0.35)
	Matched	0.091 (0.12)	−0.143 (0.07)	0.279 (0.45)
Poor quality work vs. retirement	Unmatched	0.531*** (0.16)	0.092 (0.11)	0.407 (0.59)
	Matched	0.447* (0.22)	−0.067 (0.14)	−0.349 (0.71)
Good quality work vs. retirement	Unmatched	−0.202* (0.07)	−0.226*** (0.07)	1.788*** (0.38)
	Matched	−0.133 (0.09)	−0.201** (0.09)	0.456 (0.47)

Note: ***p < 0.001, **p < 0.01, *p < 0.05.
Source: Matthews et al. 2014b, based on English Longitudinal Study of Ageing, waves 1–5.

in both instances, the matched results are non-significant. This suggests that the initial differences observed are actually the result of individual differences in wealth, social class and education between workers and retirees, rather than the result of working as opposed to taking retirement.

The second panel of results shows the effect of continued poor quality work on health as opposed to retirement. In this instance, those who continue to work in poor quality employment see higher depression scores than those in retirement (but no differences in health and cognitive function), and this significant difference remains after matching has been implemented. Similarly, the third panel of results shows self-rated health to be significantly better among those who continue to work in good quality work than those who retire.

In light of the analysis of all workers simultaneously, these results highlight the importance of accounting for different types of work when considering the effects of continued working on older populations. If all workers are examined as if they were one homogenous group, there would be no negative effects on health observed and treating the population 'as one' masks group-specific patterns. The consideration of working circumstances demonstrates that continued work is actually detrimental to the psychological health of those in poor quality jobs, and of benefit for those in good quality jobs. This is particularly important from a policy perspective, where encouragement to work for longer periods of time could potentially result in adverse (and non-cost-effective) outcomes. Indeed, increasing pension ages may well force those who cannot afford to retire before or at SPA into longer periods of working, and many of these will be people in poor quality employment. In turn, these people are placed at risk of the onset of poor health as well as exacerbation of existing (especially work-related) mental and physical health problems.

12.4 The effect of retirement circumstances on health

The analysis of working beyond SPA shows that if working conditions are optimal, continued working can be better for health than retirement, but where working conditions are suboptimal, individuals may experience better health outcomes if they instead take retirement. The research clearly shows the importance of accounting for the heterogeneous nature of employment in later life. However, it is equally important to consider the heterogeneous nature of the retirement process and its implications for health post-retirement. The ELSA data can also be used to examine the impact of retirement on health taking into account both the route into retirement and the individual's economic well-being (as indexed by level of wealth) at the time of retirement.

While the analysis of working beyond SPA used a sample inclusive only of those who worked until retirement age, this analysis uses a sample inclusive of those who were working prior to stating they had retired as well as those who classed themselves as unemployed, semi-retired and permanently sick or disabled. Those who stated absence from work because they looked after the home or family were excluded from the analysis. Routes into retirement are described as 'routine', 'involuntary' or 'voluntary'. 'Routine' retirement relates to those who retired because they reached retirement age and 'involuntary' to those who retired for reasons beyond their control, such as the onset of ill health, the ill health of a family member or friend, redundancy or unemployment and subsequent inability to find another job. 'Voluntary' retirement includes those who retired in order to spend more time with family, enjoy life while still young and physically able, experience change from a job of which they were bored, to give the younger generation a chance in employment, to retire at the same time as their partner or spouse, or because they were offered reasonable financial terms. The analysis

Table 12.2 Effects of route into retirement and retirement wealth on post-retirement depression (linear regression coefficients)

Routes into retirement
(ref. remaining employed)

	Model 1		Model 2	
Routine	0.05	(–0.12, 0.21)	–0.02	(–0.15, 0.19)
Involuntary	0.56**	(0.18, 0.94)	0.42*	(0.02, 0.81)
Voluntary	–0.21	(–0.58, 0.16)	–0.10	(–0.48, 0.27)
Model controls for	sex, baseline age, baseline CES-D score		sex, baseline age, baseline CES-D score, wealth quintile, occupational class, education	

Wealth circumstances in retirement
(ref. remaining employed)

	Model 3		Model 4	
Lowest wealth	0.53**	(0.19, 0.86)	0.43*	(0.08, 0.77)
Middle wealth	0.01	(–0.13, 0.16)	–0.01	(–0.15, 0.14)
Highest wealth	–0.24**	(–0.41, –0.06)	–0.17	(–0.35, 0.003)
Model controls for	sex, baseline age, baseline CES-D score		sex, baseline age, baseline CES-D score, occupational class, education	

Notes: Linear regression coefficients reflecting change in depression score in retirement compared with those still working; 95%-confidence intervals shown in brackets; **$p < 0.01$, *$p < 0.05$.
Source: Own calculations based on English Longitudinal Study of Ageing, waves 2–4.

examines the impact of routes into retirement and wealth in retirement on change in the level of depression, using the same eight-item version of the CES-D scale used in the analysis of working beyond SPA. The analysis used a stepwise linear regression approach to modelling the difference between pre- and post-retirement depression scores, taken from the two waves before and after the respondent reported entering retirement. Longitudinal weights are applied in order to account for issues of survey attrition. Results are presented in Table 12.2.

Compared to remaining in the workforce, involuntary retirement was associated with an increase in depression scores of half a point (0.56), and this association remained significant in the fully adjusted model. Voluntary retirement was associated with a reduction in depression scores, although this was not significant. In terms of wealth in retirement, those in the lowest wealth category had the highest level of increase in depression score, by just over half a point (0.53), and this result remained significant after controlling for all factors. Belonging to the highest retired wealth group is associated with significantly lower depression than working, although this result is no longer significant after controlling for occupational class, which is, of course, strongly correlated with wealth and has corresponding significant effects on depression.

The finding that lower wealth in retirement is associated with a large increase in depression is particularly important to consider in light of the research concerning working beyond SPA. Extended working in poor quality employment, which is likely to be characterized by inadequate income for the tasks carried out, was associated with an increase of almost half a point in depression (0.45). Similarly, retiring in poorer circumstances is also associated with an increase in depression of a similar magnitude (0.43). Subsequently, although policy needs to be aimed at improving working circumstances to encourage a wider variety of workers to delay retirement, measures must also be taken to ensure that those who do retire from poor quality work are not left in circumstances which further deteriorate mental well-being.

12.5 Improving the circumstances of retirement

The analyses discussed so far have shown that working in later life, as opposed to retirement, may be beneficial to health as long as it is of good quality, but it may be detrimental when working conditions are not ideal. Similarly, retirement is only beneficial to mental health when the pathways into retirement and the circumstances within it are optimal. As lower levels of income and lower class positions are likely to be associated with participation in poorer quality work as well as with poorer retirement circumstances, it is particularly important to focus on the well-being of those in such circumstances. One interesting avenue to explore here is whether unpaid activities and the quality of those activities post-retirement impact on well-being, much as the quality of employment activities do pre-retirement. One such activity that lends itself to such an enquiry is voluntary work.

In fact, evidence suggests that later-life volunteering on its own has a protective effect on mortality (Smith 2004), self-rated health (van Willigen 2000; Morrow-Howell et al. 2003) and depression (Morrow-Howell et al. 2003; Hao 2008), and increases life satisfaction (van Willigen 2000). These beneficial effects might be especially pronounced among older people (Musick and Wilson 2003). The finding that volunteering, both in conjunction with paid work and on its own (Hao 2008), has beneficial impacts on well-being suits ideas of continuity and role theory, whereby individuals nearing the end of employment benefit from already having roles established which may be as rewarding as paid work in terms of remaining socially productive, socially integrated and maintaining purposeful and valued roles (Moen 1996).

A report produced by Nazroo and Matthews (2012) used waves 2 to 4 of the ELSA data to examine the impact of volunteering in later life on several dimensions of well-being. Depression was measured using the same eight-item CES-D scale discussed in relation to working beyond SPA. Quality of life was measured using the Control, Autonomy, Self-realisation and Pleasure (CASP)-19 score, which is a 19-item measure specifically designed for use among older populations and has a range from 0 to 57 (representing the highest quality of life). Life satisfaction is measured using the Diener five-item scale (Diener et al. 1985), with scores ranging from 1 to 15 (representing the highest life satisfaction). In addition, a measure of social isolation was included to explicitly examine whether volunteering might help those post-retirement to remain socially integrated. The measure used for this was a five-item scale covering companionship, feeling left out, feeling isolated, feeling 'out of tune' with others and loneliness, and which has scores ranging from 1 to 15 (highest feelings of social isolation). In this instance, respondents were included in the study if they were above SPA and had provided information at each of the three waves of interest to the analysis (3,632 cases).

Basic descriptive results showed rates of volunteering were much higher among those who were younger, wealthier and in better health. Among those in the highest wealth quintile, 42.5 per cent volunteered, compared to just 13.2 per cent of those in the lowest. Similarly, 41.3 per cent of those who reported their general health to be 'excellent' volunteered, compared with 8.5 per cent of those who classed their health as 'poor' (Nazroo and Matthews 2012).

In order to examine the longitudinal effects of volunteering, linear regression models were run to show changes in well-being scores between waves 2 and 3 for respondents volunteering at wave 2, between waves 3 and 4 for those volunteering at wave 3 but not wave 2 and between waves 2 and 3 for respondents who did not volunteer at either wave. The models controlled for baseline well-being. Differences in volunteers and non-volunteers were then examined on the basis of how appreciated volunteers felt for the work they provided (based on an according item indicating the extent to which they felt appreciated for the voluntary work they carried out), the number of voluntary activities participated in and the frequency of volunteering. The analysis also aimed to establish whether the effects of volunteering persisted after stopping participation in voluntary work, by comparing the change in well-being between waves 2 and 3 for those who volunteer at both waves and those who volunteer at wave 2 only, relative to those who do not volunteer at all (those who started volunteering at wave 3 were included in the analysis, but their results were similar to those who continuously volunteered, and so are not included in the table of results). Fully adjusted models controlled for age, gender and marital status, wealth quintiles, perceived own social status, self-reported health and difficulties with physical mobility, activities of daily living and instrumental activities of daily living. The results of the analysis are presented in Table 12.3. The results represent the change in well-being scores of volunteers compared to non-volunteers over a two-year period. So, for example, the coefficient of –0.21 for depression and all volunteering means the depression score for volunteers decreased by 0.21 points relative to those who did not volunteer, over a two-year period.

All significant results presented in Table 12.3 suggest a beneficial effect of volunteering on well-being. Overall, volunteers consistently see greater decreases in depression and social isolation, and greater increases in quality of life and levels of life satisfaction. In more detail, when scores of all volunteers are analysed simultaneously, volunteering is associated with significantly better quality of life and life satisfaction scores than not volunteering. It also shows a non-significant link with reduced depression and social isolation. However, previous research has suggested the greatest effects of volunteering are observable among those who feel their input is appreciated and well reciprocated (Siegrist et al. 2004a; McMunn et al. 2009; Siegrist and Wahrendorf 2009). Similarly, in this instance, volunteering for which the individual feels appreciated is associated with significantly lower levels of depression, and better quality of life and life satisfaction, while volunteering for which the individual does not feel appreciated has no significant effect on well-being and the coefficients are typically much smaller in magnitude. This finding reflects those we reported for employment earlier in the chapter, although that also showed that those in poorly reciprocated work saw significantly higher rates of depression than retirees. It may be that the lack of detrimental effects of poorly reciprocated voluntary

Table 12.3 Change in well-being over two years: Volunteers compared with non-volunteers (linear regression coefficients)

Type of volunteering (ref. non-volunteers)	Depression	Quality of life	Life satisfaction	Social isolation
All volunteering	−0.21	0.66**	0.69**	−0.09
Model controls for	age, sex, marital status, wealth quintile, self-perceived social status, self-reported health, physical mobility, difficulties with activities of daily living and instrumental activities of daily living, employment status, caregiver status			
Appreciated	−0.22**	0.74**	0.76**	−0.11
Unappreciated	−0.16	−0.08	0.17	0.07
Model controls for	age, sex, marital status, wealth quintile, self-perceived social status, self-reported health, physical mobility, difficulties with activities of daily living and instrumental activities of daily living, employment status, caregiver status			
Volunteers continuously	−0.16*	0.43	0.48*	−0.09
Stops volunteering	0.04	0.27	0.40	0.00
Model controls for	age, sex, marital status, wealth quintile, self-perceived social status, self-reported health, physical mobility, difficulties with activities of daily living and instrumental activities of daily living, employment status, caregiver status			
1–2 activities (all volunteering)	−0.41**	1.88**	0.49*	−0.12
3+ activities (all volunteering)	−0.67**	4.03**	1.15**	−0.40**
Model controls for	age, sex, marital status			
Volunteers < once a month	−0.39**	2.35**	0.75	−0.32*
Volunteers once a month	−0.57**	4.21**	1.73**	−0.29
Volunteers > once a month	−0.68**	3.72**	1.33**	−0.40**
Model controls for	age, sex, marital status			

Note: *p < 0.05; **p < 0.01.
Source: Nazroo and Matthews (2012), based on English Longitudinal Study of Ageing, waves 2–4.

work is a result of the active choice to participate in subjectively meaningful voluntary work, whereas this is unlikely to be the case among those who feel trapped to stay in poor quality work due to financial constraints, so-called job lock (Ekerdt et al. 1983).

The next analysis aims to show whether the effects of volunteering persist even after the respondent no longer participates in it. Table 12.3 shows a noticeable difference between the coefficients of those who volunteered across two waves of data and those who had stopped volunteering by the second wave. Those who participated in continuous volunteering saw significantly better depression and life satisfaction scores than non-volunteers, but those who had stopped volunteering saw no significant differences in any well-being scores compared to non-volunteers.

There is a notable difference in coefficients on the basis of the number of voluntary activities carried out. Those who participate in one or two activities of any volunteering type see significant improvements in their levels of depression, quality of life and life satisfaction compared to non-volunteers, but these effects are even greater among those who participate in three or more activities. Additionally, participating in three or more activities of any type is associated with significantly lower social isolation. Frequency of volunteering shares a similar pattern of results to number of activities carried out. Volunteering more than once a month has a strong significant association with all well-being outcomes. Volunteering once a month is associated with significantly lower levels of depression, higher quality of life and life satisfaction, and volunteering less than once a month is associated with less depression and higher quality of life. Coefficients are larger for those who volunteer once a month or more than once a month, suggesting that higher frequencies of volunteering are more beneficial than lower frequencies. Social isolation is the outcome which sees the most infrequent significant coefficients, yet it is worth noting that it is most significantly reduced among those who participate in the highest numbers of activities or who volunteer on the most frequent basis. This finding ties in with research by Smith (2004) who found effects of volunteering to be greatest among those who had low levels of social interaction.

12.6 Discussion

This chapter aimed to strengthen an understanding of the role of social and health inequalities across both the extension of working life and the retirement process. The analysis of working beyond SPA demonstrated that the effect of prolonging employment as a means of dealing with an ageing population will only be beneficial if inequalities persisting across the life course, such as those due to education, social status and employment type, are accounted for. While those in favourable occupations might see benefits to health of working beyond the traditional SPA, those who are forced to remain in poorer working conditions due to unaffordability of retirement are likely to see deterioration in health, and particularly in mental well-being.

The second piece of analysis considered the effects of differing pathways through retirement on inequalities in risk of depression. Involuntary retirement, including retirement on the basis of redundancy, illness or another's illness, was associated with higher levels of depression. Also retirement among those with lower levels of wealth was associated with an increase in depression of a similar magnitude to continued working in poor conditions. Both higher rates of disability and lower wealth are associated with poorer quality work, and so the importance of considering this group of people at risk of poorer mental health regardless of whether they continue to work or retire was highlighted.

The final piece of analysis considered the effects of volunteering after retirement on well-being among older people. Volunteering was consistently shown to be beneficial to well-being, especially when respondents self-reported feeling adequately appreciated for it, and when a higher number of activities were participated in. Consequently, it was considered that volunteering might be a potential way to improve the retirement experience among retirees by means of maintaining social networks and remaining socially productive and active. Participation in volunteering might be especially beneficial in improving the retirement circumstances of those with lower levels of wealth who also experience poorer mental well-being.

Research conducted over the past few years has provided mixed results when examining the effects of extended work and retirement on the health and well-being of older populations. The first two analyses discussed in this chapter demonstrated the importance of accounting for individual inequalities when considering how both extended working and retirement might affect health outcomes. The analysis of people who continued to work beyond retirement age demonstrated that prolonged working lives are only beneficial among those in high quality work, and that extending the working lives of those in poorer quality employment is likely to lead to higher rates of poorer mental health among older people. This finding supports the two key ideas proposed by role theory. On the one hand, continuation of roles which provide a sense of importance and social productivity can lead to better well-being than the loss of such roles. Conversely, however, the continuation of roles which are not appreciated or meaningful can be detrimental and better health would be achieved among those in this group who stop such activities (Kim and Moen 2001).

If policies to extend working lives are to be successful, the retirement period thereafter also has to be a time which is fulfilling and which one can expect to spend in good health. Previous research has demonstrated links to exist between wealth and the ability to enjoy retirement, with people with higher levels of wealth able to afford to participate in enjoyable leisure activities and therefore experiencing a higher sense of personal control and autonomy during retirement (Ross and Drentea 1998). Research has shown that not only are inequalities in wealth and health in retirement the product of inequalities across the life course (Blane et al. 1993; Marmot and Nazroo 2001), class inequalities also increase post retirement age (Marmot and Shipley 1996; Chandola et al. 2007). This suggests those individuals who participated in poor quality work, characterized by lower incomes, are likely to also be the individuals who continue to see poorer mental well-being after leaving the workforce, even when unfavourable roles have been exited. This complements the ideas of both the 'honeymoon phase' which immediately follows retirement (Atchley 1982: 127) and role strain theory (Ekerdt et al. 1983): The relief of stressful roles immediately following retirement provides a temporary increase in mental well-being, which subsides as the length of retirement increases and the individual adjusts to circumstances which are less than ideal in terms of being able to afford participation in preferred activities. Subsequently, policy needs to focus on dealing with those groups at risk of exposure to poorer circumstances, both prior to workforce exit and thereafter.

The finding that involuntary retirement is associated with a significant increase in depression corresponds with previous research which has found that early retirement is associated with increased depression when it is involuntary on the basis of health (Buxton et al. 2005; Butterworth et al. 2006), long-term unemployment, job loss or redundancy (Gallo et al. 2006; van Solinge and Henkens 2007; Mandal and Roe 2008). Again, the potential negative changes to well-being following job loss are likely to be especially prevalent when exit from the workforce occurs suddenly, earlier than expected and for unfavourable reasons (Herzog et al. 1991). The loss of personal control and positive roles gained through employment in such circumstances has been shown to have significant detrimental effects on a range of well-being outcomes across the retirement process (Kim and Moen 2002; Calvo et al. 2009), while voluntary retirement has shown links with higher well-being (Herzog et al. 1991; Reitzes and Mutran 2004).

The research on work and retirement discussed in this chapter highlights several key issues which must be considered in terms of changes to policy affecting later life. While changes to pension age might not be detrimental to the health of those working in good quality employment, evidence suggests those who continue to work in poor quality work will see poorer mental health. Additionally, the research highlights the importance of considering that it is likely to be those who have suffered poorer quality employment into later life who subsequently retire into lower wealth circumstances and, as a result, suffer further declines in mental well-being in the long run, even if the initial relief from unrewarding employment serves to improve health. Conversely, those who have enjoyed well-paid work and who felt their employment was well reciprocated in terms of appreciation, salary, prospects and security are those who are likely to retire with higher levels of wealth and be able to maintain good well-being through participation in preferred activities.

The final piece of analysis presented in this chapter focused on the effect of voluntary work among retirees. Participation in meaningful and rewarding social activities, such as voluntary work, provides a potentially useful means of improving the retirement experience, which might be particularly important for those in poorer retirement circumstances. Feeling integrated in society, remaining active and productive and maintaining good social networks, especially with tasks which the individual enjoys and feels appreciated for, are potentially a way to limit the negative effects on mental well-being which might be brought about by retirement, regardless of the route into it. Additionally, engaging in activities which promote a sense of autonomy and personal control offers further potential well-being benefits. These benefits might be continuous for those who have exited good quality employment beforehand, or new for those whose prior work was unrewarding. The greater involvement an individual has in voluntary work, the more likely they are to be performing roles with greater responsibility and importance as well as replicating the structure of employment. This may be especially relevant among those who retired involuntarily and may have suffered a sudden or unexpected loss of job roles. Additionally, volunteering may provide a particularly useful means of increasing well-being among those who suffer poorer mental health due to both poor quality employment and poorer retirement circumstances. However, it must be remembered that as inequalities in wealth are associated with inequalities in health and disability across both the working and retirement periods of life, we need to be careful not to exclude from activities such as volunteering those who might benefit the most from them. Special attention should be paid to ensuring those who suffer disadvantages in both wealth and health are as able to access relevant opportunities, services, networks and resources as readily as those who are in better social positions.

References

Adam, S., Bonsang, E., Germain, S. and Perelman, S. (2007), *Retirement and cognitive reserve: A stochastic frontier approach applied to survey data*, CREPP (Center of Research in Public Economics and Population Economics) Working Paper No. 2007/4, Liège: University of Liège.

Atchley, R. C. (1982), 'Retirement: Leaving the world of work', *The Annals of the American Academy of Political and Social Science*, 464 (1), 120–31.

Behncke, S. (2012), 'Does retirement trigger ill health?', *Health Economics*, 21 (3), 282–300.

Blane, D., Smith, G. D. and Bartley, M. (1993), 'Social selection: What does it contribute to social class differences in health?', *Sociology of Health & Illness*, 15 (1), 1–15.

Bonsang, E., Adam, S. and Perelman, S. (2012), 'Does retirement affect cognitive functioning?', *Journal of Health Economics*, 31 (3), 490–501.

Butterworth, P., Gill, S. C., Rodgers, B., Anstey, K. J., Villamil, E. and Melzer, D. (2006), 'Retirement and mental health: Analysis of the Australian national survey of mental health and well-being', *Social Science & Medicine*, 62 (5), 1179–91.

Buxton, J. W., Singleton, N. and Melzer, D. (2005), 'The mental health of early retirees', *Social Psychiatry and Psychiatric Epidemiology*, 40 (2), 99–105.

Calvo, E., Haverstick, K. and Sass, S. A. (2009), 'Gradual retirement, sense of control, and retirees' happiness', *Research on Aging*, 31 (1), 112–35.

Calvo, E., Sarkisian, N. and Tamborini, C. R. (2013), 'Causal effects of retirement timing on subjective physical and emotional health', *The Journals of Gerontology Series B: Psychological Sciences and Social Sciences*, 68 (1), 73–84.

Chandola, T., Ferrie, J., Sacker, A. and Marmot, M. (2007), 'Social inequalities in self-reported health in early old age: Follow-up of prospective cohort study', *BMJ*, 334 (7601), 990.

Coe, N. B. and Lindeboom, M. (2008), *Does retirement kill you? Evidence from early retirement windows*, CentER Discussion Paper Series No. 2008-93, Tilburg: Center for Economic Research, Tilburg University.

Coe, N. B. and Zamaarro, G. (2011), 'Retirement effects on health in Europe', *Journal of Health Economics*, 30 (1), 77–86.

Dannefer, D. (2003), 'Cumulative advantage/disadvantage and the life course: Cross-fertilizing age and social science theory', *The Journals of Gerontology Series B: Psychological Sciences and Social Sciences*, 58 (6), S327–37.

Diener, E. D., Emmons, R. A., Larsen, R. J. and Griffin, S. (1985), 'The satisfaction with life scale', *Journal of Personality Assessment*, 49 (1), 71–5.
Ekerdt, D. J., Bosse, R. and LoCastro, J. S. (1983), 'Claims that retirement improves health', *Journal of Gerontology*, 38 (2), 231–6.
Fries, J. F. (2002), 'Aging, natural death, and the compression of morbidity', *Bulletin of the World Health Organization*, 80 (3), 245–50.
Gallo, W. T., Bradley, E. H., Dubin, J. A., Jones, R. N., Falba, T. A., Teng, H. M. and Kasl, S. V. (2006), 'The persistence of depressive symptoms in older workers who experience involuntary job loss: Results from the health and retirement survey', *The Journals of Gerontology Series B: Psychological Sciences and Social Sciences*, 61 (4), 221–8.
Hao, Y. (2008), 'Productive activities and psychological well-being among older adults', *The Journals of Gerontology Series B: Psychological Sciences and Social Sciences*, 63 (2), S64–72.
Herzog, A., House, J. S. and Morgan, J. N. (1991), 'Relation of work and retirement to health and well-being in older age', *Psychology and Aging*, 6 (2), 202–11.
Kim, J. E. and Moen, P. (2001), 'Is retirement good or bad for subjective well-being?', *Current Directions in Psychological Science*, 10 (3), 83–6.
Kim, J. E. and Moen, P. (2002), 'Retirement transitions, gender, and psychological well-being: A life-course, ecological model', *The Journals of Gerontology Series B: Psychological Sciences and Social Sciences*, 57 (3), P212–22.
Mandal, B. and Roe, B. (2008). 'Job loss, retirement and the mental health of older Americans', *The Journal of Mental Health Policy and Economics*, 11 (4), 167–176.
Marmot, M. G. and Nazroo, J. Y. (2001), 'Social inequalities in health in an ageing population', *European Review*, 9 (4), 445–60.
Marmot, M. G. and Shipley, M. J. (1996), 'Do socioeconomic differences in mortality persist after retirement? 25 year follow up of civil servants from the first Whitehall study', *BMJ*, 313 (7066), 1177.
Matthews, K. (2014), *Is working beyond State Pension Age beneficial for health? Evidence from the English Longitudinal Study of Ageing*, Manchester: University of Manchester, https://www.escholar.manchester.ac.uk/api/datastream?publicationPid=uk-ac-man-scw:227703&datastreamId=FULL-TEXT.PDF, date accessed 23 June 2015.
Matthews, K., Chandola, T., Nazroo, J. and Pendleton, N. (forthcoming a), 'The effects of later-life employment in depression and self-rated health: A meta-analysis'. Currently under review with the *Journals of Gerontology, Series B*.
Matthews, K., Chandola, T., Nazroo, J. and Pendleton, N. (forthcoming b), 'The health effects of later-life working: Evidence from the English Longitudinal Study of Ageing'. Currently under review with *The Lancet*.
McMunn, A., Nazroo, J., Wahrendorf, M., Breeze, E. and Zaninotto, P. (2009), 'Participation in socially-productive activities, reciprocity and wellbeing in later life: Baseline results in England', *Ageing and Society*, 29 (5), 765–82.
Moen, P. (1996), 'A life course perspective on retirement, gender, and well-being', *Journal of Occupational Health Psychology*, 1 (2), 131–44.
Morrow-Howell, N., Hinterlong, J., Rozario, P. A. and Tang, F. (2003), 'Effects of volunteering on the well-being of older adults', *The Journals of Gerontology Series B: Psychological Sciences and Social Sciences*, 58 (3), S137–45.

Musick, M. A. and Wilson, J. (2003), 'Volunteering and depression: The role of psychological and social resources in different age groups', *Social Science & Medicine*, 56 (2), 259–69.
Nazroo, J. and Matthews, K. (2012), *The impact of volunteering on well-being in later life, WRVS report*, Cardiff, UK: Women's Voluntary Royal Service.
Office for National Statistics (2012), *Average age of retirement rises as people work longer*, Newport: ONS, http://www.ons.gov.uk/ons/dcp29904_256641.pdf, date accessed 3 November 2014.
Phillipson, C. and Smith, A. (2005), '*Extending working life: A review of the research literature, Research Report No. 299*, Leeds, UK: Corporate Document Services.
Radloff, L. (1977), 'The CES-D Scale: A self-report depression scale for research in the general population', *Applied Psychological Measures*, 1 (3), 385–401.
Reitzes, D. C. and Mutran, E. J. (2004), 'The transition to retirement: Stages and factors that influence retirement adjustment', *The International Journal of Aging and Human Development*, 59 (1), 63–84.
Rohwedder, S. and Willis, R. J. (2010), 'Mental retirement', *Journal of Economic Perspectives*, 24 (1), 119–38.
Ross, C. E. and Drentea, P. (1998), 'Consequences of retirement activities for distress and the sense of personal control', *Journal of Health and Social Behavior*, 39 (4), 317–34.
Rosenbaum, P. R. and Rubin, D. (1983), 'The central role of the propensity score in observational studies for causal effects', *Biometrika*, 70 (1), 41–5.
Siegrist, J. (1996), 'Adverse health effects of high-effort/low-reward conditions', *Journal of Occupational Health Psychology*, 1 (1), 27–41.
Siegrist, J., Knesebeck, O. von dem and Pollack, C. E. (2004a), 'Social productivity and well-being of older people: A sociological exploration', *Social Theory & Health*, 2 (1), 1–17.
Siegrist, J., Starke, D., Chandola, T., Godin, I., Marmot, M., Niedhammer, I. and Peter, R. (2004b), 'The measurement of effort-reward imbalance at work: European comparisons', *Social Science & Medicine*, 58 (8), 1483–99.
Siegrist, J. and Wahrendorf, M. (2009), 'Participation in socially productive activities and quality of life in early old age: Findings from SHARE', *Journal of European Social Policy*, 19 (4), 317–26.
Smith, D. B. (2004), 'Volunteering in retirement: Perceptions of midlife workers', *Nonprofit and Voluntary Sector Quarterly*, 33 (1), 55–73.
Steptoe, A., Breeze, E., Banks, J. and Nazroo, J. (2013), 'Cohort profile: The English longitudinal study of ageing (ELSA)', *International Journal of Epidemiology*, 42, 1640–8.
van Solinge, H. and Henkens, K. (2007), 'Involuntary retirement: The role of restrictive circumstances, timing, and social embeddedness', *The Journals of Gerontology Series B: Psychological Sciences and Social Sciences*, 62 (5), 295–303.
van Willigen, M. (2000), 'Differential benefits of volunteering across the life course', *The Journals of Gerontology Series B: Psychological Sciences and Social Sciences*, 55 (5), S308–18.

13
The Decline of 'Late Freedom'? Work, Retirement and Activation – Comparative Insights from Germany and the USA

Silke van Dyk

13.1 Introduction

When it comes to work and retirement, we establish three topics that are currently up for debate in several countries: the rising employment rates of older workers, increases in pension age and the growing prevalence of work beyond retirement. There are several drivers of these developments. First, they are part of fundamental welfare state reforms, as well as of new employment strategies that pursue full employment, including even those sections of the population who had traditionally been willingly excluded. Second, the rejuvenation of old age in medical and social terms has been ongoing for some time, with rising life expectancy and new images of ageing in media, advertising and everyday life. Finally, new generations of elderly have entered the stage with the post-war cohorts of the rather wealthy, healthy and better educated 'young-old' currently moving into retirement age.

Against this backdrop, current developments with regard to later life go far beyond the extension of working life. We are witnessing a fundamental renegotiation of old age that challenges the institutional and moral economy of retirement as deserved rest in a categorical way – and not just the concrete transition point from work to retirement. There is a wide coalition of international organizations, national governments and academic experts promoting new concepts of age and ageing. Back in 1999, which was declared the 'International Year of Older Persons' by the United Nations, the European Commission urged its member states to change 'outmoded practices' in relation to older persons: 'Both within labour markets and after retirement, there is the potential to facilitate the making of greater contributions from people in the second half of their lives' (European Commission 1999: 21).

The message is quite clear: old age is not about sitting in the wing chair, but it is an active phase of life and a time to give something back in return for pension benefits received and (future) care provisions. There is a variety of (partly) overlapping labels that describe this new perspective, with active ageing, productive ageing and successful ageing being the most popular ones. Additionally to the increase in state pension age and the rise of employment rates of older employees, the discussion is about broadening so-called voluntary duties, mostly past retirement, such as care work, mutual aid in neighbourhoods, civic engagement and a reasonably healthy lifestyle and health-conscious behaviour. Even though there is widespread agreement on a general tendency towards old-age activation in Western industrialized countries (Morrow-Howell et al. 2001; van Dyk and Lessenich 2009; Moulaert and Biggs 2012), country-specifics persist. Notably the significance of paid work beyond retirement age plays out very differently – empirically as well as normatively (Scherger et al. 2012; Boudiny 2013; see also the contributions in Part I of this book).

Against this backdrop, the aim of this chapter is twofold: First, I discuss the paradigm shift towards active and productive ageing for the German case, paying special attention to the role of paid work (before and beyond retirement age) within the wider context of activities and newly defined duties. An empirical research project dealing with discourses and politics of old age and retirement in Germany, covering the time period from 1983 to 2011 (van Dyk et al. 2013; Denninger et al. 2014), forms the basis of this discussion.[1] The reconstruction of change and transformation in Germany is contrasted with comparative insights from the US case based on literature studies and expert interviews. After critically reviewing the productive ageing paradigm and briefly discussing major blanks and pitfalls of the public discourse, this macro-perspective is finally set against the narratives of the target group of this discourse – the 'young-old' in Germany. What do they think about the claims of active and productive ageing and how do they live up to them?

13.2 Active and productive ageing in Germany – and comparative insights from the USA

In the above-mentioned research project, we analyse old age as a social dispositive (Foucault 1990). Dispositives can be understood as a composition of heterogeneous elements, namely discourses, institutions, objects/artefacts, practices and bodies – which together, as a complex ensemble, make up the powerful order of ageing and old age. Starting from the early 1980s, we identified three historically distinct age dispositives that are mutually overlapping and interfering (Denninger et al. 2014: 63–180): 'retired old age', 'restless old age' and 'productive old age'. Thus, there are actually two distinct dispositives – 'restless age' and 'productive age' – that in dialectic interplay constitute the 'active and productive ageing paradigm'.

13.2.1 Dispositives of old age: Retirement, restlessness and productivity

We were able to identify a classical retirement dispositive that can be found from the beginning of the period under study and serves as an omnipresent and contrasting background to the two historically younger dispositives. The guiding idea of the retirement dispositive is that of the older person effectively disengaging not only from employment but also from any sort of occupation. This idea implies two different aspects: On the one hand, it suggests that the old person is legitimately liberated from work or any other public responsibility and thus enjoying life's last chapter as a time of deserved rest financed by status-based public pension schemes (Kohli 1987). In German, *Ruhestand* (for retirement) has the double meaning of an empirical, factual *state* as well as a legitimate social *status* of rest after employment. The 1957 pension reform ended old age as the poverty-prone phase of life that it had been until the 1950s and has been the institutional foundation for the emergence of a retirement culture in (Western) Germany, based on a moral economy of 'late freedom' (Rosenmayr 1983). On the other hand, retirement connotes elderly people as being detached from society, and a whole range of negative stereotypes commonly associated with old age often accompanies this. The short form of the retirement story then reads *retreat – provision – freedom – inactivity – decay – emptiness*, with the typical old person leading a relatively inanimate life in the shadow of the 'real' social world (Denninger et al. 2014: 86–93).

Starting from this point, the 'restless age'[2] dispositive emerges in the late 1980s and becomes dominant by the mid-1990s at the latest. In line with the EU (European Union) active ageing formula – 'adding life to years' (European Commission 1999: 21), in the 'restlessness dispositive' the prototypical older person is depicted as a surprisingly juvenile, active and competent senior citizen integrated into the social fabric of an active society. Far removed from the imagery of retirement, a new notion of old age and of the elderly as 'best agers' being in permanent motion emerges in the public sphere. Without questioning the legitimacy of material security and freedom from productive duties in later life, what is effectively delegitimized is physical and mental immobility. With the bodies and brains of the elderly explicitly conceived of as being plastic and amenable to continuous exercise, keeping busy means actively avoiding degradation, frailty and the need for long-term care. Gerontological knowledge and its dissemination by mass media played an important role in this turn away from the former deficit perspective on people in retirement age (for example Baltes and Baltes 1989). Thus, in the context of the plausible conjunction *plasticity – self-initiative – healthy lifestyle – competencies*, active ageing stands for an auto-productive agency of older people motivated by the will to prevent or postpone personal dependency. At the same time, this dispositive also implies the image of the well-off, open-minded and cosmopolitan pensioner engaging in expensive and time-consuming leisure activities for the sake of individual fulfilment (Denninger et al. 2014: 93–126).

However, the 'restless age' dispositive has itself become overwritten, beginning in the late 1990s, by the emerging 'productive ageing' narrative, which has been dominating public discourses since the middle of the 2000s. Here, the guiding idea is that older people obviously could – and actually should – make use of the multiplicity of resources attributed to them in the context of the 'restless age' story, not only for themselves but also in hetero-productive ways. The 'productive ageing' story is no longer about *any* activity in old age being approvable. What really matters now is that older people's activities are useful to others – which implies activities from grandparenting and civic engagement to intragenerational care and extended working life. In 2005, the German government's influential fifth report on the situation of the older generation stated that old age has to be more than waiting for pension payments and enjoying the free time (BMFSFJ 2006). It is the ageing society itself which is said to have a legitimate claim on elderly people and their potentials. *Resources – potentials – responsibility – engagement – (public) benefit*: this is how the 'productive ageing' story goes. Compared to the two other dispositives, productive ageing is an *emerging* dispositive, which is not fully established yet. Discursive strategies to promote these new duties of the elderly – for example political campaigns, declarations, expert reports – are currently dominating this dispositive order, while corresponding institutions and well-established practices are still in their infancy. Interestingly, the ongoing political problematization of retirement as 'late freedom', which is characteristic of the productivity dispositive, reliably indicates that the moral economy of 'late freedom' is still well established in everyday life (Denninger et al. 2014: 127–50). In fact, this enduring influence accounts for the specifics of the German case when it comes to productive ageing and paid work.

13.2.2 Paid work and productive ageing

Several authors have analysed the conceptual development of active and productive ageing in documents of the World Health Organization (WHO), the United Nations (UN), the Organisation for Economic Cooperation and Development (OECD) and the European Commission. They have described a shift from a rather broad concept of active ageing to an increasingly productive notion 'restricting the social contribution of older adults to work and work-like activities' (Moulaert and Biggs 2012: 23–4), with 'first priority to the labour-market objective' (Boudiny 2013: 1081). Even though the labour market participation of older workers plays an important role in Germany too, we do not find a comparably restrictive notion of active and productive ageing.

On the contrary, the most prominent and controversial institutional reform, the gradual increase of the regular pension age to 67, is usually not discussed within the context of active and productive ageing. Whereas productive ageing is a paradigm of intended, paradigmatic change, the increase in pension age is instead presented as economic necessity that does not greatly impact on people's lives because they live so much longer nowadays. So, with regard to retirement age we rather find a discourse of continuity (by institutional adaption) than an ideological renegotiation of what old age means: the moral economy of late freedom *from work duties* is not challenged in principle, but politically promised to be preserved for all by harmonizing life expectancy and the length of working life. Nevertheless, raising the pension age to 67 has been and still is highly controversial and unpopular (for critical arguments see Baecker 2011). Diverse corrections and the extension of exemption rules since the gradual increase of pension age started in 2012 have to be read as political reactions to the immense social unpopularity of the reform.[3]

Therefore, it is not surprising that work beyond retirement age does not play a major role for active and productive ageing claims in Germany. At the same time, there is a striking discrepancy: even though the employment rates are indeed still low (see Hokema and Lux in Chapter 3 of this book) and even though only a minority strives for continued work,[4] there is a rising – though still small – number of (particularly) older women who remain employed due to their financially precarious situation (Butterwegge and Hansen 2012). More often than not, this is about work under non-regular, precarious conditions. While there is rising awareness of the upcoming problem of old-age poverty in general (Vogel and Motel-Klingebiel 2013), the situation of these women (and of course some men, too) is not taken into account when it comes to demands for productive ageing under voluntary conditions. However, lacking financial resources, on the one hand, and time constraints due to continued work, on the other, are major impediments to unpaid, 'voluntary' productive activities, which are in the focus of the German productivity dispositive.

Contrary to the increase in the pension age to 67 and work beyond pension age, the abolition of early retirement options and the general rise of older workers' employment rates before retirement age form an important part of the dispositives relating to restlessness and productivity. The restless age dispositive first emerged in the early 1980s in the context of growing numbers of early retirees which gave rise to concerns about what these healthy young elderly could do with their free time (Denninger et al. 2014: 86–7). So, initially restlessness linked to autoproductive, nonwork activities was promoted as an option for retirees in their 50s and early 60s in order to prevent an individual 'pension shock' or even 'pension death' (FAZ 1984). However, at that time, neither early retirement itself nor the encompassing disengagement beyond regular retirement age had been challenged yet (Kohli et al. 1988). Moderate attempts to reduce the number of pensioners in relation to contributors (Barkholdt 2001) were made from the late 1980s onwards. At the same time, however, and as a consequence of the reunification process, almost a whole generation aged 55 and older in the new Eastern federal states had been sent into early retirement (Zähle and Möhring 2010: 335–6). Against this backdrop, and with the majority of employees and employers backing the early retirement practice (Naegele 2002), the successive abolition of early retirement options remained highly controversial (for example BMFS 1993: 97–9) and far from effective until the early 2000s. It was not before the Lisbon strategy of the European Union on full employment, launched in 2000, that the employment opportunities of older workers and the continuing institutional incentives to retire early became a popular subject in politics and academia. Again, this development is not about the *abolition* of 'late freedom' as freedom from work, but – as the popular argument runs – about ceasing early retirement rules, in order to *preserve* freedom from work beyond regular pension age. This turns out to be very different in Anglo-Saxon countries, which I will demonstrate in the following section, using the example of the USA.

13.2.3 Comparative insights from the USA

Starting out from the pension reform in 1957, the German discourse moved from 'old age as [a] problem' (Leisering 1993) to retirement as 'late freedom'. As outlined above, this retirement culture had been successively transformed within the context of the restlessness dispositive during the 1980s and 1990s. Although this was the starting point for increasingly linking retirement with activity, the basic principles of financial security and liberation from productive duties remained unchallenged. The US discourse, by contrast, transferred the deficit perspective of 'old age as a problem' to a focus on old age as a productive resource as early as the late 1970s and early 1980s (Butler and Gleason 1985).

Even though the USA had been quite successful during the 1970s in reducing the poverty risk in old age, Social Security, the American public pension, was never meant to be the only source of income, and retirement was not institutionalized as a phase of freedom and legitimate rest. Nevertheless, with regard to generally low levels of public welfare provision in the USA, the elderly still turned out to be a privileged group within the welfare system, which gave rise to an opinion shift towards a more critical view of elderly: 'Until the early 1980s, the elderly had enjoyed a privileged status among welfare-state beneficiaries – built on the image of older people as poor, frail, and dependent. But as the generational equity campaign portrayed them as politically powerful, selfish, and potentially dangerous, the dynamics of interest-group liberalism were turning against them' (Cole 1997: 234). Though contested, this reading converged with other developments that brought forward productive ageing. First, the growing amount of gerontological research that dismissed the deficit perspective increasingly questioned the legitimacy of an (indeed never realized) adequately funded retirement free of any obligation (Donicht-Fluck 1989: 241–2). At the same time, the institution of (mandatory) retirement was challenged by neo-liberal actors who had gained influence after Ronald Reagan had won the presidential election in 1980. And interestingly, grass-roots initiatives like the Gray Panthers Movement and left-wing gerontological scholars, who had established the Marxian *Political Economy of Aging* (Townsend 1981), joined the fight, although from a different perspective than neo-liberal actors and without an equivalent in Germany: they considered 'forced exclusion from work [...] as a major source of inequality in the North American context', whereas European pensioners' movements and unions have tended 'to focus on opposition to work and the need for state-funded economic support following retirement' instead (Estes et al. 2003: 76). So, back in the 1970s and 1980s, the abolition of mandatory retirement, finally enacted by the *Age Discrimination in Employment Act* in 1986, was a major issue for social movements and lobby groups. In contrast to the moral economy of 'late freedom' (from work) in Germany, a wide coalition of actors stressed the 'enforced idleness' (Select Committee on Aging, US House of Representatives 1981) of retirement, which was seen as exclusion and not as liberation. To this day the Act is undisputedly considered a key step against the discrimination of old age (Nelson 2007: 58–9), and the concept of productive ageing became inextricably linked to anti-discrimination and the acceptance of seniors as equal citizens in the USA (Butler 1975; Achenbaum 2009). Unsurprisingly, there is no true equivalent to the German restlessness dispositive with its neglect of productive activities to be found in the USA, as there was no cultural legitimation of self-sufficient, autoproductive leisure pursuits in old age possible without generating social expectations towards productive commitments.

Compared to the USA, the rise of the productivity dispositive in Germany has not just been 'delayed' by over 20 years, but it stems from a different cultural and institutional background: It is not linked to anti-discrimination issues, which are still rarely addressed in German politics and academia (see as an exception Brauer and Clemens 2010), but exclusively rooted in the crisis discourse of demographic ageing and welfare state exhaustion. The German productivity dispositive comes as a government programme 'from above', backed up by some gerontologists and other academic experts, whereas grass-roots movements do not play a role.

The German political discourse focuses on the problem of supposedly 'false' negative images of ageing – for instrumental reasons: the main political message is that these images have to be turned round and brightened, not because of their oppressive character, but in order to make the productive resources of elderly people accessible. The latest government report on the situation of older people states that 'a new concept of old age is required that stresses the capabilities and strengths of older people, in order to push their productive contribution to economy and society' (BMFSFJ 2010: V, own translation). According to this view, age discrimination is not a structural problem but a kind of misunderstanding, based on wrong information about capable elderly people. What Bill Bytheway has called the confusion 'to equate an anti-ageist stance with thinking positive' (Bytheway 1995: 128) can thus also be seen for the German context. Thinking (exclusively) positively about old age, however, generates pressure and discriminatory actions towards those who do not comply with the positive images based on capability, independence and productivity.

What do we learn from the comparative perspective? The focus on paid work (beyond retirement age) and the sensitivity for ageism is much stronger in the USA than in Germany. Different concepts of welfare and public services – individualistic, liberal and residual in the USA, more comprehensive, corporatist and status related in Germany (Esping-Andersen 1990) – explain these differences. And while productive ageing is a quite recent 'invention' of politicians and academics in Germany, the history of the paradigm is considerably different in the USA, with strong grass-roots influences. At the same time and despite the instrumental and paternalistic top-down activation, older people in Germany have more (not least financial) room to reject demanded activities – as I will show in Section 13.3. Even though the US history of the productive ageing paradigm is inextricably linked to the rising awareness of ageism and exclusion, the targeted inclusion into the labour market is very ambivalent in its consequences (see also Lain in Chapter 2 of this book), with high numbers of people aged 65 and older who simply cannot afford to retire:

> The late twentieth-century 'discovery' of age discrimination and consequent legislative action may signal a new recognition of the rights of older people. On the other hand, it may – paradoxically – herald an era in which age protection is diminished, state pension ages will rise, and there will be workfarist policies designed to force older people into the kinds of new low-grade, poorly remunerated jobs [...]. That is the dilemma.
>
> (Macnicol 2009: 246)

Despite downward pressure on old-age pensions and rising old-age poverty rates in Germany (Becker 2012), financial security as well as care provision and health care in particular are still much more comprehensive than in the USA with its traditionally low replacement rates (ILO 2010) and less generous health-care provision for older people in Medicare. Poverty rates of older adults in the USA are about twice as high as those in Germany (ILO 2011), and insecurity about future pensions has considerably increased with the shift from defined benefit to defined contribution plans (Rix 2009). Consequently, besides the high employment rates before retirement age, work beyond retirement age was never uncommon and has further risen (see Marshall in Chapter 9 of this book). According to recent polls, 35 per cent of adult workers nowadays believe that they will continue to work after retirement age for financial reasons (Gallup 2013). At the same time, however, high numbers (about 40 per cent) express their general *willingness and wish* to work beyond retirement age – though under favourable conditions with reduced hours, challenging tasks and great(er) flexibility (Schulz and Binstock 2006: 153–4). This expressed desire to work mirrors a culture of later life that rather lacks the moral economy of late freedom (from work).

13.2.4 Productive ageing revisited – critical remarks

In a sociological account, a number of pitfalls and problems characterize the policies of active and productive ageing. First, it improperly homogenizes the very heterogeneous group of 'the elderly', suggesting every single one of them has high potential and is just waiting to be discovered as a productive citizen. Furthermore, it effectively produces a new norm of ageing; a single standard which is not achieved by all those who are less resourceful than the prototypical 'best ager'.[5] The success or – more importantly – failure of the elderly to keep up with the programmatic expectations is attributed to them individually, converting not only personal well-being but also the 'public good' into a matter of individual behaviour (Rudman 2006; van Dyk 2009). In fact, policies of productive ageing go along with major welfare state changes and cutbacks, taking up and implementing the neo-liberal claim of individual responsibility and the individualization of (social) risks. The paradigm follows the idea that productive elderly are not just capable but responsible 'to fill gaping holes in the safety net' (Minkler and Holstein 2008: 197).

Whereas there is a strong Anglo-Saxon tradition of Critical Gerontology that has been addressing these problems for almost three decades now (Estes et al. 2003), these works are rarely received in German-based gerontology. Even though German social gerontology has always been sensitive towards social inequalities and diversities among the elderly, the critique of active and productive ageing remains rather weak – exceptions as always proving the rule (see Schroeter 2000; van Dyk and Lessenich 2009). Active and productive ageing is instead approved *in principle* and acknowledged as a promising *win-win* situation in times of demographic ageing (even more so by economists and psychologists, for example Grabka 2013), and only criticized with regard to its application in increasingly homogenizing and utilitarian terms. There are 'warnings' that the elderly should not be exploited and that the productivist predominance at the expense of well-being in later life should be prevented (for example Backes 2006). The paradigm, however, is not challenged in categorical ways with regard to its socio-economic context conditions of active society and flexible capitalism. Anglo-Saxon critical gerontologists, on the contrary, counter that the exploitation is not the accident but the rule under these conditions, deeply rooted in the concept itself (Estes et al. 2003: 74–5).

Because of this weak critique, additionally overshadowed by the strong individualist influence of psychology-based gerontologists, structural impediments to productivity, social inequalities or structural contradictions such as productive activities that are detrimental to health do not play a major role in current public discourses in Germany. Connected to the absence of a strong critical gerontological tradition, the blank space ageism plays a major role for this rather uncritical stance. Whereas even moderate critics and Anglo-Saxon gerontologists who do welcome productive ageing in principle discuss possible conflicts 'between ageism and productive aging' (Walker and Taylor 1993: 77), the German-based political discourse tends to promise the revaluation of old age in exchange for the productive performance of the elderly (BMFSFJ 2010). As the influential German government reports on the situation of older people, which are based on academic expertise, demonstrate, this perspective is backed rather than challenged by academics. Apart from the fact that a true anti-discriminatory stance has to be unconditional by definition, this promise is misguided, since the deficit perspective on those who are considered to be really old (that is, frail, sick, dependent) remains unchallenged (Holstein and Minkler 2003: 793–4).

Moreover, even the most positive images of the capable 'young-old' are not just unduly homogenizing but far from being as positive as it might appear at first sight. A good deal of the German media, academic and political attention focuses on older workers aged about 50 to 65. Interestingly, these indeed very young old are not characterized as 'normal' adults in pre-retirement age, which might have been the expectation according to the continuity claim with regard to the extension of working life. Instead, they are introduced as 'the others' who have important skills and potentials: These 'young-old' are considered to possess wisdom and experience, to be loyal to employers, conscientious and reliable in all respects. These qualities could – as the discourse suggests – compensate for the loss of creativity, spontaneity and flexibility.[6] Despite all the fuss about efforts to overcome the deficit perspective on old age, these losses are taken for granted. The main part of the message '... slow, non-creative, inflexible (and the like), but... loyal, experienced...' is that people aged 55 or even 50 and older are exactly this: rather slow, inflexible and non-creative. So, the asserted revaluation at the same time bears a stereotyped devaluation, since the praised experience has to compensate for a long list of deficits. If we take into account the wider social context of active and productive ageing – flexible capitalism and active society, based on flexibility, mobility and creativity (Sennett 1998) – it is obvious what this means for those who are – at least implicitly – addressed as quite the opposite.

Since productive ageing is framed as a win-win situation that benefits both society and the elderly at the same time, it suggests, amongst others, expressing the opinions, preferences and critiques of the addressees: the young-old. However, as 'expectations around active ageing are typically defined by policy makers, service planners and allied researchers' (Stenner et al. 2011: 469), there is little information so far on what older people think themselves – at least beyond standardized answers to questionnaires. The following section is dedicated to giving a voice to the young-old.[7] I will focus on commonalities of the interviewees in relation to the old-age dispositives, thereby deliberately neglecting differences among them.

13.3 'My own piece of life' – the young-olds' perspectives on late freedom

Despite the growing influence of the productivity dispositive, the biographical interviews reveal the ongoing influence of the normative world of retirement (Denninger et al. 2014: 217–47). In particular, the legitimized and institutionalized liberation from employment duties proves to be widely unchallenged; contrary to the Anglo-Saxon context, the strong norm of retirement (age) is not up for discussion except for a very small and privileged minority, mostly self-employed. The vast majority tends to regard retirement as 'late freedom' (Rosenmayr 1983), with the term 'freedom' being prevalent in most of the interviews, and they appreciate being liberated from stress and pressure. 'Late freedom' still means 'freedom *from* work' and not 'freedom *to* work'.[8] Without explicitly speaking about alienation and heteronomous working conditions, such experiences of working serve as the background against which the freedom of retirement and the chance for self-determined time use are valued.[9] This is true not just for lower educated people and blue-collar workers, who have suffered the effects of physically strenuous or monotonous work, but indeed for many professionals as well.

Being free is cherished, but it is not regarded as being free for the purpose of doing nothing or resting all the time. The general appreciation of institutionalized freedom goes hand in hand with a very negative perspective on what is thought to be a standard – passive and empty – retirement life, from which the interviewees distance themselves emphatically. They think that the majority of their fellow retirees are passive and portray themselves (as well as their partners and friends) as major exceptions. It is not retirement as such, then, which is delegitimized or challenged by the young-old, but, more concretely, domestic passivity. In fact, there is a 'little restless ager' in most of our interviewees, since emphasizing self-responsibility in health issues, claiming to be 'younger' than former generations were at the same age, and expanding one's activities (from the domestic sphere into the public) have become important parts of everyday life of most elderly people. Nevertheless, many interviewees scoff not only at the supposedly passive retirees but also at those who rush around as if they were still working.

In accordance with other research, the interviews demonstrate that 'being active [is] universally regarded as desirable and even essential' (Venn and Arber 2011: 203). However, this attitude is not the proof of the 'active and productive ageing' agenda actually influencing people's concept of (active) retirement. Instead, we find a more general recourse to activity as a *modus vivendi* in our sample of interviews, as a simple matter of being (and keeping) alive, nourished by the counter-image of dependent very old age. The interviewees talk about their (highly negative) vision of very old age very vividly, often even equating life close to death with a state of post-humanity. This abysmal imagery helps to understand that people's activity claims are not so much related to active ageing specifically but are meant in a much more basic, vitalist sense – as a tribute to 'life as the highest good' (Arendt 1998: 399). This perspective is close to the concept of 'vital aging', initially advocated by the US National Council on Aging 'as a way to underline the capacity of every individual to celebrate and engage in the gift of life' (Achenbaum 2009: 49).

At the same time, this vital perspective that denigrates passivity is easily linked to the hegemonial interpretation and renegotiation of old age in productive notions: 'Active ageing' works as an 'empty signifier' (Laclau 1996: 36), effectively incorporating different notions of activity that eventually, however, could be reframed according to productivity standards. Consequently, vital people of a certain age are discursively turned into active elderly who seem to behave in accordance with the public call for productive ageing.

If productive ageing is mentioned at all, the interviews reveal that this dispositive is much less entrenched in older people's narratives than the 'retirement' or 'restless age' dispositives. Many interviewees have never heard of productivity claims addressed to elderly people. Core elements of the 'productive ageing' dispositive, like the state-led interest in elderly people's resources and experiences, are of marginal relevance in the interviews or are explicitly rejected: roughly half of the interviewees believe (without having been asked directly) that the life experience of the elderly is nowadays not valued at all.[10] The popular claim of the productivity dispositive that there is a new appreciation of the elderly's resources and capabilities does obviously not coincide with these 'young-old' people's everyday experience. And those who mention productivist images of old age exclusively speak about calls for voluntary work and civic engagement, not about paid work, and tend to be critical about the new expectations: 'These permanent calls for voluntary work, based on the claim that there is nobody else to do it, since the state is stepping back – if you ask me, that's no good, I feel like I'm having to replace the state' (Mr. Liebig).

In spite of this, we find several interviewees who actually live up to the productivity claims, as they are very engaged in voluntary work or care duties. However, the most productive interviewees are particularly critical of activation policies, and often consider their own commitment as being self-evident and/or as a matter of personal preference, without expecting others to do the same. Many of the rather privileged 'productive agers' in our sample emphasize that a special commitment to the public good should not be expected from those with restricted financial and educational resources, who would be overburdened by corresponding obligations.

13.4 Conclusions

The interviews show clearly that the simple win-win narrative that productive ageing is 'good for everyone' (Walker 2002: 137) does not hold against empirical evidence. Hence, the productivity claim is not just problematic with regard to its political implications, but it falls short of the real lives of older persons. In particular, the popular promise that productive commitments involve a new appreciation of the elderly's virtues does not hold true in the view of many interviewees. On the contrary, many of them report the opposite, namely discriminatory experiences at the workplace as well as in civic engagement. However, and differently to their US contemporaries, they do not problematize these experiences as ageism, but tend to regard them as normal or inevitable. Except for some interviewees from the Eastern states of Germany, who characterize current policies as age-discriminatory compared to the former German Democratic Republic, ageism seems to be 'unspeakable' within the German context. Whereas many interviewees are sensitive and critical about sexism and discrimination based on differences between East and West Germany, ageism is effectively absent from public debates and thus not a meaningful concept for many.

Apart from this, the analysis of the German case exemplifies that researchers should not equate prevalent public discourses with their immediate success on the level of individual actors. Foucauldian-style *Governmentality Studies* in particular, which instructively aim to explore the complex power/knowledge arrangements that 'govern' the self-formation of late-modern subjects (Dean 2010), are prone to this shortcut: at least implicitly, they take it for granted that social discourses and dispositives actually also work on the micro level, guiding people's self-conceptions and practices. The presented results, however, indicate that beliefs and practices incompatible with productive ageing persist. In fact, they are nurtured by a deeply rooted retirement culture along with sufficient financial resources that allow for a modernized retirement life with elements of restlessness – at least for those who desire it. This retirement culture goes along with a concept of 'vita activa' that is not identical with productivity claims.

Contradicting the popular demand for the elderly's productive commitment to the public good – either by voluntary or by paid work – I conclude that there are several reasons to defend retirement as 'late freedom' from duty. Biographical interviews with the 'young-old' as well as studies of the views of adults in their 40s and 50s make it obvious that the vast majority welcomes the freedom from work and duty as well as the chance for autonomy in later life. Interestingly, these wants and hopes do not decline for younger cohorts, quite the contrary, as recent research suggests: major changes in the labour market, the increasing insecurity and flexibility of employment as well as the remaining leisure time being penetrated by work issues rather reinforce than diminish the desire for the 'late freedom' of retirement (Behr and Hänel 2013). Alienation, exploitation, insecurity and heteronomy, however, are absent in the debate on productive ageing that exclusively stresses the positive effects of productive duties – for instrumental reasons. Without a doubt it is a less than ideal solution to use fixed age limits in order to regulate when one needs to be productive and when one is liberated from the need to sell one's labour. In light of the universal trend towards radicalized marketization and commodification, however, this means of social protection is better than nothing.

Notes

1. In addition to concrete policies and their promotion, more than 2,200 text sources in four daily newspapers, several general and special interest magazines, as well as administrative and expert reports and political party programmes have been analysed.
2. In German, the term being used and often repeated to denote what is to be told and known by way of this story is 'Unruhestand', a neologism untranslatable into English (literally meaning 'state of unrest').
3. According to a recent survey, 64 per cent of the German population are in favour of suspending the raising of the pension age to 67 (Forsa 2012).
4. It has to be mentioned that quantitative data on the willingness or wish to continue paid work beyond retirement age vary widely for Germany, probably dependent on the exact question asked and the age groups under investigation. The share of those wanting to work longer ranges from less than 10 per cent (for example, Hertie-Stiftung 2013; IG-Metall 2013) to more than 40 per cent, with the latter share, however, being an exception (Dorbritz and Micheel 2010). In spring 2014, the Institut für Demoskopie Allensbach found that 19.7 per cent of respondents aged 60 and older are willing to continue work beyond retirement age, whereas the rates for those aged 60 and younger are less than 10 per cent (Institut für Demoskopie Allensbach 2014).
5. Of course, as is always the case with political programmes, the paradigm itself is fluctuating: recently, attempts have been made to redefine 'active ageing' in post-productivist terms, suggesting a multidimensional concept of activity that is not out of reach for very old and frail people (BMFSFJ 2010; Moulaert and Biggs 2012).
6. These results are backed by a recent survey among employers who were asked to name typical qualities of younger and older employees (IAB company panel, quoted from: Burkert and Sproß 2010).
7. The empirical basis are 55 qualitative, problem-centred interviews with people aged between 60 and 72 who had been retired for at least one year and who were not dependent on care at the time of the interview in 2010 and 2011.
8. In contrast, ten biographical interviews with pensioners aged 70 and older in New York, conducted by the author in spring 2010, reveal that the word 'retirement' has highly negative connotations for the US interviewees and does not represent the 'freedom' notion that is so important in Germany. Even a retirement with adequate resources tends to be regarded as institutionalized discrimination. And whereas age discrimination and ageism take centre stage in the narratives of these US elderly, this perspective remains rather implicit in the German-based interviews.

9. At the same time, many talk about the challenge to structure the day all by themselves since they never learned to do so during their working life: 'The great freedom is actually what is most difficult' (Ms. Altenberger).
10. Many elderly from the former German Democratic Republic openly doubt that there could be a public interest in their potentials and resources since they experienced the opposite after 1989 when almost a whole generation of those aged 50 and older were sent into early retirement or long-term unemployment.

References

Achenbaum, W. A. (2009), 'A history of productive aging and the boomers', in: R. B. Hudson (ed.), *Boomer bust? Economic and political issues of the graying society*, Westport/London: Praeger, 47–60.

Arendt, H. (1998 [1958]), *The human condition*, Chicago: University of Chicago Press.

Backes, G. (2006), 'Widersprüche und Ambivalenzen ehrenamtlicher und freiwilliger Arbeit im Alter', in: K. Schroeter and P. Zängl (eds.) *Altern und bürgerschaftliches Engagement*, Wiesbaden: VS Verlag, 63–94.

Baecker, G. (2011), *Rente mit 67? Argumente und Gegenargumente*, Bonn: Friedrich-Ebert-Stiftung.

Baltes, P. B. and Baltes, M. M. (1989), 'Optimierung durch Selektion und Kompensation. Ein psychologisches Modell erfolgreichen Alterns', *Zeitschrift für Pädagogik*, 35 (1), 85–105.

Barkholdt, C. (ed.) (2001), *Prekärer Übergang in den Ruhestand*, Opladen: Leske + Budrich.

Becker, I. (2012), 'Finanzielle Mindestsicherung und Bedürftigkeit im Alter', *Zeitschrift für Sozialreform*, 58 (2), 123–48.

Behr, M. and Hänel, A. (2013), 'Höher qualifizierte Angestellte als Lebenskraftkalkulierer', *WSI Mitteilungen*, 66 (2), 98–106.

BMFS (Bundesministerium für Familie und Senioren) (1993), *Erster Altenbericht der Bundesregierung. Die Lebenssituation älterer Menschen in Deutschland*, Bonn: BMFS.

BMFSFJ (Bundesministerium für Familie, Senioren, Frauen und Jugend) (2006), *Fünfter Bericht zur Lage der älteren Generation. Potenziale des Alters in Wirtschaft und Gesellschaft*, Berlin: BMFSFJ.

BMFSFJ (2010), *Sechster Bericht zur Lage der älteren Generation. Altersbilder in der Gesellschaft*, Berlin: BMFSFJ.

Boudiny, K. (2013), 'Active ageing: from empty rhetoric to effective policy tool', *Ageing and Society*, 33 (6), 1077–98.

Brauer, K. and Clemens, W. (eds.) (2010), *Zu alt? 'Ageism' und Altersdiskriminierung auf Arbeitsmärkten*, Wiesbaden: VS Verlag.

Burkert, C. and Sproß C. (2010), 'Früher oder später: Altersbilder auf Arbeitsmärkten im europäischen Vergleich', in: K. Brauer and W. Clemens (eds.), *Zu alt? 'Ageism' und Altersdiskriminierung auf Arbeitsmärkten*, Wiesbaden: VS Verlag, 149–70.

Butler, R. N. (1975), *Why survive? Being old in America*, Baltimore/London: Johns Hopkins University Press.

Butler, R. N. and Gleason, H. P. (eds.) (1985), *Productive aging: Enhancing vitality in later life*, New York: Springer.
Butterwegge, C. and Hansen, D. (2012), 'Altersarmut ist überwiegend weiblich', in: C. Butterwegge, G. Bosbach and M. W. Birkwald (eds.), *Armut im Alter*, Frankfurt/New York: Campus, 111–29.
Bytheway, B. (1995), *Ageism*, Buckingham and Philadelphia: Open University Press.
Cole, T. R. (1997), *The journey of life. A cultural history of aging in America*, Cambridge: Cambridge University Press.
Dean, M. (2010), *Governmentality. Power and rule in modern society*, Los Angeles: Sage.
Denninger, T., van Dyk, S., Lessenich, S. and Richter, A. (2014), *Leben im Ruhestand. Zur Neuverhandlung des Alters in der Aktivgesellschaft*, Bielefeld: Transcript.
Dorbritz, J. and Micheel, F. (2010), 'Weiterbeschäftigung im Rentenalter', *Bevölkerungsforschung Aktuell*, 31 (3), 2–7.
Donicht-Fluck, B. (1989) 'Neue Alte in den USA', in: D. Knopf, O. Schäffter and R. Schmidt (eds.) *Produktivität des Alters*, Berlin: DZA, 232–56.
Esping-Andersen, G. (1990), *The three worlds of welfare capitalism*, Cambridge: Polity Press.
Estes, C. L., Biggs, S. and Phillipson, C. (2003), *Social theory, social policy and ageing*, Berkshire: Open University Press.
European Commission (1999), *Towards a Europe for all ages. Promoting prosperity and intergenerational solidarity, COM 221 final*, Brussels: Commission of the European Communities.
FAZ (Frankfurter Allgemeine Zeitung) (1984), *'Und dann steht man da.' Wenn Arbeit nicht mehr das Leben ist*, 15 December 1984.
Forsa (2012), *Meinung zur Rente mit 67*, Berlin: Forsa, http://de.statista.com/statistik/daten/studie/12996/umfrage/beibehaltung-oder-ruecknahme-der-rente-ab-67, date accessed 2 August 2013.
Foucault, M. (1990), *The history of sexuality. Vol. 1: The will to knowledge*, London: Penguin Books.
Gallup (2013), *Three in four U.S. workers plan to work past retirement age*, by L. Saad, 23 May 2013, www.gallup.com/poll/162758/three-four-workers-plan-work-past-retirement-age.aspx, date accessed 10 March 2015.
Grabka, M. (2013), 'Aktives Altern – Erwerbstätigkeit und bürgerschaftliches Engagement', *WSI Mitteilungen*, 66 (5), 329–37.
Hertie-Stiftung (2013), *Arbeit und Alter. Unternehmens- und Beschäftigtenumfrage*, Frankfurt am Main: Hertie-Stiftung.
Holstein, M. B. and Minkler, M. (2003), 'Self, society and the "New Gerontology"', *The Gerontologist*, 43 (6), 787–96.
IG Metall (Industriegewerkschaft Metall) (2013): *Ergebnisse einer Umfrage zum Renteneintrittsalter: Jeder Zweite will vor 60 in Rente*, Frankfurt am Main: IG Metall, http://www.igmetall.de/ergebnisse-einer-umfrage-zum-renteneintrittsalter-12134.htm, date accessed 10 March 2015.
ILO (International Labour Organization) (2010), *World social security report 2010/2011: providing coverage in times of crisis and beyond*, Geneva: ILO.
ILO (2011), *Social security for social justice and a fair globalization*, Geneva: ILO.

Institut für Demoskopie Allensbach (2014): *Umfrage: Viele wollen länger arbeiten*, Archiv-Nr., 11021, Allensbach: IfDA.
Kohli, M. (1987), 'Ruhestand und Moralökonomie', in: K. Heinemann (ed.), *Soziologie wirtschaftlichen Handelns*, Opladen: Westdeutscher Verlag, 393–416.
Kohli, M., Gather, C., Künemund, H., Mücke, B., Schürkmann, M., Voges, W. and Wolf, J. (1988), *Leben im Vorruhestand*, Düsseldorf: Hans-Böckler-Stiftung.
Laclau, E. (1996), *Emancipation(s)*, London: Verso.
Leisering, L. (1993), 'Zwischen Verdrängung und Dramatisierung. Zur Wissenssoziologie der Armut in der bundesrepublikanischen Gesellschaft', *Soziale Welt*, 44 (4), 486–511.
Macnicol, J. (2009), 'Differential treatment by age: Age discrimination or age affirmation?', in: R. B. Hudson (ed.), *Boomer bust? Economic and political issues of the graying society*, Westport/London: Praeger, 41–51.
Minkler, M., and Holstein, M. (2008), 'From civil rights to...civic engagement? Concerns of two older critical gerontologists about a "new social movement" and what it portends', *Journal of Aging Studies*, 22 (2), 196–204.
Morrow-Howell, N., Hinterlong, J. and Sherraden, M. (eds.) (2001), *Productive aging. Concepts and challenges*, Baltimore/London: Johns Hopkins University Press.
Moulaert, T. and Biggs, S. (2012), 'International and European policy on work and retirement: Reinventing critical perspectives on active ageing and mature subjectivity', *Human Relations*, 66 (1), 23–43.
Naegele, G. (2002), 'Active strategies for older workers in Germany', in: M. Jespen, D. Foden and M. Hutsebaut (eds.), *Active strategies for older workers*, Brüssel: ETUI, 207–43.
Nelson, T. D. (2007), 'Ageism and discrimination', in: J. E. Birren (ed.), *Encyclopedia of gerontology*, Amsterdam: Elsevier, 57–64.
Rix, W. E. (2009), 'Will the boomers revolutionize work and retirement?', in: R. B. Hudson (ed.), *Boomer bust? Economic and political issues of the graying society*, Westport/London: Praeger, 77–94.
Rosenmayr, L. (1983), *Die späte Freiheit. Das Alter, ein Stück bewusst gelebten Lebens*, Berlin: Severin und Siedler.
Rudman, D. L. (2006), 'Shaping the active, autonomous and responsible modern retiree', *Ageing and Society*, 26 (2), 181–201.
Scherger, S., Hagemann, S., Hokema, A., Lux, T. (2012), *Between privilege and burden. Work past retirement age in Germany and the UK, ZeS-Working Paper No. 4/2012*, Bremen: Centre for Social Policy Research.
Schroeter, K. R. (2000), 'Die Lebenslage älterer Menschen im Spannungsfeld zwischen "Freiheit" und "sozialer Disziplinierung"', in: G. Backes and W. Clemens (eds.), *Lebenslagen im Alter*, Opladen: Leske + Budrich, 31–52.
Schulz, J. H. and Binstock, R. H. (2006), *Aging nation. The economics and politics of growing older in America*, Westport/London: Praeger.
Select Committee on Aging, U.S. House of Representatives (1981), 'Mandatory retirement: The social and human cost of enforced idleness' in: C. S. Kart and B. B. Manard (eds.), *Aging in America*, Sherman Oaks: Alfred Publishing, 260–75.
Sennett, R. (1998), *The corrosion of character*, New York: W.W. Norton.
Stenner, P., McFarquhar, T. and Bowling, A. (2011), 'Older people and "active ageing": Subjective aspects of ageing actively', *Journal of Health Psychology*, 16 (3), 467–77.

Townsend, P. (1981), 'The structured dependency of the elderly', *Ageing and Society*, 1 (1), 5–28.

van Dyk, S. (2009), 'Das Alter: adressiert, aktiviert, diskriminiert', *Berliner Journal für Soziologie*, 19 (4), 601–25.

van Dyk, S. and Lessenich, S. (2009), 'Ambivalenzen der (De-)Aktivierung: Altwerden im flexiblen Kapitalismus', *WSI-Mitteilungen*, 62 (10), 540–6.

van Dyk, S., Lessenich, S., Denninger, T. and Richter, A. (2013), 'The many meanings of active ageing. Confronting public discourse with older people's stories', *Recherches Sociologiques et Anthropologiques*, 44 (1), 97–115.

Venn, S. and Arber, S. (2011), 'Day-time sleep and active ageing in later life', *Ageing and Society*, 31 (2), 197–216.

Vogel, C. and Motel-Klingebiel, A. (eds.) (2013), *Altern im sozialen Wandel: die Rückkehr der Altersarmut?*, Wiesbaden: Springer VS.

Walker, A. (2002), 'A strategy for active ageing', *International Social Security Review*, 55 (1), 121–39.

Walker, A. and Taylor, P. (1993), 'Ageism versus productive aging: The challenge of age discrimination in the labor market', in: S. A. Bass, F. G. Caro and Y.-P. Chen (eds.), *Achieving a productive aging society*, Westport/London: Auburn House, 61–80.

Zähle, T. and Möhring, K. (2010), 'Berufliche Übergangssequenzen in den Ruhestand', in: P. Krause and I. Ostner (eds.), *Leben in Ost- und Westdeutschland*, Frankfurt/New York: Campus, 331–46.

14
Open Questions and Future Prospects: Towards New Balances Between Work and Retirement?

Harald Künemund and Simone Scherger

14.1 Introduction

The contributions in Part I of the book provide an overview of the manifold forms of work beyond pension age, its conditions and reasons, drivers and barriers. Parts II and III of the book, but also some of the country studies, further depict the institutional and cultural contexts in which the shifts between working and retirement unfold, and the consequences that these shifts possibly have for inequalities, individual and collective well-being, and for our conceptions of old age and retirement. While this structure of the book spans the scope of the questions that can be asked and the issues that are at stake, it is of course not within the realm of possibility to *answer* all these questions here, let alone to foresee future developments. Given these limitations, our final conclusions serve a threefold aim: first, to identify important gaps and blind spots in existing research on work beyond pension age and related issues; second, to reflect on some of the results presented above, their underlying concepts of the life course and their normative implications; and third, to challenge dominating views on the life course and retirement by pointing out possible alternatives. In this way, we hope to be able to stimulate further discussions on the future relationship between work and retirement.

14.2 Gaps and challenges in research on work after pension age

14.2.1 Multiple work and pension arrangements and the complexity of the retirement transition

The contributions to this book reveal the multiplicity of forms of working after pension age or while receiving a pension. Although they do not address completely identical target groups, the majority of the contributions concentrate their observations on the time after regular statutory pension age and thus on prolonging working careers into an age when most other people no longer work. Those who work mostly do so part-time, and many among them are self-employed. Furthermore, both highly qualified jobs and low qualified (service) jobs seem to be over-represented in many countries and in comparison with the population of main working age; by contrast, manufacturing is not the typical area in which post-retirement workers engage, except in China.

Two contributions in this book, Principi et al. in Chapter 4 (for Italy) and Yu and Schömann in Chapter 7 (for China), also include early pensioners whose share is large in both these countries, due to (formerly) generous rules for early pensions. In Italy (Principi et al.), many among these early pensioners work, indicating that in people's 50s or early 60s, a hybrid phase between retirement and work, a kind of semi-retirement with a combination of incomes from working and pensions, is a frequent path out of the main career and into retirement. This is also underlined by the fact that a third of these early pensioners work full-time. In China, those who continue working while receiving a pension, whose share is large among pensioners in their late 40s and early 50s, mostly continue working full-time and for very long hours. Together with the fact that a huge majority of people (mostly rural residents) do not receive a pension at all, this shows that in China's underdeveloped pension system a work-free retirement with a pension is not the rule, although general employment rates seem to be dropping quickly after the age of 60. Another country where regular pension ages do not seem to be of great salience for the actual end of employment is Russia (see Radl and Gerber in Chapter 6 in this book).

Although the other country studies in the book, on the UK (Lain in Chapter 2; Hokema and Lux in Chapter 3), the USA (Lain; see also Marshall in Chapter 9) and Sweden (Halleröd in Chapter 5), focus on paid work *after* state pension age, early or gradual retirement before pension age also exists there, and is still common for example in Germany. Both working while receiving a pension before and after pension age underline once again that retirement is in many cases rather a process (Loretto and Vickerstaff 2013) than a transition consisting of a single step. This applies more in multi-pillar pension systems (where pensions from different sources might start at different times), in pension systems with generous early retirement regulations (especially when earning extra in addition to receiving a pension is possible) and – increasingly – in systems where partial pension receipt or stepwise retirement are institutionalized and supported.

Corresponding to the many kinds of pension receipt, there are also many ways to engage in paid work, from full-time dependent employment to marginal freelancing or undeclared work in the shadow economy. Some of these forms (especially work off the books) are not well covered in the available data, partly because asking at the same time about receipt of benefits (including pensions) and income from work (possibly from participation in the shadow economy) is sensitive and might result in lower response rates and panel attrition. However, we should try to develop non-intrusive but still accurate ways to collect this kind of information. It is only on this basis that we are able to describe and examine the multiple forms of combining work and retirement precisely, only some of which are well described by the term 'bridge employment' (Alcover et al. 2014), and to relate them systematically to the surrounding institutions. Apart from pension payments as a crucial criterion for the beginning of retirement, another important aspect of this is to determine the relationship between the current job and the 'main' career, and whether the employer and/or the job have been changed. Radl and Gerber in Chapter 6 of this book give a fine example of the value of such an approach, even though confined to the time around state pension age. Job or employer changes will at least sometimes be connected to downward occupational mobility into low qualified (service) jobs. This also touches the question as to whether older unemployed, faced with the difficulties of ever finding a job again, 'escape' to self-employment or less qualified jobs in their 50s or even after retirement. Describing (and possibly classifying or clustering) trajectories from full-time 'career' jobs, via 'bridge employment' before or after regular pension age, to complete retirement would be the ideal approach to trace these 'messy' processes of retirement (Loretto and Vickerstaff 2013), and would require comprehensive longitudinal data on the individual level.

On a related note, for a large part of the population the transition to retirement has always been less distinct as their careers were more discontinuous and interrupted: for women, especially those with children.

While the normative idea of a single-step retirement – which is also at the core of the social policy regulations surrounding pensions and retirement – is a somewhat simplifying (but parsimonious) 'measure' to describe male retirement transitions, it has frequently been inadequate and insufficient for the factual transitions of women to retirement (see also Loretto and Vickerstaff 2013). Many mothers among current pensioners used to have longer career interruptions, and in some pension systems, they started work again for a few years before retirement in order to fulfil the basic requirements for pension entitlement. The critique of an oversimplified conception of retirement extends the feminist critique of traditional welfare state research (starting with Lewis 1992; Orloff 1993) into research on retirement and pensions. Both theoretical concepts and empirical observation have to be further adapted to take into account the larger complexity of women's career trajectories and their obligations in family and care, and perhaps also the increasing complexity of men's careers, and their 'fit' to the institutionalized pension entitlement rules. A starting point for this could be to relate different models of sharing household responsibilities as they have, for example, been developed by Lewis (2001) (male breadwinner, dual breadwinner in different variations and so on) to the variety of retirement transitions or typical work patterns around pension age.

14.2.2 Individual drivers of post-retirement work

At the individual level, the country studies above reveal at least some basic individual preconditions of work after pension age to be similar across all countries. Examples of these preconditions are good health and a good educational qualification, which increase the probability of someone working – and their lack creates barriers to working longer, as it also does before retirement. Whereas there are indications that additional economic resources as a *general* driver or motive for working play some role in every country, the manner and degree to which they do so differs considerably. How much people are 'pushed' into work by poverty (and perhaps regardless of mediocre health) is closely related to the resources that are available. These are in part determined by the pension system, but also by the rules for receiving means-tested benefits, the role of private savings for old-age provision, as well as of home ownership. That a lack of income or wealth cannot always be established as increasing the probability of working (see, for example, Principi et al. who in Chapter 4 of this book find the opposite for Italy, or, Lain in Chapter 2 for the UK and the USA) has both substantial and technical reasons. Data on income and wealth are notorious for their poor quality and missing information; this problem can be exacerbated if the number of people working after pension age is low in the respective samples. Furthermore, it is difficult to capture all relevant sources of income and wealth correctly. There is no agreement on whether wealth or which types of (pension) incomes are the right approach to measure financial resources, and the adequacy of these indicators will depend on the degree to which private pension savings play a role in financing old age (compare the contributions of Lain in Chapter 2 and Hokema and Lux in Chapter 3 of this book). Home ownership (and, conversely, renting) is an additional important factor to take into account, whose importance differs considerably between countries. Moreover, regarding the sources of income, it would not only be paramount to distinguish different kinds of pensions, but also to be able to distinguish people of pension age without any pension claims based on their own employment record from those without pension income who only *defer* pension receipt. The latter form a growing group in some countries, who might seem poor if mainly or only income is used as a measure of financial resources. At best, we would need data measuring both pension entitlements and income across the life course at the household level, which is rarely available at the moment. Linking panel survey data and pension system records seems a promising way to approach this challenge (see, for example, Rasner et al. 2013).

While poor financial resources might be a driver for working, many poorer pensioners, however, will at the same time display characteristics which pose barriers to working: they are, on average, less healthy, less well educated and thus may have poorer labour market opportunities. In particular in purely descriptive quantitative approaches they may therefore not appear as being more likely to work, and only when these barriers are (sufficiently) controlled for in multivariate models, will a lack of financial resources emerge as a driver of working. Further, means-tested benefits might pose a disincentive for this group to working (see introduction).

In summary, people who combine characteristics favourable to working with a certain financial need should have the highest probability of working – that is, those who have good health, more education, a long career, on the one hand, *and* a lack of financial resources, on the other, be it due to the wish to pay off a mortgage, to keep one's standard of living, or old-age pensions that are (or are deemed to be) insufficient. Qualitative research indicates that it is a *subjective* need for financial resources which goes together with the desire to work (Hagemann et al. forthcoming). This subjective need builds on comparisons to others and to oneself over time; hence, taking into account *relative* income changes around retirement would be a good way of measuring financial needs over and above poverty.

Many of the contributions in this book also mention social and intrinsic reasons for working and cite them as an explanation for the high work propensity of professionals and the self-employed. Generally, qualitative research finds these 'pull' motivations for working to be important. However, these factors are rarely accounted for directly quantitatively, which could be done by including work satisfaction with the job before retirement, the individual importance of work or the status and recognition connected to a job. Qualitative methods should be used to develop reliable scales that allow for the measurement of both push and pull factors.

More *qualitative* studies on the motivations and experiences of working after pension age could also help to further disentangle the complexities around the financial motives for working. There are, for example, indications that some poorer people prefer working instead of claiming means-tested benefits. So far, we know very little about how this kind of decision is taken, how much knowledge people have about the alternatives to working, and the experience and the 'symbolic' value of being financially independent and earning one's own money. Such an approach would connect post-retirement work to more general research on non-take-up of benefits. This is only one example of many speculative points which could be better understood by applying interpretative methods – others are the gender dynamics and the subjective interpretation of welfare regulations, topics which would also benefit from qualitative comparisons across societies. Hokema and Lux, in Chapter 3 of this book, furthermore give a first impression of the value of systematically integrating and connecting quantitative and qualitative approaches which may help to overcome the limits of the specific perspective.

These scattered remarks on individual drivers and barriers, and on motives for post-retirement work boil down amongst others to a call for better data, especially those of a *longitudinal* nature. Individual longitudinal in-depth data – *both quantitative and qualitative* – would be crucial to describe employment pathways around and after retirement and to distinguish continued employment in the same job and/or for the same employer from 'pensioner jobs' possibly connected to downward mobility. Not only the relationship of late careers, on the one hand, and health-related or family trajectories, on the other hand, could be studied in this way. Plans for post-retirement work which people make in their 50s could be followed up too, and ideally be reconstructed based on qualitative data, or by deriving a thorough quantitative approach on a qualitative basis. In this way, concrete drivers and barriers involved in the realization of work preferences and plans for old age could be established. Longitudinal data could finally also serve to observe income trajectories in late careers and around retirement, in order to get a more detailed picture of potential financial motivations for working beyond strict poverty. Of course, tracing careers, motivations, plans and their realization would be an ambitious enterprise, as individual interpretations will often change with changing life situations. This has to be taken into account in quantitative analyses and in qualitative approaches where careful reconstructive interpretations are essential.

14.2.3 The role of contexts: Macro-level factors and the workplace

The individual drivers and barriers form of course only half of the story; macro- and meso-level influences, the latter in particular workplace related, need to be taken into account to get a fuller picture. The six country studies presented in Part I of the book do not even begin to cover the huge range of such conditions for post-retirement work, and Part II of the book on contexts of work beyond pension age only offers some first ideas on the systematization by context. Crucially, Central and Eastern European countries are not included at all, which have undergone comprehensive reforms of their framing (pension) institutions. The example of China (Yu and Schömann in Chapter 7 of this book) underlines how the study of emerging pension systems offers the possibility of thought-provoking structural comparisons to countries undergoing pension-retrenching reforms. In a similar vein, the historical scope of comparison should be broadened as well, as some critical observers suggest that current pension policies might, in the end and at least for some, lead to the return of endless working careers, characterized by dequalification processes and rising inequalities. Focusing on how to exploit the favourable conditions that clearly set apart the present situation from the past (in most countries), such as prosperity, improved health and longevity, expanded education and tertiarization, will help to develop alternative scenarios for the future (see below).

Closely related to the individual influences on working beyond pension age, labour market opportunities and factors of labour demand seem to be hard to grasp directly. While a working pensioner's education and class of former or current job appear to be reasonable approximations for job opportunities, the impact of education and class can, at the same time, indicate the above-mentioned intrinsic drivers for working. Anecdotal evidence from qualitative research illustrates that in some cases pursuing paid work is clearly facilitated by demand.[1] However, this is rarely investigated directly by including the demand for specific jobs into the equation, or by looking at jobs with known skills shortages. At the same time, post-retirement work might be a possible measure against skills shortages. Including the unemployment rate (of countries or regions) into the explanation of post-retirement work forms the most general and easy to realize approach to this question (see Hokema and Lux in Chapter 3 of this book).

Similarly, quantitative models related to the macro level and explaining the employment rates of older people in a large number of countries are an additional way of better understanding the interplay between pension policies, labour market policies and old-age employment. Especially the institutional framing of post-retirement work, in the form of rules for earning extra money, for combining work and the receipt of benefits, or for partial retirement, is worth a more detailed analysis, comparing countries or changes over time. Studying the composition of working pensioners over time also sheds light on the underlying dynamics of inequality (see Halleröd in Chapter 5 of this book).

Finally, in between the individual level and the macro level, workplace characteristics and policies and their impact on the probability of working past retirement age have not yet been investigated often (see Schmitz in Chapter 10 of this book), although some factors, such as company size or the sectors of companies with many post-retirement workers, are well known (see, for example, the introduction to this book and Lain in Chapter 2). Still, the role of companies in facilitating or impeding work post retirement can barely be overestimated because it is there where recruitment decisions are made, as the contribution by Schmitz demonstrates for employment before retirement.

14.2.4 Consequences of post-retirement work

Part III of the book offers some initial evidence on what the individual and social consequences of work beyond pension age will or might be. To grasp and describe these consequences poses complex methodological challenges, as the exact causal relationship between working and its consequences can only be established by using longitudinal data and adequate methods of statistical analysis, such as matching methods (which account for the selectivity of the group in focus) or fixed-effects models (focusing on individual changes). As Matthews and

Nazroo show in chapter 12, post-retirement work does not necessarily have negative effects on individual well-being – on the contrary. However, if the rewards of the work are not (subjectively) in balance with the efforts made, working can go together with a decrease in well-being. More large-scale research is needed to explore the complex institutional and structural lines along which these consequences for well-being unfold; again, this should be complemented by qualitative evidence exploring what exactly it is about working that can improve or impair well-being.

Large-scale longitudinal data are also needed to find out which of the general inequality-related scenarios presented in the introduction is an appropriate description of reality. The exact answer is of course highly context-dependent and will differ according to country. Halleröd's analyses in Chapter 5 suggest – for the favourable context of Sweden – that extended work after pension age has not led to increasing, but potentially decreasing inequalities between classes. At the same time, inequalities *within* classes between those working and those not working seem to be increasing. So, in a sense and abstracting from the Swedish case, variables on class might not (or only very incompletely) capture the crucial differentiation between those who can work longer and those who cannot. There might be people in every class who have sufficient pensions in the first place or who can 'compensate' low pensions by working, while the most disadvantaged have low pensions, for example due to interrupted careers, and cannot compensate this by working longer. Labour market-related segmentation theories, the idea of 'insider-outsider' or 'dualized' labour markets (see, for example, Emenegger et al. 2012) are perspectives worthwhile extending into pension age. However, the consequences of extended work after pension age are of course not limited to inequality-related shifts. They will probably affect many other fundamental aspects of social life. One very important example which needs further exploration are intergenerational relationships on labour markets, in companies and in families.

Comparative perspectives on both the individual and socio-structural consequences of working in retirement – which so far barely exist (but see Lux and Scherger forthcoming) – would further help to carve out *which* contextual factors explain the consequences of prolonged working lives. The same applies to the observation of long-term changes in these consequences, changes which will go together with changed meanings of post-retirement work and of retirement in general. These meanings are, as Hagemann and Scherger show in Chapter 11 of this book, embedded into welfare regimes, which imply not only concrete regulations but also ideas and ideals, for example regarding individual life courses and old age. Welfare regimes and their cultures provide specific (and always temporary) answers to fundamental conflicts regarding the relationship between individual lives and societies, such as the conflict between individual freedom of choice and collective regulations which allow for strategic planning on the individual, the company and the state levels. Discourses around work and retirement, in turn, are expressions of the symbolic struggles around these basic conflicts. A good example is the discussion on intergenerational justice and solidarity. Working longer might be regarded as a contribution to both intergenerational justice and solidarity and is often discussed in this way. However, before introducing unequal treatment of age groups or cohorts – for example by reducing pension levels or raising the age boundaries of pension systems – one should first ask about social inequalities *within* cohorts, for example, who is able to and going to work longer (and who is not), who contributes to the pension system (and who does not), who benefits disproportionately from existing institutional rules (and who suffers systematic disadvantage) or who in further ways acts non-solidarily, non-sustainably or unjustly. Many controversies today, but especially that surrounding intergenerational justice, strategically or unwillingly distract attention from the underlying, still predominantly class-based inequalities and conflicts (see already Berger 1984).

Another example of symbolic struggles around welfare and old age are the discourses on the 'activation' of old age, which are discussed by van Dyk in Chapter 13 of this book. They imply not only a very general shift in the meaning of old age and retirement but also a shift to a (more) individualized concept of the old or ageing individual actor. Although the questions as to whether and how these sometimes detached political discourses are translated into everyday life – that is, how exactly they play out in concrete practices, whether they are adopted or met by resistance by individual actors, and by which ones – have already been explored (see, for example, Denninger et al. 2014), they are worth further exploration, especially in comparative perspective and for different welfare contexts. The same applies to both the dissemination and the consequences of these discourses, which seem to be at least ambivalent, ranging from the successful fight against ageist stereotypes to new tendencies to exclude those who cannot be 'activated'.

14.3 Rethinking the life course: Alternatives to tripartition

To some extent it is or might be the tripartite life course (Kohli 1986) itself that is in question in the renegotiation of old age and retirement, that is, a specific form of regulating individual life courses, building on a fixed set of institutions of social policy, labour markets and education. Apart from demographic change, dampened economic growth and transformed labour markets, changing gender relations and rapid technological change, to name just a few trends, combine and exert pressure on this normative model of the life course to change.

There are many indications that the current strategy of raising pension ages as well as attempts to abolish age boundaries will result in increasing social inequalities. These increasing inequalities are probably unintended side effects of reforms due to simplified, overly optimistic and sometimes biased projections of policy outcomes. In general, predictions assume a continuing increase in life expectancy by about three years per decade – 'half of the children alive today in countries with high life expectancies may celebrate their 100th birthday' (Vaupel 2010: 539). Medical progress, healthier lifestyles and technical advances have contributed to this development. These predictions have fuelled worries about how to finance the growing longevity of increasing parts of the population. These worries are usually based on rising dependency ratios whose calculation, however, rests on crude assumptions, leading to oversimplifying if not flawed interpretations. The simple quantitative comparison between the number of people aged 65 and older (the 'dependent', in the numerator of the dependency ratio) and the number of those aged 18 to 64 (in the denominator) neglects the just mentioned changes: improved health due to medical progress, healthier nutrition, less arduous working and living conditions, better education and technological progress are reshaping biographies, so that future cohorts of people aged 65 and over will most probably be healthier (Vaupel 2010), presumably also have more resources for active and productive lifestyles, and will retire later compared to older people today. Furthermore, in many countries, future pensioners will also have lower pension entitlements due to more unstable working careers and atypical employment, and face reduced public pension levels. On the whole, all these factors will contribute to less 'dependency' in these future cohorts of older people, but they are not taken into account in the calculation of the dependency ratio. With regard to the denominator of the ratio, the actual number of those financing the 'dependent' older people through their contributions tends to grow as well, as the employment participation of women is increasing in most countries and more older people keep on working until pension age. It follows that the expected 'burden' is considerably overestimated.

Even if the number of the 'dependents' in such calculations were to increase by 100 per cent within 50 years (a realistic scenario in some countries), this only translates into an increase in the dependency ratio of about 2 per cent per year; financing such an increase seems feasible since technological change, better health, better education and improved working conditions will result in increasing productivity. In pay-as-you-go systems, this would require, however, that all forms of labour market participation, including marginal or atypical (self-)employment, involve contributions to an ideally unified (and not fragmented) pension system. If, for example, the increasing labour force participation of women is mainly in marginal employment without contributions to the pension system, financing increasing numbers of pensions would be difficult. On a different level of redistribution, this also applies to other kinds of pension systems, where redistribution would have to be financed via taxes.

Despite these misleading figures which measures like the dependency ratio and the related forecasts provide, many European governments have decided to increase pension age, reduce public pension levels and increase the role of private pensions (including occupational schemes); the shift from defined benefit towards defined contribution pension plans in both public and private pensions follows a similar logic (for all of these changes see also Anderson in Chapter 8 of this book). In many countries, these reforms and changes will result in (more) insecure and instable old-age incomes, increasing old-age poverty and social inequality among older people (see Fachinger et al. 2013), possibly leading to social exclusion and more strained or even troubled intergenerational relations within the family (see Künemund 2008).

The rise in pension ages itself already implies increasing inequalities: it affects individuals with different life expectancies unevenly as it goes together with unequal losses in work-free years; and some of those with low life expectancies do not even live until pension age. Poorer health care, strenuous or stressful working and living conditions, unhealthy lifestyles and so on contribute to lower life expectancies of disadvantaged people, whose longevity also rises much more slowly, if at all, than that of the more privileged. In such a situation of unequally rising life expectancies combined with generally rising pension ages, these disadvantaged groups contribute, relatively seen, more and more to the pensions of those with higher life expectancies. Consequently, the current developments go together with a real danger of moving towards a systematic redistribution of resources within the pension system that contradicts the idea of solidarity in welfare states.

One of the alternatives to the reforms might be individualization – that is, to completely abolish age boundaries, for example for withdrawing from work and for pension receipt. This idea is by no means new (see, for example, Best and Stern 1977), but increasingly becoming popular in the context of 'ageism' and age discrimination discourses. Releasing individuals from the strict (age) norms of an institutionalized life course could be interpreted as a continuation of the historical process of individualization. In this view, the institutionalized tripartite life course with its three clear-cut stages of preparation (education), activity (labour participation) and leisure (retirement) (Kohli 1986) might have been be a transitional period.

However, as argued by Kohli and Künemund (2002), age boundaries are an essential part of current work societies and serve a number of important purposes: They allow for the timing of retirement for individuals and thus the planning of succession in families and companies, and they permit the calculation of pension expenses in pay-as-you-go systems. Furthermore, age boundaries legitimate exit (or dismissal) from work, and in this way protect not only individual workers themselves but also society from the potential consequences of being old – the latter can be illustrated by the example of pilots. Abolishing all age boundaries would most likely result in (even further) increasing social inequalities as some individuals would have the resources to – for example – retire from work, while others would not be able to give up work. Given these arguments, the immediate abolition of age boundaries might have unintended side effects and negative consequences for society and welfare. At the same time, even if age boundaries are generally accepted, their exact matter (work exit, pension receipt or more specific transitions), their bindingness and possible exceptions are worthwhile objects of dispute.

Some gerontologists – especially Riley and Riley (1994) – have argued in favour of an 'age-integrated' society, where education, work and leisure are significant parts of life at all ages (see Figure 14.1). Although this suggestion can be seen as overly optimistic and simplistic (as it would, for example, also imply the re-introduction of child labour), the example makes clear that there are alternatives to the current life-course regime to consider (see also Phillipson 2002 for a similar conclusion).

Starting from the question of what to do with and how to 'distribute' the years gained through rising life expectancy across the (tripartite)

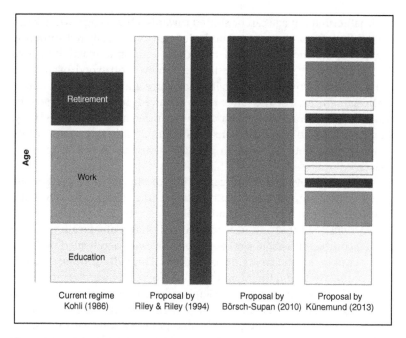

Figure 14.1 Life-course models
Source: Own figure on the basis of Künemund (2014).

life course, Börsch-Supan (2010) proposes adding two-thirds of this additional time to the phase of employment and one-third to final retirement (see Figure 14.1). This, however, would result in the above-mentioned injustices and rising inequalities. Therefore, we would like to point to an alternative suggested earlier (Künemund 2013, 2014): a reorganization of the life course that keeps the separation of preparation, activity and leisure, but alters the duration of these periods and multiplies their sequence (see Figure 14.1).

After the first essential phase of education similar to the current regime, we may think of a first phase of employment of about 15 years, followed by one year of temporary withdrawal from paid employment (which could be time for the family, for example) and one year of additional education. This further education could build on the first occupation, for example update the relevant knowledge and skills in order to improve productivity and employability; it could also be used as an opportunity to change from an unwanted or outdated job into a new occupation. After another 15 years, another year of temporary economic inactivity and another year of further education may serve to prepare for a third phase of activity of 10 or 12 years. In this way, the final retirement age can be shifted towards a higher age, while keeping the ageing workforce rested and educated as a rule. By redistributing phases of education, work and temporary economic inactivity across the life course, it would be possible to achieve both good health and increased productivity in the ageing workforce and reduced inequalities at the same time. For the latter to be achieved, it would, however, be crucial that the nonworking phases are organized and financed publicly, for example by an expanded pension system, as only this would ensure that everyone, not only those in privileged jobs, can benefit from these phases.

The benefits of such a reorganization of the institutionalized life course, which implies the organized 'loosening' of its tripartition, are manifold and can be found on all levels of society: With respect to *lifelong learning*, repeated and institutionalized phases of education during the life course will enable workers to keep up with accelerated progress, new technologies and changing job profiles. This will contribute to both their productivity and their continued employability. Furthermore, people may more easily adapt to societal and technological changes even in old age and after having finished their working careers, so that, for example, caregivers are possibly able to better cope with the strain and pressure of providing care by making use of new care-supporting technologies.

Previous experiences with temporary phases of economic inactivity in the life course will also help to solve one of the oldest problems in social gerontology (see Cavan et al. 1949), the *adjustment to retirement*: finding meaningful activities and roles in retirement and in that sense 'successful ageing' can be a challenge for many retirees which they can meet more easily based on prior experiences with nonwork phases. Especially in conjunction with the effects of lifelong learning, this will probably also contribute to 'productive ageing'.

In the earlier and middle phases of the life course, the discussed model would broaden the scope for *healthy and 'active' ageing*: having regular times of rest from work will improve health, lead to a well-rested workforce and hence increased productivity across the whole life course. Together with the points mentioned so far, the alternative life-course regime can thus be interpreted as a growth-friendly way of organizing social protection. The phases outside paid employment and spent with the family may also contribute to reducing the stress and burden of the 'rush hour' phase in life (that is, the frequent coincidence of family formation and important early career steps) and of periods of providing care for elderly parents, and the related potentially negative consequences for health.

On a structural level, the suggested reorganization even has the potential to *reduce inequalities*. In comparison to a strategy which only raises the pension age, having phases of temporary economic inactivity during the whole life course will allow social groups with lower life expectancies to at least experience some periods without work, which could probably mitigate the increasing social inequalities described above. This also applies to *gender inequality*. Having a life-course regime that supports, honours or even promotes phases without labour market participation and with subsequent education for all would most likely result in improved re-entry opportunities into the labour market for mothers compared to the current situation: regular interludes in employment careers become the institutionally rewarded life course. This would be more compatible with demands from families and might even allow for stabilized fertility rates. As men would also follow this pattern of a reorganized life course, interrupted careers would become the normal 'benchmark' of social security, reducing their relative negative consequences (of difficult labour market re-entry, wage penalties and so on) and thus gender inequality.

If all these arguments are correct, the economic advantages following from the suggested reorganization of the life course are obvious: Companies allowing for these life courses will have – on average – a better rested, better educated and healthier workforce. The same applies to nations, which will be better able to succeed in international competition due to more flexibility, better education and increased productivity. Of course, this brief list of advantages is one-sided and probably simplifying, as it does not discuss the costs and the problems that may arise from such a reorganized life course or the many practical difficulties its realization might encounter. However, the aim of this exercise is to show that alternatives exist to raising retirement ages or removing age boundaries, prolonging working lives and accepting the consequently increasing social inequalities.

From our point of view, it is nonetheless crucial that a reorganized life course is not completely flexibilized, 'released' from social regulation and protection, and subjected to organization through market principles. Age boundaries and their advantages can and should be preserved as an important normative orientation. Work past retirement is only one of many 'deviations' from the 'fiction' of the tripartite life course which serves as standard model of the normal life course and guides large parts of social policy regulations. Such deviations have become (much) more prevalent, and demographic ageing is only one among many factors driving this development. Systematically reorganizing education, work, family obligations and phases of economic inactivity across the life course would imply acknowledging the fictional character of the institutionalized tripartite life course, while at the same time preserving its benefits in the field of social protection. It may also remind us of visions of the future that are not constructed around working longer and even in retirement age, but allow for both more freedom and welfare for all. These visions might inspire policies for better lives and fewer inequalities not only in industrialized countries but even more so in countries with so far underdeveloped or non-existent welfare states and pension systems.

Note

1. In the qualitative project forming the basis of Hokema and Lux in Chapter 3 of this book, for example, engineering seemed to be a classical occupation among post-retirement workers in Germany. Many of the respondents or potential respondents worked for their former employers (often as project-related freelancers) and in direct reaction to being approached by them.

References

Alcover, C.-M., Topa, G., Parry, E., Fraccaroli, F. and Depolo, M. (eds.) (2014), *Bridge employment. A research handbook*, New York: Routledge.
Berger, B. M. (1984), 'The resonance of the generation concept', in: V. Garms-Homolowá, E. M. Hoerning and D. Schaeffer (eds.), *Intergenerational relationships*, Lewiston: Hogrefe, 219–27.
Best, F. and Stern, B. (1977), 'Education, work, retirement – must they come in that order?', *Monthly Labor Review*, 100 (7), 3–10.
Börsch-Supan, A. (2010), 'Generationengerechtigkeit in der Alterssicherung', in: Deutsche Rentenversicherung Bund (ed.), *Gerechtigkeitskonzepte und Verteilungsströme in der gesetzlichen Alterssicherung*, Berlin: DRV.
Cavan, R. S., Burgess, E. W., Havighurst, R. J. and Goldhammer, H. (1949), *Personal adjustment in old age*, Chicago: Science Research Associates.
Denninger, T., van Dyk, S., Lessenich, S. and Richter, A. (2014), *Leben im Ruhestand: Zur Neuverhandlung des Alters in der Aktivgesellschaft*, Bielefeld: transcript.
Emenegger, P., Häusermann, S., Palier, B. and Seeleib-Kaiser, M. (eds.) (2012), *The age of dualization. The changing face of inequality in deindustrializing societies*, Oxford: Oxford University Press.
Fachinger, U., Künemund, H., Schulz, M. F. and Unger, K. (2013), 'Dynamisierung kapitalgedeckter Altersvorsorge', *Wirtschaftsdienst*, 93 (10), 686–94.
Hagemann, S., Hokema, A. and Scherger, S. (forthcoming), 'Erwerbstätigkeit jenseits der Rentengrenze. Erfahrung und Deutung erwerbsbezogener Handlungsspielräume im Alter' (journal article).

Kohli, M. (1986), 'The world we forgot: A historical review of the life course', in: V. W. Marshall (ed.), *Later life: The social psychology of ageing*, Beverly Hills: Sage, 271–303.

Kohli, M. and Künemund, H. (2002), 'La fin de carrière et la transition vers la retraite. Les limites d' âge chronologiques sont-elles un anachronisme?', *Retraite et Société*, 36, 84–107.

Künemund, H. (2008), 'Intergenerational relations within the family and the state', in: C. Saraceno (ed.), *Families, ageing and social policy – intergenerational solidarity in European welfare states*, Cheltenham: Edward Elgar, 105–22.

Künemund, H. (2013), 'Demografie, Politik und Generationenbeziehungen', in: M. Hüther and G. Naegele (eds.), *Demografiepolitik*, Wiesbaden: Springer VS, 164–76.

Künemund, H. (2014), 'Soziologische Perspektiven', in: V. Schumpelick and B. Vogel (eds.), *Demographischer Wandel und Gesundheit. Lösungsansätze und Perspektiven*, Freiburg: Herder, 189–206.

Lewis, J. (1992), 'Gender and development of welfare regimes', *Journal of European Social Policy*, 2 (3), 159–73.

Lewis, J. (2001), 'The decline of the male breadwinner model: Implications for work and care', *Social Politics*, 8 (2), 152–69.

Loretto, W. and Vickerstaff, S. (2013), 'The domestic and gendered context for retirement', *Human Relations*, 66 (1), 65–86.

Lux, T. and Scherger, S. (forthcoming), 'In the sweat of their brow? The effects of starting work again after pension age on life satisfaction in Germany and the UK' (journal article).

Orloff, A. S. (1993), 'Gender and the social rights of citizenship: The comparative analysis of gender relations and welfare states', *American Sociological Review*, 58 (3), 303–28.

Phillipson, C. (2002), *Transitions from work to retirement. Developing a new social contract*, Bristol: Policy Press.

Riley, M. W. and Riley, J. W. (1994), 'Structural lag: Past and future', in: M. W. Riley, R. L. Kahn and A. Foner (eds.), *Age and structural lag. Society's failure to provide meaningful opportunities in work, family, and leisure*, New York: Wiley, 15–36.

Rasner, A., Frick, J. R. and Grabka, M. M. (2013), 'Statistical matching of administrative and survey data, an application to wealth inequality analyses', *Sociological Methods & Research*, 42 (2), 192–22.

Vaupel, J. V. (2010), 'Biodemography of human ageing', *Nature*, 464, 536–42.

Index

active ageing, 18, 21, 22, 81, 101, 130, 169, 278–93, 312
active labour market policies, 15, 85, 111, 133, 147, 217, 220, 224, 282
activities of daily living, 37, 40, 47, 264, 269–70
age boundaries, 8, 220, 247, 248, 252, 293, 307, 308, 310, 313
age discrimination, 12, 21, 34, 85, 145, 169, 170, 206, 220, 227, 241, 285–6, 293
ageism, *see* age discrimination
age management, 12, 226, 231, 245
age norms, 9, 12, 15, 60, 74, 310
age-related legislation, 9, 12, 14–15, 18, 32, 34, 35, 85, 220, 222
age stereotypes, 12, 222, 225, 228, 253, 280, 289, 307
attitudes/perceptions towards old age, 18, 129, 221, 222, 240, 246, 249, 252
atypical employment, *see* non-standard employment

benefits
  disability, 4, 156, 220
  housing, 33
  incapacity, 13
  means-tested, 10, 13, 14, 20, 32–4, 38, 77, 188, 241, 242, 302, 303
  sickness, 111
  social, 195, 205
  unemployment, 85, 111, 133, 224
  see also pensions, health care provision
Beveridge system (of pensions), 179–80, 241
Bismarck system (of pensions), 179–80, 182–7, 190, 240–1, 242, 251
blue-collar workers, *see* class

breadwinner model, 68, 100, 136, 177, 301
bridge employment, 6, 13, 21, 41, 203, 207, 209, 300

Canada, 208–9
care, 183, 250, 270, 279, 281, 286, 291, 301, 312
  caregiver credits, 184
  for family and children, 100, 102, 200
  for grandchildren, 153, 159, 169
  for older relatives, 93
  for partner, 52
career management, 12, 209, 230–1
Central and Eastern Europe, 179, 189–91, 304
China, 2, 17, 22, 151–7, 299, 304
civil service, 3, 59, 86, 96, 154, 157, 167, 170, 178, 179, 204
  see also public sector
class, 7, 10–11, 65, 100, 102, 107–2, 139–42, 178, 204, 262–7, 273, 306–7
class mobility, *see* occupational mobility
collective actors, 21, 180, 190, 237–54, 284, 285
collective agreements, 14, 15, 85, 111, 133, 143, 187, 220, 243
companies and older workers, 12, 22, 198–212, 217–32
conservative welfare regime, 16, 59, 180, 182, 186, 191, 286
continuity theory, 7, 81, 268
crisis, *see* economic crisis
culture, 9
  culture of ageing/later life, 18, 287
  see also ideas, retirement culture, welfare culture

debt crisis, *see* economic crisis
debts, 10, 39–40, 52, 302

Index   317

decommodification, 13, 15, 240, 241, 242
default retirement age, *see* mandatory retirement age
deferral of pension, 4–5, 13, 67, 110, 115, 155, 302
defined benefit (DB)/defined contribution (DC) pensions, *see* pensions
Denmark, 3, 100, 181, 182, 187, 192, 193, 194, 220, 222, 227
dependency ratio, *see* old age dependency ratio
depression, 203, 261, 263–4, 266–73
disability, 4, 40, 148, 156, 193, 205, 220, 259, 272, 275
see also benefits
discourse, 23, 244, 278–93
divorce, 11, 44, 66, 67, 93, 94–5, 98–100

Eastern Europe, 2, 131, 179, 189–91, 219
economic crisis, 14, 85, 110, 133, 218, 229
see also financial crisis
economic development, 16, 190, 204, 208, 211
economic growth, 17, 21, 87, 134, 177, 199, 246, 308
economic recession, 85, 108, 133, 146, 203, 211, 221
electoral politics, 177, 180, 181, 182
employer organizations, 109, 246–50
employers, *see* companies and older workers
employment after pension age, *see* work beyond pension age
England, 31–53, 57–78, 237–54, 262
Estonia, 191, 192, 193
ethnicity, 43–4, 222
euro crisis, 177, 186, 194
European Central Bank (ECB), 186
European countries, *see under single countries*
European Union (EU) policy, 13, 15, 21, 88, 181, 186, 218, 220, 226, 253, 280, 283

exclusion, 102, 253, 285, 286, 307, 309

fertility, 131, 147, 177–8, 181, 200, 313
financial crisis, 59, 134, 145, 177, 186, 189, 191, 192
financial markets, 59, 181, 186, 189, 192, 240
financial need, *see* poverty
Finland, 182, 192, 193
flexibility, *see* labour market flexibility
flexible retirement, 1, 12, 13, 110, 130, 147, 220, 225, 230, 249, 300, 305
Fordism, 199, 206–9, 211
France, 2, 185–6, 192, 193, 222
further training, *see* training

gatekeeper, 205, 221–3
Germany, 2, 16, 57–78, 84, 153, 180, 184–5, 193, 218–19, 220, 225, 237–54, 278–94, 299
gerontology, 81, 284–5, 287–8, 310, 312
globalization, 13, 198–212
gradual retirement, *see* flexible retirement
grandchildren, 11, 153, 157, 159, 164–5, 166–7, 168, 169, 281
Greece, 2, 186, 192, 193, 194, 219

health, 8, 9, 10, 17–20, 259–75
see also partner's health
health care provision, 14, 49–51, 132, 201, 203, 204, 207, 286, 308
health management, 12, 228, 230
history of pensions/retirement, 3, 177–9
hours worked, 17, 50, 51, 64, 96–7, 139, 146, 161–2, 203, 228
see also part-time work
housing, 10, 39–40, 48, 53, 93–4, 98, 102, 179, 301–2
see also benefits
Hungary, 2, 190, 192, 193

ideas/ideology, 15, 18, 237–54, 281–2, 287
immigrants, 33, 131, 162, 169, 208

incapacity, 13, 40, 220, 228
individualization, 21, 200, 208, 238, 249, 287, 307, 310
inequality in old age (incomes), 60, 75, 112, 124, 127, 170, 177, 193, 306, 309
informal work, 16, 61, 102, 133–4, 136, 137, 146, 204, 300
interest organizations, *see* non-profit organizations
intergenerational justice, 284, 307, 309
intergenerational relationships, 306, 309
International Monetary Fund (IMF), 186
Ireland, 66, 186, 187, 193
irregular work, *see* informal work
Italy, 2, 81–103, 181, 183, 185, 192, 193, 194, 222, 299, 301

job mobility, 14, 60, 132, 135, 142, 228, 229

labour market flexibility, 14, 15, 64, 84, 205, 228–30, 289
labour market policies for older workers, 133, 169, 220, 243
labour market theories, 7–8, 134, 222–3, 227, 306
labour shortages, 18, 101, 132, 147, 169, 178, 221
leisure, 52, 70, 273, 281, 285, 292, 310, 311
liberal welfare regime, 15, 16, 31, 59, 180, 182, 187, 241, 250, 286
life course regime, 3, 8, 20, 151, 198–200, 238, 308–14
life expectancy, 131, 189, 206, 245, 259, 278, 282, 308–10
    life expectancy links pension age, 31, 87, 110, 184, 186, 188, 194
lifelong learning, 169, 248, 312
    see also training
loneliness, 229, 268–71

mandatory retirement age, 14, 15, 31–5, 152, 155, 220, 243, 248, 284

manufacturing sector, 6, 97–8, 133, 141, 161, 163, 165, 207, 209, 230, 299
marginal employment, *see* non-standard employment
marital status, 11, 44, 69, 76, 94, 98–100
means-tested benefits, *see* benefits
mobility, *see* occupational mobility, job mobility
moral economy, 237–9, 278, 280, 282
mortgage, *see* debts
multi-pillar pension systems, 5, 13, 179–80, 182, 187–91, 194, 300

Netherlands, 2, 181, 187, 188, 192, 193, 194, 222
networks, 9, 169, 272, 274–5
non-profit organizations, 237–54
non-standard employment, 61, 89, 182, 194, 283, 300, 308, 309
Northern Europe, 2, 178
Norway, 182, 192

occupational class, *see* class
occupational mobility, 3, 6, 14, 20, 139–41, 229, 300, 303
OECD (Organisation for Economic Co-operation and Development), 226, 282
old age dependency ratio, 178, 308–9
old age interest organizations, 230, 243, 250–1, 284, 285
old age stereotypes, *see* age stereotypes

partial retirement/pensions, *see* flexible retirement
partner, *see* divorce, marital status, widowhood
partner's employment/retirement, 11, 37, 42–4, 46, 153, 159, 164, 166–7, 264, 265
partner's health, 11, 53, 159, 162, 166–7
part-time employment, 6, 61, 89, 96, 101, 138–9, 202, 209, 229, 230, 299
    see also hours worked
pension reform, 177–95

pensions, *see also* inequality, poverty
  defined benefit (DB)/defined
    contribution (DC), 14, 39,
    49–51, 86–7, 110, 154, 183,
    194, 207, 242, 286, 309
  financing, 179–81
  privatization, 13, 182, 190
  regulation of private/occupational
    pensions, 187–8
  social insurance, 59, 73, 248
phased retirement, *see* flexible
  retirement
physical job demands, 40, 50–2, 87,
  162, 170, 228, 259, 290
planned economies, 151–2, 154, 182
Poland, 2, 192, 193, 194, 219, 220, 222
Portugal, 2, 186–7, 192, 193
post-retirement work, *see* work
  beyond pension age
post-socialist countries, 16, 134, 179,
  182, 189–91
poverty, 11, 13, 18, 100, 177, 192–3,
  242, 286, 301, 309
precarious employment, *see*
  non-standard employment
psychological approaches, 7, 210, 288
psychosocial work environment, 228
public sector, 72, 87–8, 96, 100, 171
  see also civil service

race, *see* ethnicity
recession, *see* economic crisis
recruitment of older workers, 12, 85,
  89, 96, 100, 143, 206,
  223–4
regions, 14, 62, 66–8, 102, 131, 132,
  133, 205, 207, 226
retention of older workers, 12, 14,
  101, 222, 225–6
retirement
  attitudes towards, 129, 136, 143
  concept of, 237–54
  definition of, 5, 102, 113–14, 130,
    174, 238
  involuntary, 40, 143–4, 265–6,
    272–3
  reasons for, 265, 273
  retirement culture, 280, 284, 292
risk society, 198–9, 208, 211

role theory, 7, 268, 272–4
Russia, 2, 17, 129–48, 299

seniority wages, 14, 223, 299
service sector, 6, 14, 70, 97, 163, 299,
  300
skills shortages, 14, 18, 131, 226, 305
social contacts, 7, 9, 11, 48, 71, 228,
  245
social-democratic welfare regimes, 15,
  16, 180, 182, 187, 188
social insurance pension systems, *see*
  Bismarck systems
social integration, 74, 132, 262, 268,
  274, 280
socialism, *see* post-socialist countries,
  China, planned economies
solidarity, 244, 307, 310
Southern Europe, 2, 16, 153, 181, 219
Spain, 2, 186, 192, 193
spouse, *see* partner, marital status
stress, 12, 228, 273, 289, 309, 312
successful ageing, 130, 279, 312
Sweden, 2, 20, 84, 107–2, 180, 182,
  183, 184, 192, 193, 194, 218, 222,
  306
Switzerland, 187, 188, 189, 192, 194

tenure, *see* housing
theoretical perspectives on
  post-retirement work, 7–8
  see also under specific theoretical
    approaches
training, 12, 60, 85, 133, 169, 226–7,
  229, 248

UK, 1, 14, 31–53, 57–78, 84, 95, 180,
  187–8, 194, 219, 237–54, 262, 301
undeclared work, *see* informal work
unemployment
  general rate, 12, 14, 60, 68, 86, 169,
    178, 182, 199, 205
  of older workers, 8, 60, 101, 136–7,
    170, 209, 226, 265, 275
  of younger workers, 85, 101
  see also benefits
unions, 109, 133, 204, 206, 223, 243,
  244–6, 285
United Nations, 278, 282

USA, 2, 4, 14, 15, 31–53, 84, 108, 198–212, 241, 278–94

varieties of capitalism, 15, 31, 59, 60, 84
volunteering, 7, 20, 22, 244, 250, 261, 267–71, 272, 274–5

wealth, 10, 36–9, 48–9, 52, 76, 90, 95, 265–70, 272–5, 278, 301–2
welfare culture, 9, 15, 21, 237–41, 250–3, 307
welfare regimes (Esping-Andersen), 15–16, 22, 31, 59, 180, 182
well-being, 19, 112, 259–75, 287, 298, 306
see also health
Western Europe, 2, 6, 13, 16, 179, 182, 189
white-collar workers, *see* class
WHO (World Health Organization), 282

widowhood, 42, 44, 66, 94–5, 98–100
work ability, 9, 12, 17, 111, 220, 227, 228–9
work beyond pension age
 definition, 4–5, 61, 90, 157
 subjective experience, 8, 61, 69–76
 types, 4–6
working conditions, 14, 20, 74, 95, 101, 126, 220, 261, 262, 265, 272, 289, 309
work off the books, *see* informal work
workplace influences, 9, 12, 49–51, 164, 167, 217–32, 304
 size of employer, 12, 51, 224, 225
workplace mobility, *see* job mobility
workplace training, *see* training
World Bank, 179, 190

younger workers, 7, 12, 42, 51, 85, 86, 132, 134, 203, 223, 225, 240, 244, 246

Printed and bound by CPI Group (UK) Ltd, Croydon, CR0 4YY